HAWKS, EAGLES, & FALCONS
OF NORTH AMERICA

Stooping peregrine falcon and bufflehead

HAWKS, EAGLES, & FALCONS
OF NORTH AMERICA

Biology and Natural History

PAUL A. JOHNSGARD

Smithsonian Institution Press
Washington and London

This book was designed by Lisa Buck Vann
and edited by Matthew Abbate.

Cover: Crested caracara (*Polyborus plancus*). Photo by
Kenneth W. Fink.

Library of Congress Cataloging-in-Publication Data
Johnsgard, Paul A.
Hawks, eagles & falcons of North America : biology
and natural history / Paul A. Johnsgard.
p.   cm.
Includes bibliographical references.
ISBN 0-87474-682-5 (alk. paper)
1. Falconiformes—North America.
I. Title.
II. Title: Hawks, eagles, and falcons of North
America
QL696.F3J6 1990
598'.916'097—dc20   89-48558

British Library Cataloguing-in-Publication Data
available

A paperback reissue (ISBN 1-56098-946-7) of the
original cloth edition

⊚ The paper used in this publication meets the
minimum requirements of the American National
Standard for Information Sciences—Permanence of
Paper for Printed Library Materials ANSI Z39.48-1984.

Manufactured in the United States of America

Color illustrations and paper cover were printed in
Hong Kong by the South China Printing Company.

10 9 8 7 6 5 4 3 2 1
08 07 06 05 04 03 02 01

For permission to reproduce individual illustrations in
this book, please correspond directly with the owners of
the works, as listed in the caption. The Smithsonian
Institution Press does not retain reproduction rights for
these illustrations individually or maintain a file of
addresses for photo sources.

Dedicated to those who, like King Solomon,

would know the way of an eagle in the sky

Contents

Preface

In 1986, as I was writing my *North American Owls: Biology and Natural History* for Smithsonian Institution Press, I necessarily had to read extensively in the hawk and eagle literature, and was struck by the fact that no up-to-date book-length survey of the North American falconiform species was yet available. As with the owls, the best available reference was still Bent's *Life Histories of North American Birds of Prey.* This two-volume set in Bent's classic series was published nearly a half-century ago, but had not been effectively updated in spite of the enormous amount of published research that had subsequently appeared on the falconiform birds of North America. To be sure, the falcons of the world had been beautifully treated in a monograph by Tom Cade (1982), and a more general monograph of all the Falconiformes of the world had been produced earlier by Leslie Brown and Dean Amadon (1968). However, the former reference is limited in its coverage to the true falcons of the genus *Falco* and the latter has become dated and has long been out of print.

Recognizing the existence of a broad literature data base and still fascinated with the biology of raptors, I suggested to Ted Rivinus at Smithsonian Institution Press that a book on North

American hawks and their comparative biology would make a useful companion volume to the one I was then preparing on owls. He quickly agreed, and during the fall of 1986 I set about the considerable task of surveying the appropriate ornithological literature.

Knowing that a two-volume reference (Palmer, 1988) dealing with the North American falconiform birds would almost certainly be published before I could complete my own book, I decided that I should at least try to finish my initial draft of the manuscript prior to publication of those volumes (volumes 4 and 5 of the *Handbook of North American Birds* series). I thereby was assured an independent approach to the group, in spite of reviewing essentially the same literature. My initial draft was completed during the spring of 1988, just before the appearance of the *Handbook*. In revising the manuscript since then, I have incorporated some information from the *Handbook* as has seemed appropriate. I trust that my somewhat different organization of the species accounts will offer an alternative source of information to interested readers, and that the preliminary comparative chapters in Part One may offer some insights that might not be so apparent or so easily extracted either from the *Handbook* volumes or from the individual species accounts in Part Two.

Some idea of the magnitude of the literature base that had accumulated by the late 1980s is indicated by the extensive bibliography on falconry and falconiform birds produced by Olendorff and Olendorff (1968–70), which contains about 7,500 references to mostly English-language literature published through the late 1960s. Palmer (1988) included over 2,500 references, of which about half postdate those included in the Olendorffs' bibliography, suggesting a reference base of at least 10,000 English-language titles available as of the mid-1980s. The North American literature is particularly large (probably collectively approaching 4,000 references) with respect to three relatively rare species, the bald eagle, peregrine falcon, and osprey. At least in my own citation files the number of references relating to these three species nearly equals the total number for all of the other 28 species of North American falconiform birds included in this book. Among the latter are such rare and primarily tropical American species as the gray hawk, short-tailed hawk, white-tailed hawk, crested caracara, and aplomado falcon, all of which are still very inadequately studied and collectively have fewer than 100 significant citations in the ornithological literature.

The chapters on comparative hawk biology and the individual species accounts are similarly organized to those of the owl book, but I have included a considerably greater number of literature citations (over 1,000, or twice the number in the owl book). Nevertheless, this total represents perhaps only about ten percent of the currently available English-language technical literature on the group. I have, however, concentrated on the most recently published relevant literature (with about a third of the citations postdating 1980, as compared with about a sixth of Palmer's citations), and have included in my references nearly all (about 60) of the relevant unpublished theses and dissertations that I located, which are often overlooked during literature searches.

As in my earlier Smithsonian Press books, I have tried to present an adequate if far from exhaustive survey of the general biology, ecology, and behavior of all the included species, written so as to be understandable to the interested layman as well as useful to the biologist who might be looking for relevant literature citations or trying to deal with a specific question without resorting to extensive library searches. The larger number of species of North American hawklike birds (31, compared to 19 owl species) has required that I keep the individual species accounts substantially shorter. I have excluded the New World vultures from consideration, primarily because it is now increasingly apparent that these birds are not directly related to any of the other falconiform groups. I have similarly excluded a few vagrant Eurasian or Mexican species that, although they may rarely have occurred on mainland North America north of Mexico, are not yet (as of 1988) known to have nested within these limits. For species whose ranges extend to the United States, however, I have included all the generally accepted races occurring in Mexico, the Greater Antilles, and/or other nearby islands.

In common with my earlier books, I have provided anatomical drawings, measurements, keys, and plumage descriptions (these mainly adapted from those of Friedmann, 1950) to facilitate in-hand species identification and further to assist in aging and sexing living or dead specimens. Except for a few generalities, units of measurement have been presented metrically, or converted if necessary to that system. I have provided North American distribution maps and provided range descriptions (mainly adapted from the American Ornithologists' Union *Check-list of North American Birds*, but with substantial modifications) for all the included 31 species and their generally accepted subspecies. The taxonomic sequence, species-level taxonomy, and English vernacular names of birds are primarily those most recently (1983) adopted by the AOU. The listed subspecies are those of the fifth edition (1957) of the AOU *Check-list*, with additions and modifications based on more recent literature or as needed for my somewhat expanded geographic coverage. The nomenclature of nonavian terrestrial vertebrates follows Banks, McDiarmid, and Gardner (1987); that of other organisms follows various standard references.

Field identification criteria and a limited reference series of field-guide-type drawings have been included in this book, although it is not intended to substitute for a good field guide. Indeed, the field identification of many hawk species is one of the more difficult but most fascinating aspects of studying them, owing to their remarkable individual variability associated, for example, with age changes and plumage phases. However, a very good field guide to the North American falconiform birds has recently been published (Clark and Wheeler, 1987) that has an unusually complete illustrative coverage of such plumage variations.

I have not provided any specific rationale for the current legal protection of our raptors (all of which are now fully protected by federal and state statutes, as well as by comparable Canadian laws), which in the view of some persons might represent threats to livestock, poultry, and the like. It should be evident from the information included on each species's foods that such a threat is largely imaginary, and that the occasional "valuable" or domesticated animal that raptors may take is of infinitesimally small significance compared to their enormous value as biological controls of rodents, insects, and other economic "pests." Nor have I mentioned raptors' obvious aesthetic roles as palpable symbols of wilderness, freedom, or any of the other things that might come to mind upon seeing for the first time, for example, talon locking in flight by territorial red-tailed hawks or a peregrine falcon stooping on its prey. To describe such glories adequately is extremely difficult, and a poor substitute for the extraordinary sense of sheer wonder and individual discovery upon first experiencing such unforgettable events personally.

As usual, I have had the welcome assistance of a variety of people during the preparation of this book. Betsy Hancock and her many volunteers at the Raptor Recovery Center of Lincoln, Nebraska, allowed me to photograph various raptors in their care, and Betsy additionally helped and advised me in a variety of other ways, including the loan of reference materials. The librarians of the University of Nebraska, the Bureau of Land Management's Denver Service Center, Oxford University's Edward Grey Institute, and the Nebraska Game and Parks Commission were likewise extremely helpful. Similarly, Bill Andelt, David Busch, Dirk Derksen, Nancy Green, Charles Henny, Scott Johnsgard, Mike Kaspari, Mark Linderholm, Ross Lock, Kenneth Meyer, James Rodgers, Jr., Ronald Ryder, Dwight Smith, and Paul

Sykes, Jr., provided me with information or reference materials. As he has in the past, Lloyd Kiff very kindly extracted a great deal of useful egg and nest data for me from the files of the Western Foundation of Vertebrate Zoology. I must also thank many people who have offered the use of their photographs, including Robert Behrstock, Rob Bennetts, William Clark, Larry Ditto, Ed Dutch, Craig Farquahar, Kenneth Fink, Kenneth Hollinga, Jon Judson, Philip Kahl, Mark Kopeny, H. P. Langridge, William Lemburg, Thomas Mangelsen, Jack Murray, Alan Nelson, K. Neuman, B. J. Rose, William Shuster, Kathy Watkins, and Brian Wheeler. The wonderful paintings by John Felsing, Jr., were done at my request, and somehow manage to catch the beauty of these rare and magnificent birds. As usual, Ken Fink was more than generous with my use of his splendid photos, and also kindly read an early draft of the manuscript. The final manuscript was also critically read by William Clark. As was also true of my earlier book on owls, Matthew Abbate's careful and thoughtful editing has operated as efficiently as natural selection in weeding out undesirable typographic mutations and semantic aberrations, leaving a substantially improved "hopeful monster" for the reader to contemplate.

Although I finished intensive work on the manuscript during the summer of 1988, I had an opportunity to make some late changes and additions while attending the annual meeting of the Raptor Research Foundation during October of that year, and greatly appreciate the wealth of advice and assistance accorded me at that time by the Foundation's participating members. The raptorial birds of North America are lucky to have among their defenders such enthusiastic and dedicated people, whose efforts represent the forefront of raptor research and whose quarterly *Journal of Raptor Research* (formerly *Raptor Research*) has been of enormous assistance to me in preparing this book. Individuals interested in learning more about raptors could do no better than join this organization of nearly 800 raptor enthusiasts from at least 25 countries; it holds its annual meetings at varying locations across North America to discuss recent research findings and to deal with conservation issues that affect raptors, and lists almost 100 raptor-related organizations in its most recent membership list.

Other nonprofit foundations with similar if somewhat narrower purposes are the Eagle Foundation (formerly the Eagle Valley Environmentalists), sponsor of the annual "Bald Eagle Days" symposia, the International Osprey Foundation, sponsor in recent years of several regional symposia on the osprey, and The Peregrine Fund, a foundation devoted to the study and preservation of falcons and other raptors, which has been especially active in recent peregrine reintroduction efforts. Additionally, the Hawk Mountain Sanctuary Association and the more geographically inclusive Hawk Migration Association of North America sponsor periodic (five through 1988) Hawk Migration Conferences, and the latter association publishes the *Journal of the Hawk Migration Association*. The National Wildlife Federation's Institute for Wildlife Research has published several bibliographies of various raptors, publishes a semiannual periodical *The Eyas*, and sponsors annual surveys of wintering bald eagle populations. The Institute has also sponsored a recent series of five regional symposia on raptor management, as well as the publication of a raptor management techniques manual (Pendleton et al., 1987). Finally, the Society for the Preservation of Birds of Prey and the North American Falconers' Association attract their own groups of interested persons and support various raptor-related activities, including the latter's publication of the *North American Falconers' Association Journal*.

This profusion of raptor-related organizations and their activities represents a welcome change of public attitudes since I was a child in a small North Dakota town during the 1930s, when all hawks and owls were considered to be vermin. I can still vividly remember gazing in mixed awe and repulsion at the bloody carcass of a golden eagle (the first eagle I had ever seen) that had been killed and placed conspicuously in a neighbor's front yard, presumably for all to appreciate the efforts of the man who had killed such a terrible predator. Today the killing or even the harassing of eagles is a federal crime, and it is increasingly common for people to go out eagle-watching during winter months, reveling in the incredible sight of dozens or even hundreds of eagles gathered in such favored wintering areas as the upper Mississippi River or the Pacific Northwest Coast.

Perhaps the day has come when, like the Native Americans before us, we can look to the eagle (or, in its most mystical and symbolic form, the thunderbird) as a kind of sacred symbol, secure in our knowledge that in order to appreciate any living thing to the fullest we must be able to cherish its freedom, and grateful to be able to share it with everyone.

Color Plates

Plate 1 (left). Osprey, adult at nest. Photo by Kenneth Fink.
Plate 2 (above). Hook-billed kite, adult female. Photo by Ed Dutch.

Plate 3 (opposite). Swallow-tailed kite. Photo by Jeff Palmer.

Plate 4 (left). Black-shouldered kite, adult. Photo by Kenneth Fink.

Plate 5 (below). Snail kite, adult male. Photo by Rob Bennetts.

Plate 6 (above). Snail kite, adult female. Photo by Rob Bennetts.
Plate 7 (right). Mississippi kite, immature. Photo by author.
Plate 8 (opposite). Bald eagle, adult. Photo by Kenneth Fink.

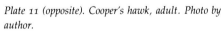

Plate 11 (opposite). Cooper's hawk, adult. Photo by author.
Plate 12 (above). Northern goshawk, adult of atricapillus. *Photo by author.*
Plate 13 (right). Common black-hawk, adult. Photo by Brian Wheeler.

Plate 14 (opposite). Harris' hawk, adult. Photo by author.
Plate 15 (above). Gray hawk, adult. Photo by Kenneth
Fink.

Plate 16 (left). Red-shouldered hawk, adult of elegans.
Photo by Kenneth Fink.
Plate 17 (below). Broad-winged hawk, adult. Photo by
author.
Plate 18 (opposite). Short-tailed hawk (dark morph).
Photo by Brian Wheeler.

Plate 19 (overleaf). Swainson's hawk, adult. Photo by Kenneth Fink.
Plate 20 (overleaf). White-tailed hawk, adult. Photo by Larry Ditto.

Plate 21 (above). Zone-tailed hawk, adult. Photo by Kenneth Fink.
Plate 22 (right). Red-tailed hawk, adult of calurus. Photo by Kenneth Fink.

Plate 23 (right). Red-tailed hawk, adult of harlani. *Photo by author.*
Plate 24 (below). Red-tailed hawk, immature of borealis. *Photo by author.*

Plate 25 (opposite). Ferruginous hawk, adult (typical morph). Photo by Alan Nelson.
Plate 26 (left). Ferruginous hawk, adults (dark and typical morphs). Photo by author.
Plate 27 (below). Rough-legged hawk, adult (dark morph). Photo by author.

Plate 28 (left). Rough-legged hawk, adult (typical morph). Photo by Alan Nelson.

Plate 29 (below). Golden eagle, adult. Photo by Kenneth Fink.

Plate 30 (opposite). Crested caracara, adult. Photo by Kenneth Fink.

Plate 31 (right). American kestrel, pair. Photo by Kenneth Fink.
Plate 32 (below). Merlin, adult female of richardsonii. *Photo by Alan Nelson.*
Plate 33 (opposite). Peregrine, adult of anatum. *Photo by Alan Nelson.*

*Plate 34 (above). Prairie falcon, adult male. Photo by
Kenneth Fink.*
*Plate 35 (opposite). Gyrfalcon, adult (white morph).
Photo by Kenneth Fink.*

Plate 36 (above). Gyrfalcon, adult (gray morph). Photo by Kenneth Fink.
Plate 37 (opposite). Aplomado falcon, adult female. Photo by Kenneth Fink.
Plate 38 (overleaf). Aplomado falcon, adult. Painting by John Felsing, Jr.

PART ONE

Comparative Biology

Ferruginous hawk

Evolution, Classification, and Zoogeography

EVOLUTION AND CLASSIFICATION

It seems highly probable that the falconiform raptors evolved very early while the ancestral bird groups were undergoing adaptive radiation in late Mesozoic and early Cenozoic times; yet their fossil record is extremely limited and puzzling. Feduccia (1980) states that "the fossil record tells us almost nothing about the evolution of raptorial birds," and that the relationships of the falconiform subgroups to each other and to other avian orders "remains one of the major challenges of avian systematics," a point that had been made earlier by Sibley and Ahlquist (1972).

A few general comments on the falconiform fossils will be useful. Fossil hawks or eagles have been found that have been dated from as far back in time as the late (upper) Eocene or early (lower) Oligocene of France, and from the early to middle Oligocene in the Americas (Feduccia, 1980), or roughly from the period of about 35 to 50 million years ago. Contemporary North American genera extend back as far as the middle and upper Oligocene (or about 25–30 million years ago) in the case of *Buteo*, and to the upper Miocene (or about 10 million years ago) in the case of *Aquila* and *Haliaeetus*, but many of the other broadly ranging contemporary genera still remain unknown in the older fossil record (Brodkorb, 1964). As a result, one must rely almost totally on nonfossil evidence for reconstructing the probable phylogeny of the hawks.

In Elliott Coues's *Key to the North American Birds*, first published in 1872 and one of the landmark books in North American ornithological history, many important taxonomic decisions were made that to some extent are still widely adopted today. Coues included all the diurnal birds of prey in his suborder Accipitres (the owls and New World vultures comprising the other two suborders Striges and Cathartides, respectively, of his order Raptores). At least in the later editions of this work (e.g., 1903), Coues recognized only two families within the suborder Accipitres, the osprey as the sole member of the family Pandionidae and all the remaining species of vultures, falcons, hawks, eagles, and kites included in the large family Falconidae. He regarded the osprey to be "buteonine" rather than "falconine" in most of its structural characteristics, but possessing many peculiarities that at least in part were related to the "semi-aquatic piscivorous habits" of this species. Within the Falconidae Coues recognized six subfamilies, including the harriers (Circinae), the kites (Milvinae), the accipiter hawks (Accipitrinae), the falcons (Falconinae), the caracaras (Polyborinae), and the buteo-like hawks and eagles (Buteoninae). He was "perfectly willing" to associate taxonomically the falcons and caracaras on the grounds of their sharing some anatomical similarities in the pectoral girdle (shoulder joint) and some skull similarities, in spite of the great differences in the outward appearances of these two groups. However, he offered few other opinions as to possible evolutionary relationships within this large assemblage, other than to note some similarities between harriers and owls that would now be attributed to convergent foraging adaptations. Coues later himself recognized the evolutionary affinities of the owls to be with the goatsuckers rather than with the diurnal raptors, and placed the two groups ("Striges" and "Nycticoraciae") in adjoining orders.

The now obviously convergent similarities between the currently recognized orders of owls (Strigiformes) and hawklike raptors (Falconiformes) were discussed in my earlier book on owls (Johnsgard, 1988) and need no further discussion. However, the apparently distant relationships of the New World or cathartid vultures and the other hawklike raptors has puzzled ornithologists for a very long time, and only recently has it generally become recognized that the traditional inclusion of this group of hawklike scavengers in the Falconiformes is a major taxonomic error, and that the affinities of the New World vultures are actually with the storklike birds (Rea, 1983). With this recognition the evolution of the remaining hawklike birds becomes something less of a puzzle, although by no means are all the higher-level relationships at all clear.

In reviewing the taxonomic history of the Falconiformes, Sibley and Ahlquist (1972) identified five major problems in their classification, one of which is the just-mentioned case of the New World vultures. The second involves the uncertain relationships of the long-legged African secretary-bird (*Sagittarius*), which at least superficially resembles certain cranelike (gruiform) birds, such as the cariamas of South America. The third problem concerns the relative closeness of the affinities between the falcons and caracaras on the one hand and the remainder of the hawk, eagle, and kite assemblage. The fourth concerns the proper level of taxonomic distinction and sequential placement of the osprey, and the last the general affinities of the remaining or "typical" accipitriform birds.

In their monograph on the falconiform birds of the world, Brown and Amadon (1968) advocated a classification that has since been widely adopted, and suggested the major lines of evolution that apparently have occurred within this group. Later, Stresemann and Amadon (1979) modified this basic classification slightly, and more recently Amadon (1982) provided a slightly amended version of the original accipitrid phylogeny suggested by Brown and Amadon. In the lower portion of Figure 1 the suggested classification of Stresemann and Amadon is diagrammatically presented to the subfamily level, while the "major trends" of evolution within the subfamily Accipitrinae as visualized by Brown and Amadon are shown in the upper part. They regarded the osprey as most probably a closer relative of the Accipitridae than of the Falconidae, with its greatest similarities to certain kites. They considered the kites to be perhaps the most primitive members of the family Accipitridae. This idea was based on the fact that these birds tend to feed on insects such as bee and wasp larvae and on snails, and thus have rather poorly developed associated predatory adaptations. Kites also show generally reduced sexual dimorphism and greater sociality, at least in some species, apparently both regarded as primitive traits by Brown and Amadon. The kites consist of several long-recognized subgroups, the seemingly more primitive of which (the pernine kites such as the American swallow-tailed kite) lack a bony shield above the eye and sometimes have densely feathered sides of the head, apparently as an antistinging adaptation. In the elanine kites there is a bony shelf above the eye (probably an antiglare

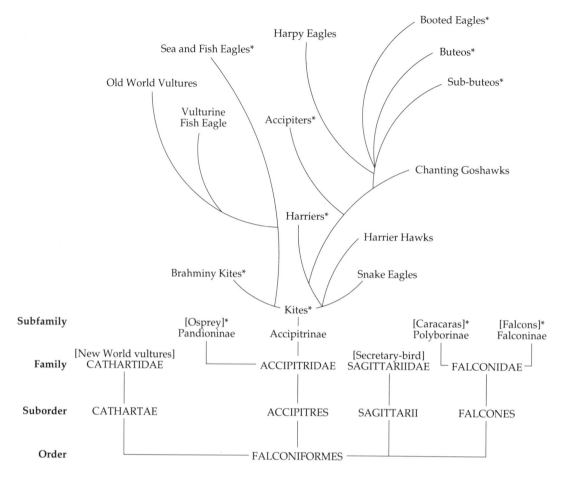

Figure 1. Classification and suggested evolutionary relationships of the Falconiformes, based mainly on Brown and Amadon (1968), but incorporating taxonomic changes suggested by Stresemann and Amadon (1979) and accepted by the American Ornithologists' Union (1983). Groups marked with asterisks are those having representative North American species included in this book.

adaptation), and unlike more advanced hawks their talons are not grooved below. The apparently most advanced of the groups of kites are the Brahminy or milvine kites, at least some of which (such as *Ictinia*) have the basal joint of the middle toe fused with the second joint, a trait of unknown adaptive significance. However, because this same trait appears in sea eagles, Brown and Amadon regarded sea eagles and fish eagles as derived from kites.

They believed the very large group of Old World vultures to be perhaps related to the sea eagles and fish eagles through the

apparently transitional vulturine fish eagle, the entire group perhaps originating from kitelike ancestors.

Also apparently derived from ancestral kites, but radiating
out separately from them, are the remaining major groups of
accipitrids. Perhaps the most primitive of these in the opinion of
Brown and Amadon are the reptile-eating and Old World snake
eagles such as the African bateleur eagle (*Terathopius ecaudatus*).
More advanced but closely related to the snake eagles are the
harrier hawks, crane hawks, and harriers, of which only the last-
named have any North American representatives; harriers are
highly diversified predators adapted to hunting during extended
and low-level searching flights.

Perhaps the most primitive of the accipiterlike hawks are the
chanting goshawks (*Melierax*) of Africa and Arabia, which often
capture slow-moving prey on the ground, while the typical goshawks and sparrowhawks of the genus *Accipiter* and related
genera are more specialized for aerial chase and capture of birds.
According to Brown and Amadon, the differences between *Accipiter* and the often mammal-eating hawks of the genus *Buteo* are
not very great, in spite of their very different hunting techniques
and flight adaptations. From a generalized type similar to the
chanting goshawk these authors would sequentially derive the
"sub-buteos," buteos, and "booted eagles." This last-named group
consists of a considerable number of mostly predatory eagles that
prey on a variety of birds and mammals and in the view of Brown
and Amadon represent the climax of one line of raptorial
evolution.

Brown and Amadon regarded the secretary-bird as being a
member of the Falconiformes, albeit a specialized one. They also
concluded that the falcons should be placed in a separate suborder Falcones, with important differences from the other
hawklike birds in their wing-molting patterns, head musculature,
a variety of skeletal features, and in some behavioral ways that
collectively suggested that a family-level distinction was inadequate. The large number of structural and behavioral differences
(and also some similarities) between the "typical" hawklike birds
or Accipitridae, and the falconlike birds or Falconidae, has been at
least partially tabulated in Table 1. Some of these differences are
of questionable phyletic significance and to varying degrees may
reflect ecological or food-related adaptations, but others are not so
easily explained on such bases.

One of the persons who examined in detail the question of
the possible relationships of the falconlike birds to the rest of the
accipitriform raptors was Jollie (1953; 1976–77), who accumulated

Table 1.

Some Comparative Traits of Accipitridae and Falconidae

Trait	Accipitridae	Falconidae[1]
Bill	Cere usually naked; no bony tubercle in nares; upper mandible rarely with a "tooth," the lower smooth	Cere often bristly; bony tubercle in nares; both mandibles often toothed and notched
Iris color	Usually varies with age	Usually permanently dark
Syrinx	First bronchial semi-ring incomplete, with small lateral and large medial tympaniform membranes	First bronchial semi-ring fused to syrinx; large medial and lateral membranes
Head feathers	Often crested; lores often bare or nearly bare, most with hidden white nape patch	Rarely crested, lores well feathered; no nape patch but malar stripe common
Secondaries	13–16 (rarely to 25)	11–14
Outer primaries	Emarginated, 7th or 8th primary the longest in most spp. (9th in some kites)	Not strongly emarginated; 9th primary the longest in most spp. (8th in caracaras)
Primary molt	Descendant (outwardly from 1st primary)	Ascendant and descendant (from 4th primary)
Carrion eating	Frequent in most species	Typical only of caracaras
Nest site	Usually in trees, on cliffs, or on ground	Usually in cavities, recesses, or on ledges; rarely on ground
Nest building	Mostly by female, nest often large, of branches	Present only in some caracaras
Aerial displays	High circling Soar and call Sky dancing (soar and dive) Talon presentation Food passing "Parachuting" "Whirling" (cartwheeling)	High circling Soar and call Sky dancing Talon presentation Food passing Bill touching Aerial chases Undulating roll
Mutual displays	Duetting Mutual billing Bill stroking	Duetting Mutual billing Allopreening
Other displays	Food begging (female) Courtship feeding (male) Perch and call (male)	Food begging (female) Courtship feeding (male) Ledge display (male) Slow landing (male) Nest scraping (female) Wailing (female)
Copulation site	Rarely at nest	Often at nest
Eggs	1–6, usually green-tinted	2–6, reddish-yellow tint
Greens in nest	Frequent in most	Lacking in most
Laying interval	2–5 days	2–3 days

(continued)

Table 1. (Continued)

Trait	Accipitridae	Falconidae[1]
Incubation	Often by female only, both sexes in some spp.	Usually by both sexes, but mainly female
Incubation period	28–59 days	25–36 days
Incubation patch	Single median patch	Usually paired patches
Hatching of eggs	Asynchronous (2–5 days apart)	Asynchronous to nearly synchronous
Clutch replacement	1–3 relayings reported	1–3 relayings reported
Fratricide	Common	Rare or absent
Brooding sex	Female, very rarely both	Female, sometimes both
Hissing by young	Absent	Present in some spp.
Defecation	Ejected over side of nest or perch	Directly below perching or nest site
Fledging period	23–125 days	25–49 days
Postfledging dependency	3–ca. 30 weeks	1–8 weeks
Juvenal plumage	Often streaked below	Sometimes streaked
Adult plumage	At 1–5 or more years	Usually at 2 years
Initial breeding	At 1–6+ years, usually 2 years	At 1–4 years, usually 2 years

[1]*Particularly with reference to Falconinae.*

evidence from a large array of osteological and other structural criteria that cast strong doubts on a close relationship of these two groups. Indeed, Jollie advocated that, in addition to removing the New World vultures and the secretary-bird from the traditional Falconiformes altogether, two separate phylogenies should be recognized within the remainder. (Figure 2 presents these in simplified form, excluding non–North American genera.) One of the two, the Falconiformes, would include the caracaras and falcons previously separated as the Falconinae. Jollie (1953) placed this group adjacent to a group that included both the owls and the caprimulgiform birds. He believed (1976–77) that the ancestors of the falcons evolved in the New World, perhaps South America, from probably weakly predaceous ancestors whose feet were used for holding prey but lacked real clenching abilities. Evolution was then mainly a radiation of foraging types, with some lines tending toward scavenging (as in typical caracaras), and others becoming increasingly efficient aerial predators (as in typical falcons). Within this latter line, including the genus *Falco*, there has apparently been a trend toward reduction in size that may in part be related to improved aerial agility. Jollie (1976–77) provided a suggested phylogeny of the 13 genera that he included in this group, two of

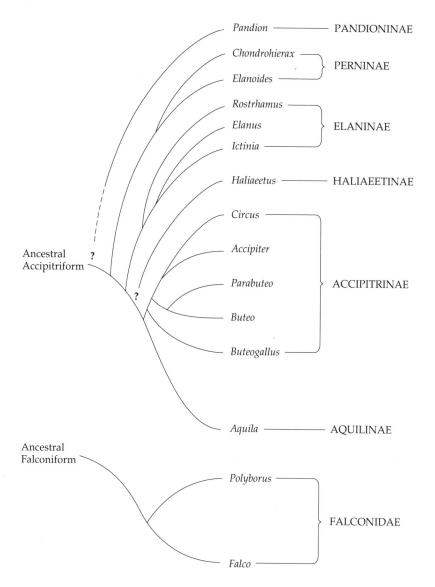

Figure 2. Possible evolutionary relationships among the genera of North American hawks, eagles, and falcons, based on diagrams presented by Jollie (1976–77), but excluding all non–North American genera.

which are represented in North America. Clearly, the genus *Falco* in Jollie's view represented one extreme type of this radiation, while such South American genera as *Micrastur* and *Herpetotheres* were regarded as old "relict" types showing a combination of primitive and specialized features.

Jollie reconstructed his "accipitriform" (here equivalent to non-falconid falconiform) adaptive radiation as proceeding from an ancestral type of terrestrial carrion eater that gradually increased its power of flight but retained its walking abilities. It also gradually developed a shorter and stronger hooked bill and grasping feet, in conjunction with tearing up food. This change allowed not only for increased vulturine efficiency but also for seizing and killing smaller animals, including birds. Thus, evolution away from a scavenging existence rapidly proceeded. Perhaps the group most closely approaching this original ancestral type is the subfamily Aegypinae, which includes a large number of Old World vulturine genera. Many of these species are quite large birds, which are mostly associated with tropical grasslands and deserts. These recently evolved species perhaps radiated during the Pliocene, when there was an abundance of large herbivores in the extensive grassland areas of Africa and elsewhere.

There are no North American representatives of this "aegypiin" assemblage recognized by Jollie, but he believed that perhaps the "haliaeetin" group of fish eagles and sea eagles (*Haliaeetus* and *Ichthyophaga*) belongs somewhere near this group of Old World vultures. All of the former are fairly closely associated with water, have massive bills, and often rely largely on fish for their primary foods.

Another core group that apparently radiated out from a common ancestral source was the assemblage Jollie called the "aquilins," which are easily identified by their feathered or booted tarsi and thus might be called "booted eagles." A number of smaller species, usually called hawk-eagles, were included as well. Among North American forms the aquilins include only the golden eagle, but in general they consist of large, efficient predators with strong bills and a powerfully constructed body skeleton. The group includes some of the most impressive mammalian predators of all the eaglelike birds, which sometimes specialize on such elusive or large prey as hares, larger monkeys, and sometimes even small antelope. They are widely distributed in the Old World and to a lesser degree in the New World, ranging from tropical rain-forest habitats to steppes and arctic tundra.

Apparently very closely related to the aquilins are the "accipitrins," a large group consisting of three subgroups. This assemblage includes all the remaining North American species included in this book except for the kites. Although the three subgroups tend to intergrade, Jollie suggested that the most primitive of them is the "buteogallin core," which includes only the genus *Buteogallus* among this book's coverage. This group is characterized by long legs, a well-marked plumage pattern, and certain other common osteological features. It intergrades with the more centrally placed "buteonin core," which consists of the extremely diversified genus *Buteo* (with 10 species in North Amer-

Table 2.

A Comparison of Some Suggested Classifications of the Falconiformes

Peters, 1931	Brown and Amadon, 1968	Jollie, 1976–77
Order Falconiformes	Order Falconiformes	Order Sagittariiformes
Fam. Cathartidae	Suborder Cathartae	Order Cathartiformes
Fam. Sagittariidae	Superfam. Neocathartoidea[1]	Order Accipitriformes
Fam. Accipitridae	Fam. Neocathartidae[1]	Fam. Accipitridae[2]
Subfam. Elaninae	Superfam. Cathartoidea	Subfam. Aegypiinae
Subfam. Perninae	Fam. Cathartidae	Subfam. Aquilinae
Subfam. Milvinae	Fam. Teratornithidae[1]	Subfam. Accipitrinae
Subfam. Accipitrinae	Suborder Accipitres	Subfam. Perninae
Subfam. Buteoninae	Superfam. Accipitroidea	Subfam. Haliaeetinae
Subfam. Aegypiinae	Fam. Pandionidae	Subfam. Elaninae
Subfam. Cercinae	Fam. Accipitridae[3]	Subfam. Pandioninae
Subfam. Circaetinae	Superfam. Sagittarioidea	Order Falconiformes
Subfam. Pandioninae	Fam. Sagittariidae	Fam. Falconidae
Fam. Falconidae	Suborder Falcones	
Subfam. Herpetotherinae	Fam. Falconidae[4]	
Subfam. Polyborinae		
Subfam. Polihieracinae		
Subfam. Falconinae		

ica) and the similar *Parabuteo* (which includes only the Harris' hawk). This group is predominantly centered in the New World but has attained a limited dispersal into the Old World; it is characterized by species that tend to be rather short-legged with relatively long and broad wings well suited to gliding and soaring. The third subgroup is the "accipitrin core," which intergrades with the previous group but consists mostly of species with longer tails and shorter, narrower wings. By far the largest component of this core is the genus *Accipiter,* with three North American species and approximately 50 species worldwide, making it one of the largest of all avian genera. Many of these are swiftly flying bird-catching hawks, including a variety of generally larger "goshawks" and smaller "sparrowhawks." Jollie includes the harrier genus *Circus* within this core group, considering it an early derivative of *Accipiter,* but with a great modification of the outer ear in conjunction with improved hearing adaptations.

From this great accipitrin assemblage, Jollie visualized the milvine kites and kitelike hawks as connecting the accipitrin group to the most aberrant of the kites, the pernine kites. He considered, for example, that the milvine kite group (which includes *Ictinia, Elanus,* and *Rostrhamus* among North American representatives) represents a transitional state between the reticulate scaled tarsal condition (as found in the pernine kites) and a more advanced scutellate type typical of most hawks.

The pernine kites include in North America only the very

Stresemann and Amadon, 1979	Sibley et al., 1988
Order Falconiformes	Order Ciconiiformes
Suborder Cathartae	Infraorder Falconides
Fam. Cathartidae	Parvorder Accipitrida
Suborder Accipitres	Fam. Accipitridae
Fam. Accipitridae	Subfam. Pandioninae
Subfam. Pandioninae	Subfam. Accipitrinae
Subfam. Accipitrinae	Fam. Sagittariidae
Suborder Sagittarii	Parvorder Falconida
Fam. Sagittaridae	Fam. Falconidae
Suborder Falcones	Infraorder Ciconiides
Fam. Falconidae	Fam. Ciconiidae
Subfam. Polyborinae	Subfam. Cathartinae
Subfam. Falconinae	

[1]*Group known from fossil taxa only.*

[2]*Jollie did not provide a formal sequence of subfamilies. Nine genera* (Gymnogenys, Gypohierax, Neophron, Gypaetus, Pithecophaga, Harpyopsis, Eutriorchis, Driotriorchis, *and* Machaerhamphus) *were left unplaced and tentatively regarded as representing monotypic subfamilies.*

[3]*Subfamilies not recognized by Brown and Amadon; Brown (1976b) later listed the following: Perninae, Machaerhamphinae, Elaninae, Milvinae, Haliaeetinae, Aegypiinae, Circaetinae, Polyboroidinae, Circinae, Accipitrinae, Buteoninae, and Aquilinae.*

[4]*Subfamilies subsequently recognized by Brown (1976b) are Polyborinae, Herpetotherinae, and Falconinae.*

distinctive swallow-tailed kite of the genus *Elanoides* and the equally remarkable hook-billed kite of the genus *Chondrohierax*. In these birds the orbital socket for the eye is very large and completely lacks an overhanging shelf. Small feathers typically cover the base of the bill and loral area in front of the eyes, and a variety of other unique traits suggest that this group has had a long and relatively independent evolutionary history.

The last of the assemblages within the accipitriform group recognized by Jollie is represented by the osprey. He thought that this species exhibits the maximum divergence that can be allowed within a family, and that it should be regarded as the sole representative of a lineage of great age and uncertain affinities.

It may be seen that, whereas Brown and Amadon regarded the kites as primitive and most like the presumed ancestral accipitriform birds, Jollie hypothesized a large, mostly ground-feeding and carrion-eating ancestral hawk not unlike modern Old World vultures. These differing assumptions lead to rather different conclusions as to which traits may be primitive and which derived, but not to many major differences in basic taxonomic groupings or sequences. A comparison of major twentieth-century suggested taxonomies (summarized to the level of subfamilies) is shown in Table 2. The first is that of Peters (1931), whose survey of the entire falconiform order is perhaps the archetypical modern taxonomy of the group. Most more recent classifications, such as those of Brown and Amadon (1968), Jollie (1976–77), and

Stresemann and Amadon (1979) involve little more than juggling
of sequences and varying degrees of elevating or lowering of the
levels of the major taxonomic groups already recognized by
Peters.

It was not until the innovative work of Sibley and Ahlquist
(1985), using the capacity for measuring degree of biochemical
hybridization of the genetic material DNA from different phyletic
lines as an index to evolutionary relationships, that a totally new
and objective basis for avian classification appeared. This revolu-
tionary technique has implications that have shaken the long-held
concepts of the avian "evolutionary tree" virtually to its roots,
forcing it to be reshaped and "pruned" in a manner undreamed
of only a decade ago. For example, Sibley and Ahlquist confirmed
the storklike affinities of the New World vultures, placing them
with the storklike birds in a supraordinal assemblage that also
includes such structurally diverse and seemingly unrelated groups
as the flamingos and pelicans. They similarly confirmed the
accipitriform affinities of the secretary-bird, placing it in a monotypic family adjacent to the Accipitridae, and concluded that its
similarities to gruiform birds such as the seriamas are indeed the
result of convergence. They judged that the Accipitridae and
Falconidae probably diverged from one another about 68 million
years ago (or at about the beginning of the Cenozoic era), thus
placing them both appropriately within the same order, but
assigning them to separate parvorders.

In a more recent classification, Sibley, Ahlquist, and Monroe
(1988) placed all the species included in this book within the
infraorder Falconides, which in turn was included in a greatly
enlarged order Ciconiiformes. They further included the New
World vultures in the stork family Ciconiidae, and reduced the
status of the osprey to that of a subfamily within the Accipitridae.

BREEDING AND ECOLOGIC DISTRIBUTIONS OF
NORTH AMERICAN HAWKS

Hawks are among those highly mobile and often behaviorally
flexible species that typically occupy rather broad ranges and have
relatively varied prey and competitive interactions. Thus, the
peregrine falcon has an almost worldwide distribution that is
among the largest of any bird's, and, although it tends to forage
selectively on birds, its specific prey organisms differ greatly in
various regions. In general, however, because of their large size,
predatory nature, and associated individual needs for large ter-

ritories or home ranges, hawks generally tend to have low population densities even in regions where they are relatively common. This attribute makes the construction of distribution maps more difficult than for such generally abundant, tame, and conspicuous species as, for example, the American robin (*Turdus migratorius*). For reasons such as these it is rarely possible to construct hawk distribution maps that illustrate varying breeding population densities in different areas, and except for a few unusually well studied species (osprey, bald eagle, and peregrine falcon), the breeding maps I have prepared for this book simply show an area as either occupied or unoccupied by a particular species, without further attempts at quantification.

On the basis of the species maps that I have prepared, it is possible to make a collective map showing the species density of breeding North American hawks north of Mexico (Figure 3). This map is similar in its mode of construction to the ones that I prepared earlier for North American owls (Johnsgard, 1988) and for North American hummingbirds (Johnsgard, 1983). These three maps are further similar in that all illustrate generally increased species density toward the western and southwestern portions of North America. The reasons for this are not altogether clear but perhaps relate to the high levels of habitat and climatic diversity typical of the western and southwestern regions. All three maps also show a generally high level of species diversity in Mexico and Central America, where a combination of topographic diversity and increasingly tropical climates seems to favor this trend.

The presence of an adequate food base is one of the factors that dictate the breeding distributions and abundance of various North American hawks. Two species, the snail kite and hook-billed kite, have extremely specialized food requirements (involving large aquatic and terrestrial snails respectively) that probably account for their highly localized distributions and densities. For several others (such as the fish-dependent osprey and bald eagle), the distributions are similarly closely associated with only slightly less restrictive foraging needs.

Another factor that may strongly influence the breeding distributions of some raptors is the availability of suitable nest sites. This is most evident among some of the larger falcons, such as peregrine falcons, gyrfalcons, and prairie falcons, that require or at least strongly prefer steep cliffs with rather inaccessible nesting ledges. All of these species require a large food base as well, and thus the presence of adequate nesting sites does not assure the presence of any of them. Similarly, the rough-legged hawk occurs widely across arctic North America, but it is most common by far along eroded river valleys with outcrops and

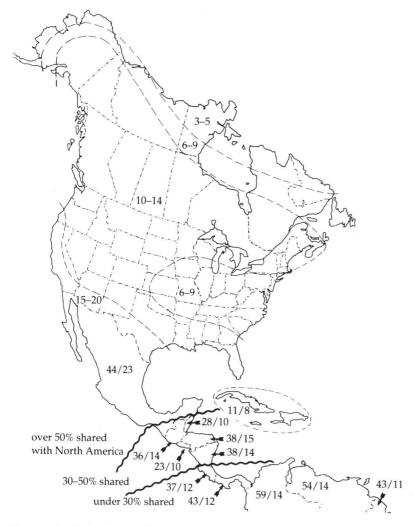

Figure 3. Species density map of North American falconiform birds, showing the number of breeding species present in indicated regions north of Mexico. From Mexico and the West Indies southward the numbers indicate each country's probable number of indigenous breeding species, followed by the number of these that extend north to at least the Mexican-U.S.A. border and are thus included in the species accounts in this book.

promontories that allow for optimum nest site placement. Increased nesting densities and increased productivity of ospreys have been obtained by supplementing natural nest sites with artificial nesting structures (Postupalsky, 1977; Rhodes, 1977; Westfall, 1983). Somewhat comparable effects have been found by introducing artificial elevated nest structures such as electrical transmission lines into the habitats of grassland-adapted buteos and the golden eagle (Howard, 1980; Steenhof and Kochert, 1987). The erection of nesting boxes for American kestrels in grassland areas lacking suitable natural cavities provides a similar example of this approach.

Suitable foraging habitats are of course a critical component of breeding habitats, and major changes in such habitats may have devastating effects on bird distributions and populations. Hector (1987) suggested that ecological changes that degraded hunting habitats of the aplomado falcon in the American southwest most probably caused its eventual extirpation from that region, and Oberholser (1974) similarly suggested that several species of hawks that once nested widely in Texas were eliminated (or nearly so) from that state early in the present century as a result of ecological changes that affected their foraging and breeding habitats. Contrariwise, the recent spread of the black-shouldered kite in the southwest as well as in much of Central America (Eisenmann, 1971) probably results from similar kinds of ecological changes that at the present time happen to favor that species's foraging requirements.

It is much harder to prove that competition with another species is responsible for the limitations or changes in a species's distribution, but such at least seems to be a fairly likely scenario in the northern Great Plains and Pacific northwest, where Swainson's hawk populations are declining as red-tailed hawk populations are increasing. This is apparently the result of red-tailed hawks being able to invade and nest in grassland habitats that had previously lacked adequate tree growth to support nesting red-tails, habitats that had historically been exploited by the more grassland-adapted Swainson's hawk. The slightly larger red-tails are able to dominate the Swainson's and thereby probably cause their local reduction, or at least place them at a competitive disadvantage (Houston and Bechard, 1983; Janes, 1987).

Several studies (Schmutz, 1977; Cottrell, 1981; Janes, 1987) of Swainson's, ferruginous, and red-tailed hawks have indicated that substantial ecological overlap exists among these three species, in some cases causing apparent reductions in productivity in situations of proximate interspecific nesting. Similarly, the red-shouldered hawk's population has been declining in eastern North America since the early 1950s, where the red-tailed hawk has gradually increased during that same period (Kiltie, 1987), and it is possible that similar competitive interactions are occurring between these two species. Thus, selective logging that reduces tree and canopy density favors the red-tailed hawk over the red-shouldered hawk (Bryant, 1986).

In a non–North American but still interesting study of seven coexisting species of accipitrine African raptors, Simmons (1986) found that those species closest in body size overlapped least in geographic range, and that the degree of ecological overlap among all these species (using criteria of similarities in each species's diet, general habitat, foraging habitat, foraging mode, and foraging period) was both clear-cut and minimal.

WINTER DISTRIBUTIONS OF NORTH AMERICAN HAWKS

Although interspecific competition is perhaps most important and most intense on breeding areas, it certainly also occurs on wintering ranges. This aspect of raptor biology is not so well studied, although a few observations such as those of Bock and Lepthien (1976) and Wilkinson and Debban (1980) suggest that habitat selection and associated ecological segregation during winter may be well developed. Thus, Bock and Lepthien concluded that, along a habitat-climate gradient from the treeless prairies and shrub steppe of central North America to the moist eastern woodlands and the Florida peninsula, the rough-legged hawk, red-tailed hawk, and red-shouldered hawk sequentially appear and ecologically replace one another. More will be said of such apparent interspecific ecological segregation in chapter 2. However, as a general introduction to the topic of wintering raptor distribution patterns, I have analyzed the 1986 Christmas Bird Counts of the National Audubon Society (published in *American Birds* 41:583–1307). This analysis encompassed all of the Canadian provinces and mainland U.S. states except Alaska, and all of the raptor species individually described in this book. For each state or province I calculated an average abundance estimate (average total falconiforms seen per count) and species diversity index (total number of falconiform species reported per state or province), as presented in Figure 4. (In a few cases two or three states or provinces were treated as single sampling units because of their small areas and/or sample sizes: New Brunswick, Nova Scotia, and Prince Edward Island; Vermont and New Hampshire; Connecticut and Rhode Island; Maryland and Delaware.) Incompletely identified raptors such as "*Accipiter* sp." and "*Buteo* sp." were excluded from these analyses. For those raptors considered sufficiently common and widespread in winter to warrant it, the individual range maps later in the book indicate relative winter abundance (average number of individuals seen per count) by state and province.

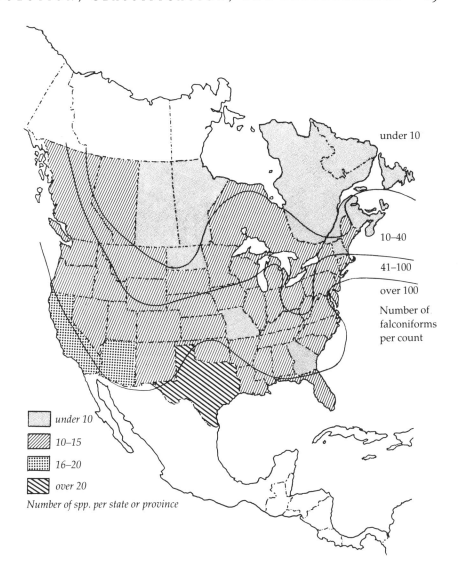

under 10

10–40

41–100

over 100

Number of
falconiforms
per count

under 10

10–15

16–20

over 20

Number of spp. per state or province

Figure 4. Winter distribution patterns of North American falconiform birds, excluding vultures. Lines separate zones having comparable average total numbers of falconiforms reported per count, and levels of shading indicate the total number of falconiform species reported per state or province on the 1986 Audubon Society's Christmas Bird Counts. Data from some smaller eastern states and provinces have been merged (see text).

These data have also been reorganized and summarized by individual species in Table 3. This table provides a highly subjective estimate of each species's overall wintering population in the Canadian provinces and the contiguous United States. These estimates were based on collective state and provincial count

Table 3.

Estimated North American Winter Populations of Falconiform Birds, 1980s

Species	Estimated population	Basis[1]
Red-tailed Hawk	350,000	Anonymous, 1986
American Kestrel	236,000	
Northern Harrier	111,500	
Bald Eagle	ca. 100,000	Table 8
Rough-legged Hawk	49,600	
Red-shouldered Hawk	34,250	
Sharp-shinned Hawk	30,100	
Cooper's Hawk	19,400	
Golden Eagle	18,520	
Black-shouldered Kite	14,120	
Merlin	11,080	
Northern Goshawk	8,500	Anonymous, 1986
Prairie Falcon	7,800	
Harris' Hawk	7,280	
Osprey	7,080	
Ferruginous Hawk	5,480	
Crested Caracara	2,280	
White-tailed Hawk	1,040	
Gyrfalcon	500	Anonymous, 1986
Snail Kite	ca. 500	Rodgers et al., 1988
Short-tailed Hawk	under 500	
Hook-billed Kite	under 100	
Swallow-tailed Kite	under 100	
Mississippi Kite	under 100	
Broad-winged Hawk	under 100	
Swainson's Hawk	under 100	

[1]*Based on 1986 Christmas Bird Counts covering Canadian provinces and 48 contiguous states of U.S.A. unless otherwise indicated. (For method of calculation, see text.)*

averages for each species (birds seen per count, each count covering a uniform area), weighted according to the relative areas of these political regions and keyed to the fact that in the early 1980s there were an estimated 350,000 red-tailed hawks present in this subcontinental region around Christmas (Anonymous, 1986). The red-tailed hawk was chosen as a population indexing species because it is an extremely abundant, widespread, conspicuous, and relatively easily recognized species. The numerous statistical problems evident in using Christmas Bird Counts as a nationwide censusing technique (variable observer skills and sample sizes, inadequate coverage of most western and northern regions, probable sampling bias toward suburban and easily accessible habitats, etc.) are readily evident. The technique is also likely to underestimate numbers of such elusive, forest-dwelling, and nonmigratory species as the northern goshawk, and to overestimate numbers of raptors that conspicuously concentrate in local areas, particularly the bald eagle. No easy methods for adjusting for such effects are apparent to me. Thus Table 3 substitutes estimates based on other sources for the northern goshawk and bald eagle population, and all species recorded in such small numbers as to produce total population estimates of less than 500 birds are estimated only in a general way.

These estimates are apparently the only ones that currently exist for nearly all the accipitriform birds that winter primarily in North America. Comparable estimates of seven species were made by the Laboratory of Ornithology during the early 1980s using Audubon Christmas Count data (but normalized by using birds seen per party-hour rather than birds seen per count) for the same general geographic area (Anonymous, 1986). Five species occurred in adequate numbers in both counts to provide cross checks on these statistics. The Laboratory's population estimate for the red-tailed hawk served as the indexing device for my calculations, leaving their estimates of the prairie falcon (13,000 individuals), merlin (8,000), Harris' hawk (5,500), and peregrine (1,300) available for comparative purposes. My own respective estimates for these four species are: prairie falcon 7,800, merlin 11,080, Harris' hawk 7,280, and peregrine 2,280, or an average estimate deviation of about 33 percent. This does not seem to be an unacceptable level of variation, and in the absence of other survey data for the region it perhaps gives additional credibility to both of these estimation techniques.

Though not attempting to provide total population estimates, a recently published analysis of Christmas Count data (Root, 1988) provides an additional interesting and independent comparison with my own results (reported on individual species' range maps) illustrating wintering hawk distribution patterns. Root's data base extended over a ten-year period beginning with the 1962–63 count, and he used contour map methods to illustrate relative abundance data (based on birds seen per party-hour) for 19 hawk species. Some distinct similarities can be seen in the results of these two methods, and readers are urged to make their own visual comparisons for species of special interest.

CHAPTER 2

Foraging Ecology and Foods

In order to survive and reproduce, raptors must obtain their foods by the most difficult method, namely catching, killing, and consuming other animals of varied size, elusiveness, and capabilities for self-defense. They must also avoid undue competition with other individuals of their own and other species, even including their own mates. As a result, a great deal of time and energy must be spent in hunting, and a good deal of experience in hunting and killing techniques must be gained if an individual is not only to avoid starvation itself but also provide efficiently for its offspring's survival.

Because of all these requirements, it should not be surprising that an important area of the study of raptor biology concerns their foods and foraging ecology, including intra- and interspecific competition, ecological energetics, and behavioral ecology. Newton's (1979) book on population ecology deals thoroughly with these and other aspects of ecology, and more general discussions of foods and foraging can be found in the books by Brown (1976b), Brown and Amadon (1968), and Grossman and Hamlet (1964). In this chapter I will limit my discussion to interspecific differences in foods and foraging ecology, associated evidence of niche segregation and competition, and the ecological implications of sexual dimorphism in hawks. Certain strictly behavioral aspects of foraging, such as territoriality, hunting techniques, and the like, will be dealt with in the following chapter.

VISUAL ADAPTATIONS FOR FORAGING

Whether predators or scavengers, the North American hawks are preeminently visually dependent hunters; their eyesight is generally regarded as being among the most acute of all birds'. Indeed, the eyesight of hawks is almost legendary; Walls (1942) reported the Eurasian buzzard (*Buteo buteo*) to be the probable "grand champion" in terms of the density of visual cells in its foveae, with an estimated million cones per square millimeter. He judged that this must provide it with a visual acuity at least eight times that of humans. This degree of visual acuity has not apparently actually been demonstrated with live birds, but there can be little doubt that the visual abilities of hawks is indeed acute. Thus, Fox, Lehmkuhle, and Westendorf (1976) estimated the visual acuity of the American kestrel as 2.6 times that of humans, although this estimate was subsequently criticized (Martin, 1985). Although the foveal structure of most hawks has apparently not been studied, it is believed that diurnal birds of prey have both lateral (or temporal) and central (or nasal) foveae, the two circular areas connected by a linear area that is believed to be related to a high degree of visual acuity and stereoscopic vision in these raptors. At least in the American kestrel, both of these foveae are located above the pecten, the central or nasal fovea near the optical center of the retina and the temporal one somewhat laterally (see Figure 5B, and Figure 7.18 in Martin, 1985). Although the actual visual projections of falcons remain unstudied, it is often assumed that the central fovea may be associated with monocular vision and the temporal fovea with stereoscopic or binocular vision (Figure 5B). There are also uncertainties as to the actual visual function of the foveae, the effects of refraction on the visual axis, and abilities of the birds to move their eyes in the sockets. While the large eyes of hawks would seem to seriously limit potential for eye movements, their generally oval shape might allow more movement than, for example, the tubular eyes of owls, and indeed close observation of captive hawks and falcons suggests that at least a limited ability for such eye adjustment does exist.

The eyes of hawks are relatively very large, as is apparent from Figure 5, and this trait plus their rather globular shape tends

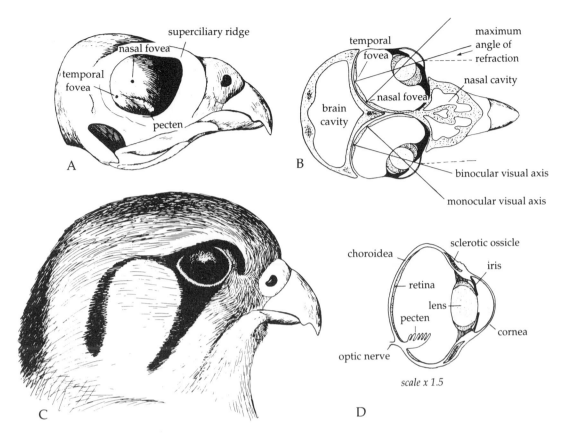

Figure 5. External (A) and cross-sectional (B) views of a falcon (Falco) skull relative to bird's appearance in life (C), all drawn to same scale. Also shown (D) is sagittal-section view of a typical hawk (Buteo) eye, drawn to somewhat larger scale. Suggested angle of refraction (of about 25 degrees) assumes perfect binocularity without movement of eye in socket (see text), and is probably excessive. A and D after Grassé (1950); B after Wood (1917).

to cause them to project somewhat beyond the contours of the head itself, presumably providing for a wide visual field. Although some hawks (especially the more primitive kites among North American species) evidently survive perfectly well without any special adjustment to this situation, most hawks, eagles, and falcons have a distinctive shelf or "eyebrow" directly above and in front of the eye, which produces a characteristic hawklike appearance (see Figures 5C and 6). This overhanging shelf results partly from a well-developed bony superciliary ridge and partly from the shape and organization of feathers in that area.

Clearly, this distinctive "eyebrow" must be of adaptive visual significance to hawks, and yet little has been written of its possible role. The most probable hypothesis is that it provides

Figure 6. Side and front views of a northern goshawk, showing the eye-shading effects of the superciliary ridge and associated feathers. From photographs.

shading from excessive overhead sun glare passing through the iris and causing a resulting dazzling of the retinal surface; such shading effects are apparent from the accompanying drawing of a northern goshawk (Figure 6). It is of course possible that this protruding shelf may provide the eye some physical protection as well, and it may possibly even serve as a kind of wind or dust deflector for fast-flying birds. Although the osprey lacks a bony superciliary ridge, its feathers above the eye do indeed provide some overhead shading, and furthermore the feathers in front of and below the eye are black, probably helping to reduce glare reflected off water surfaces. Similar black shading in front of or around the eye occurs in many aerial foraging species with white- or light-colored heads, such as the Mississippi and black-shouldered kites. Curiously, it does not occur in the swallow-tailed kite, which also does extensive soaring in sunny habitats and additionally lacks an overhanging shelf of feathers or bone.

INTERSPECIFIC DIFFERENCES IN ANATOMY AS RELATED TO FORAGING

The foods and foraging adaptations associated with the raptorial way of life have had effects on virtually all the anatomical aspects of hawks and their relatives. The most obvious examples include bill and foot anatomy (see Figures 65–70), which must in various ways be adapted to killing, tearing, and carrying food. The talons are invariably long and sharp, and in the osprey there are spicules present on the undersides of the toes that assist in grasping slippery fish. In this species too the outer front toe can swing around behind and, together with the hind toe, thus help grasp and distribute the weight of heavy prey more effectively between front and back toes. Even the nostrils may be modified, as in the osprey where they can be closed when the bird is diving into water. In the falcons distinctive bony tubercules are present in the nostrils that are believed possibly to serve as air baffles during high-speed flight. Besides being strongly hooked, the upper mandible of most falcons has a variably developed "tooth" that helps the birds to break the cervical vertebrae of their prey. Curiously, some mostly insect-eating kites have similar "teeth" that must serve quite different purposes. Two distantly related snail-eating kites, the snail kite and hook-billed kite, have greatly elongated upper mandibles that are used for extracting snails from their shells, although in quite different ways as described later in their species accounts.

The wing shapes of hawks help to predict their foraging behavior at least as much as do variations in their bills and feet. The majority of North American hawks and eagles have broad, rounded wings, with the longest primaries usually being the third and fourth from the outermost, and most of these outer primaries are strongly narrowed by notching or emarginations near their tips. As a result, when the wings are spread they appear to be slotted, with each of the outer primaries able to act as a separate fingerlike structure, bending and twisting independently of the others and thereby responding in such a way as to minimize air turbulence and drag while presumably maximizing lift capabilities. As a result, it is the larger raptors having broad wings such as the *Buteo* hawks and the eagles, with maximum wing surface areas relative to body weight (or minimum wing loadings), that are most notable for their soaring abilities.

However, by evolving such soaring wings these birds have necessarily sacrificed any great potential for high-speed flight. Thus they are ill equipped for capturing elusive prey by extended chase, and must instead largely rely on surprise. Two other groups of North American hawks have evolved wings that superbly adapt them to swift flight and associated chases of such prey as birds and mammals. These are the accipiter hawks and the falcons. In these two groups the wing shapes are quite different (see Figure 61), but both allow for the attainment of great speed.

Wing shapes in the accipiter hawks (Figure 61, top) have distinctly rounded tips, somewhat like those of *Buteo* hawks but smaller and with well-developed notching on the outer primaries. Wingbeat rates are faster than in buteos, with interspersed gliding and little or no soaring. However, great bursts of speed are possible with such wings, and with their wings and their long tails the accipiters have tremendous maneuverability in flight, often dodging branches and tree trunks as they chase their prey under woodland canopies.

By comparison, the wing shapes of falcons (Figure 61, middle and bottom) are pointed, with the outer three primaries the longest and the secondaries quite short. Little notching or emargination occurs on the primaries, so wing slots are not evident in flight. Like the accipiters, falcons have high wing loading, so the wings are beat very rapidly with interspersed gliding; soaring is common only during migration. This high wing-loading characteristic probably assists in maximizing speed during steep glides, or stoops, during which the wings are held back so that their leading edges are nearly parallel to one another, and a near-vertical angle of descent is attained. There are strong arguments as to the maximum speeds attained during such dives, which at least in theory and according to some questionable estimates might reach about 80 meters per second before air resistance prevents any further acceleration (Brown, 1976b). The tails of falcons, like those of accipiters, are fairly long, which should facilitate high maneuverability, but at least the larger falcons tend to rely more on their blinding speed in open-country chases than on erratic maneuvering for capturing prey. This is not so true of the smaller and more highly ground-foraging falcons called kestrels, which not only are quite maneuverable but also are among the few North American hawks that are adept at extended and nearly stationary hovering as they orient into the wind prior to dropping down to capture prey on the ground directly below. Similar prolonged hovering is frequent in rough-legged hawks, and much briefer hovering is not uncommon in other hawks such as ospreys, which often hover for a few seconds before plunging down to water.

Other groups of North American hawks have rather different wing shapes. Those of ospreys and harriers tend to be quite long and rounded; at least in harriers the long wings, held in steep dihedral, appear to be designed for maintaining maximum stability during low-altitude flights, where ground-level turbulence and erratic winds might otherwise make flying difficult. The wings of some typical kites such as the American swallow-tailed kite appear to be designed for "kiting," a kind of nearly stationary gliding against the wind. These kites have apparently very light wing loading, allowing them remarkable aerial maneuverability in capturing aerial insects. The species's deeply cleft tail (Figure 62, top left and center) seems at first glance to be more a decorative species-specific display structure than associated with any needs for aerial maneuvering. However, Brown (1976b) suggested that it might actually function as a long supplementary wing flap, helping to provide extra lift at slow speeds as well as allowing for high maneuverability. Interestingly, a relatively distantly related kite in Africa, the African scissor-tailed kite (*Chelictinia riocourii*), has a similarly shaped tail and is apparently quite adept at catching insects in flight.

The legs and toes of hawks are rather less diverse than the wings, a reflection of the former performing very similar functions. Most hawks walk very little and have short legs, but the only North American ground-nesting hawk, the northern harrier, has relatively long legs. Harriers are probably also the only North American hawks with such specialized hearing that they are able to locate prey by sound alone, crashing down on it from above and pinning it to the ground in an owllike manner with their talons. They are the only North American hawks with owllike facial disks of feathers and very large ear openings (see Figure 68E). Their long legs also allow harriers to reach down and pluck prey out of rank vegetation (Brown, 1976b). The crested caracara has fairly long legs that perhaps are related to the fact that it often feeds terrestrially on carrion. Its partially bare face may also reflect carrion eating (a trend that reaches its maximum in the bare-headed New World and Old World vultures), since these feathers would only become matted and bacteria-ridden when in contact with rotting meat. The caracara's talons and bill are also much less raptorial in shape than those of the similar-sized osprey (see Figure 67C,D), and its toes and tarsi are considerably weaker as well.

The golden and bald eagles both have massive tarsi, short and powerfully grasping toes, and long talons. The talon of the hind toe is especially well developed in both species (see Figure 69). Brown (1976c) stated that the presence of a long and very strong hind talon is indicative of the eagle being able to kill larger animals than would be expected, and is used for piercing vital areas while the prey is held immobile by the front toes. The hind talon is also used by large falcons in an effective raking fashion when they strike prey in full flight, as described in the next chapter. In all the North American hawks except for the osprey and black-shouldered kite the talons are not rounded in cross section but instead are flattened or even concave or grooved on their undersides. It is not clear what advantages either shape would have over the other; it is possible that effective talon penetration is easier with grooved talons, although rounded talons would seemingly be stronger and more rigid. Talons tend to grow constantly and must be correspondingly worn down by regular use, although in some cases they may become distinctly blunted by excessive wear.

Unlike owls, hawks have well-developed crops in which they can temporarily store food that they have swallowed, and they have apparently more efficient digestive capabilities for dealing with bones than do owls. The result is that, although both owls and hawks regurgitate pellets of undigested materials, bones tend to make up a substantially smaller proportion of hawk than of owl pellets, and the total proportion of ingested foods that later appears as regurgitated pellets is also substantially smaller (Duke et al., 1975).

INTERSPECIFIC DIFFERENCES IN FOODS AND FORAGING ECOLOGY

It is a generally accepted truism that each species of raptor has evolved in such a way as to be able, by behavioral and morphological adaptations, to exploit a certain spectrum of prey more effectively than any of its coexisting competitors. Some species, such as the osprey for example, are highly restricted (stenophagous) in their diet and mode of prey capture. Others such as the crested caracara are extremely opportunistic and have broadly based (euryphagous) adaptations, shifting back and forth from scavenger to predator as conditions demand and taking a wide variety of animal foods as a result.

In theory at least, animals are able to reduce interspecies competition for food in several ways, including interspecific variations (1) in food diversity (variable degrees of foraging breadth, as just noted), (2) in the specific nature of foods of the same general type (such as snakes rather than lizards), (3) in the sizes of foods taken (specializing in a certain range of prey sizes), or (4) in foraging habitats (such as feeding on open-country rodents rather than woodland rodents). Additional foraging segregation may be attained by two species of populations foraging on the same prey but at different times (temporal foraging differences) or dividing a particular kind of habitat in various multidimensional ways (such as feeding on prey in smaller trees and bushes rather than in larger, canopy-level trees).

Jaksic (1985) has suggested that community studies of raptor populations are still in their infancy as compared with those of some other groups of predators, and urged that future comparative studies of food-niche relationships be done along several lines. According to Jaksic, measurements are required of the possibility that sympatric raptors use different hunting techniques; of different habitats used by sympatric raptors while hunting; and of relative hunting success of sympatric raptors in different habitats and using different hunting techniques. Finally, efforts should be made to determine the clues that sympatric raptors use when choosing hunting habitats (relative prey availability, vulnerability, size, etc.), although this last area would seem to be so difficult to measure as to be nearly impossible.

A large number of comparative ecological studies of foraging have been undertaken with raptors, of which only a few can be summarized here. The studies of raptor predation ecology in Michigan and Wyoming by the Craigheads (1956) provide one early example of using the comparative ecological approach in studying raptor biology in North America, while even earlier population studies were done in California (Dixon and Bond, 1937; Bond, 1939). More recently, Simmons (1986) studied ecological segregation of seven coexisting accipitrine hawks in southern Africa. Using a variety of criteria, Simmons concluded that there was a minimal amount of ecological overlap among these species. Similarly, Diamond (1985) found that seven species of accipiters in New Guinea could be separated by three ecological variables: habitat, foraging methods, and prey type.

Closer to home, Janes (1984a, 1984b) examined interspecific competition and territoriality in red-tailed and Swainson's hawks, while Schmutz (1977) and Cottrell (1981) both studied resource partitioning and competition among red-tailed, Swainson's, and ferruginous hawks. Bildstein (1987) found that among four raptors wintering in Ohio, the red-tailed hawk, northern harrier, and American kestrel all exhibited at least one major niche difference from one another and also from the rough-legged hawk. Hunting activity and dietary differences appeared to be more important niche dimensions among these species than were habitat differences, and temporal differences were least important. Smith and Murphy (1973) undertook an extensive ecological study of 12 species of raptors (including three owls and the northern raven) in desert scrub and arid woodlands of north-central Utah, Olendorff (1972, 1973) studied the population ecology of 10 raptor species (including two owls) in the grasslands of eastern Colorado, and White and Cade (1971) studied the ecology of four raptors (including the northern raven) nesting along cliffs of arctic Alaska.

The studies by Smith and Murphy are particularly useful, as they evaluated interspecific differences in breeding home ranges or territories, hunting activity patterns and habitat use, and food habits. They found that although intraspecific home ranges over-lapped little if at all, interspecific home range overlap was more common, especially among relatively unrelated raptors. Territorial size was largely correlated with raptor weight (see next chapter), and the authors doubted whether territoriality either limited or regulated the raptor populations, at least in areas of low popula-tion density. As to predation, all the large raptors of the area relied heavily on lagomorphs such as jackrabbits as prey. In general, average prey size correlated positively with average rap-tor weight, suggesting but not proving substantial foraging com-petition among raptors of similar weights, the birds apparently choosing the most abundant prey species and taking the largest prey animals that they could efficiently capture and kill.

In the Alaskan studies of White and Cade, a smaller number of breeding raptor species were present (gyrfalcon and peregrine falcon, rough-legged hawk, and northern raven). The authors found no evidence for competition, or even serious overlap, for available food resources among these four species. They believed the relative availability of ptarmigans to be the prime factor regulating the distribution and density of gyrfalcons, and to a lesser degree the density of the rough-legged hawk to be prey-dependent on rodents such as lemmings. However, they found no evidence that food availability influenced the density of breed-ing peregrines or ravens. They also doubted whether sexual dimorphism in any of these species had strong implications in reducing possible intraspecific foraging competition, but believed that some other explanation must account for this dimorphism.

In a recent ten-year study of the diets of three raptor popula-tions breeding in the Snake River canyon, Steenhof and Kochert (1988) found that during this period the red-tailed hawk con-sumed at least 69 prey species, which averaged 135 grams. The golden eagle consumed at least 65 species, which averaged 690 grams and consisted in large part of jackrabbits (*Lepus*). The prairie falcon consumed at least 64 species, which averaged 97 grams and consisted mostly of ground squirrels (*Spermophilus*). Of these three raptor species, the red-tailed hawk was the most diverse and flexible in its diet, and the prairie falcon the most specialized and inflexible. During a crash in jackrabbit and ground squirrel populations the dietary diversity of all three raptor spe-cies increased, but that of the red-tailed hawk did so most markedly and that of the prairie falcon the least. These results generally fit the "optimum diet" foraging model, namely that predator preferences are probably based on relative profitability of making a particular prey choice, rather than on a prey species's relative abundance. The generalist red-tailed hawk foraged within a fairly small, defended area around the nest, spent most of its time searching for rather than actively pursuing prey, and probably encountered a wide array of potential prey during its extensive movements. The more specialized prairie falcon had a greater hunting range and spent more of its hunting time in active pursuit of prey, and its major prey species was noncyclic. The golden eagle was intermediate in various of these traits and in its estimated dietary breadth index.

INTRASPECIFIC DIFFERENCES IN FOODS AND FORAGING ECOLOGY

The condition of reversed sexual dimorphism in raptors, in which females tend to be larger and heavier than males in contrast to the usual vertebrate situation, has attracted the attention of numerous biologists who have attempted to provide adaptive explanations for this phenomenon. Thus it has long been known that among raptorial birds (owls and hawks), the extent of such reversed sexual dimorphism is related to the degree of predatory behavior, with scavengers showing little or no dimorphism and the highest degree of dimorphism typical of the most highly effective predators, the accipiters and falcons.

One of the first persons to apply actual data to this question was Storer (1966), who examined the food habits of the three species of North American accipiters from the standpoint of the frequency of prey captures, grouped according to weight categories, typical of each sex in the three species. His results, summarized in Figure 7, show a remarkable tendency for females of each species to take larger-sized prey than males, and for the three species in turn to take a spectrum of prey species ranging in average weight from about 3 grams in the case of male sharp-shinned hawks to about 1,500 grams in the case of female northern goshawks. For each raptor the largest prey species approximated or even exceeded the average weight of their respective predator, but in general the average prey weight varied from about 12 to 50 percent of the adult predator weight. In general, the largest species (northern goshawk) preyed on the largest prey relative to its own body weight, and fed on a disproportionately broader spectrum of prey sizes than did the smaller ones. Storer attributed these results to an adaptation for reducing intraspecific competition by maximizing differential prey usage between the two sexes. At the same time, overall size differences among the three species are sufficient to restrict in-

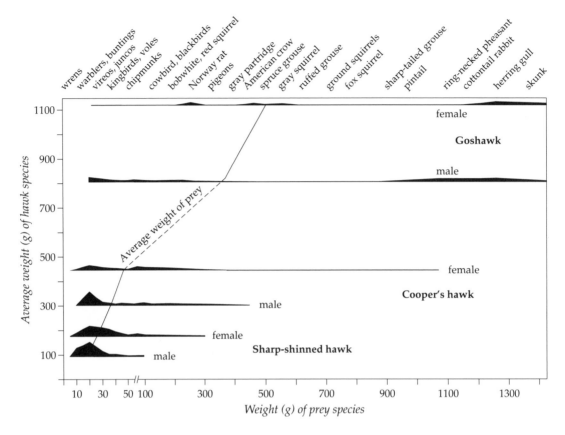

Figure 7. Relationship of average body weight in the North American accipiters to weight range of prey species and average prey weights. (Weight of each prey species based on midpoint of its weight range.) Thickness of lines represents proportion of total prey consumption for that species and sex category. Based on tabular data of Storer (1966).

terspecific competition for prey. Storer suggested that intraspecific competition is likely to be greatest among the migratory sharp-shinned hawk, where concentrations of wintering individuals might be common, whereas in the nonmigratory northern goshawk the advantages of sexual dimorphism might be least evident. At about the same time, Selander (1966) also suggested that selection for differential foraging niche utilization may account for the presence of sexual dimorphism in accipiters and falcons, although he lacked actual data to support this view. A few years later, Reynolds (1972) added a new wrinkle to this basic hypothesis by suggesting that a smaller male accipiter hawk may be better able to feed himself, his mate, and the young through the breeding season inasmuch as a smaller raptor may encounter larger numbers of smaller and more accessible prey, and further-more would himself have reduced metabolic needs from this more efficient mode of hunting. Recently, Meyer (1987) concluded that sexual dimorphism in the sharp-shinned hawk was not evolved to provide separate foraging niches.

Perhaps the best illustration of interspecific differences in raptor foraging niches comes from the work of Snyder and Wiley (1976), who compared the food habits of most North American raptors (as measured by percentage of food items taken by the species, rather than biomass estimates, using Fish and Wildlife Service stomach analysis data and literature sources) with the relative degree of reversed sexual dimorphism in adults (ex-pressed as a mean dimorphism index based on average wing, culmen, and weight measurements of the two sexes). Their data are summarized in Figure 8, with certain groups of species classified collectively as invertebrate consumers (those species whose foods numerically include more than 50 percent inverte-brates), "pursuers" (those species that spend relatively great amounts of energy in immediate prey capture activities), and bird catchers (those species having more than a third of their diets comprised of birds). It may be seen that the species that have the highest degree of reversed sexual dimorphism are indeed the bird-catching accipiters and the larger falcons, and that some other exclusively vertebrate eaters such as black-shouldered kites and ospreys exhibit almost no sexual dimorphism. Although the species classified by Snyder and Wiley as "pursuers" (as opposed to "searchers" or those with "mixed" hunting styles) do tend to exhibit considerable dimorphism, the same is true of such searchers as the zone-tailed hawk and northern harrier. Snyder and Wiley basically argued that the degree of sexual dimorphism in raptors is most closely correlated with the regularity with which a species may be stressed by food shortage during the latter part of the breeding season, when both sexes are foraging and when prey might be limited, especially for bird-hunting species. At such times sexual dimorphism may favor more effi-cient food capture of different-sized prey, in the view of these authors. They further suggested that greater numbers of prey are not necessarily available to smaller males as was argued by Reynolds, but instead suggested that it might be energetically advantageous for females to hunt close to the nest if the costs of transporting heavy prey to the nest were substantial, while the smaller and more mobile males could presumably hunt at greater distances from the nest. They also suggested that larger females are probably superior in their ability to cover and provide warmth to eggs and young.

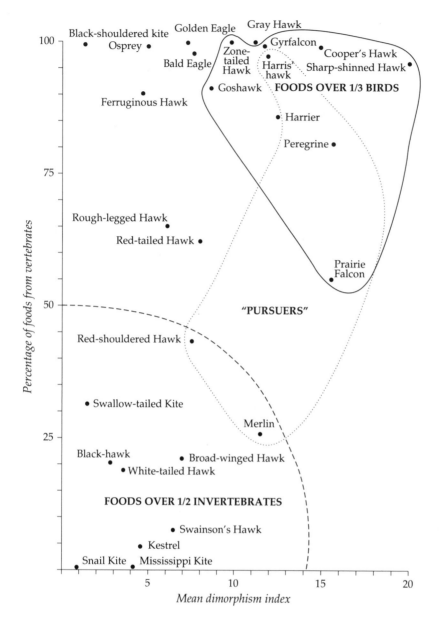

Figure 8. Relationship of sexual dimorphism to diet (stomach content analyses) in North American falconiforms, based on tabular data of Snyder and Wiley (1976).

Amadon (1975) addressed the reversed sexual dimorphism in raptors using information relative to two basic types of theoretical explanations. The first of these is basically the ecological explanation associated with foraging efficiency that was proposed by Storer, Selander, and Reynolds, and generally also supported by

Snyder and Wiley. It involves the idea that the two sexes of a dimorphic species can exploit differential foraging niches, thereby reducing intraspecific competition between the sexes while simultaneously maximizing the species's overall foraging niche. Amadon cautioned that in doing so, a species's possibilities of encountering interspecific competition are increased, and suggested that it may be more difficult for such a sexually dimorphic species to adjust to changing food resources or levels of interspecific competition.

For such reasons, Amadon urged that an alternative and more strictly behavioral explanation should be sought for the reversed sexual dimorphism in raptors. Specifically, he believed that the relationships of the sexes during pair formation and perhaps later on in the reproductive period might help explain this phenomenon. Related to this idea is the possibility that small male mates might pose less of a direct threat to the young than larger ones, and thereby might be favored by females as mates. This idea Amadon rejected as unrealistic, as he did the ideas that a larger female is better able to protect her eggs and young from other predators, or that smaller males are better at aerial maneuvers than larger ones and thereby more effective during aerial display. Finally, Amadon proposed that females are more likely to select smaller males as mates because they represent less of a threat to the females during this period and can be more easily dominated. This idea, originally suggested by Cade (1960) for the gyrfalcon and peregrine falcon, was favored by Mueller and Meyer (1985) in a recent review, who believed such dimorphism might facilitate the formation and maintenance of pair bonds; but Snyder and Wiley (1976) suggested that female domination may simply be a secondary effect rather than a cause of sexual dimorphism, and believed that more "modest" behavioral signals might achieve the same effect of sexual harmony.

In a general review of the evolution of size dimorphism in birds, Jehl and Murray (1983) commented on this problem as it relates to hawks. They believed that neither ecological segregation explanations involving food niche competition nor those that revolve around some reproductive advantage to the females (such as female domination, ability to store more energy resources, surviving cold better, etc.) offer fully acceptable answers to the question. They suggested that the possibility that small males have advantages of agility in aerial combat or are otherwise selectively chosen by females for mates should receive further attention before being rejected. In line with this view, Olsen and Olsen (1986) suggested that reproductively fit males should be considered a scarce resource for females, and that females must thus compete for experienced males holding favorable territories and nest sites. Competition for such males is greater among those species that consume more elusive prey such as birds, and since larger females probably are able to dominate rivals, increased size in females would be selected.

Temeles (1985) has recently revived the general ecological explanation involving the role of intersexual food competition and food division, by showing that average hunting success rates are highest among falconiform species that consume invertebrates, next highest for fish eaters, lower for mammal eaters, and lowest for bird eaters. This is of course exactly the opposite of the trend toward dimorphism, and Temeles suggests that among this series bird-eating raptors have the lowest net energy intake per unit of foraging time. The correspondingly reduced level of prey availability (rather than general prey abundance) selects for reduced intersexual food competition by adaptations favoring the use of different prey size classes in the two sexes. Other arguments such as those already made relative to the advantages of larger females make reversed rather than normal sexual dimorphism the most effective way of achieving this end.

Most recently, Pleasants and Pleasants (1988) analyzed the possible mechanism by which reversed sexual dimorphism in raptors might have evolved, based on the energetic costs of producing a full clutch of eggs. They concluded that at least in falconiform raptors this was achieved by the female becoming larger while still retaining her original egg size, rather than by the male becoming smaller. If true, ecological explanations such as the male becoming smaller in order to gain in aerial agility and/or increase food niche differences from the female become less attractive than the possibility that females became larger in order to produce a larger clutch without increasing the physiological cost of doing so (by having a relatively lower unit cost per egg), or in order to achieve the behavioral advantages of social dominance over males, the idea primarily favored by Mueller and Meyer (1985).

It is apparent that this fundamental question has not been answered to the satisfaction of all ornithologists, and indeed a single inclusive answer to the question of reversed sexual dimorphism in raptors may not be the most appropriate solution to anticipate. Let it suffice to say that the study of foraging ecology in hawks is still indeed an actively growing and changing one, and viewpoints firmly held today may well be discarded tomorrow as new information becomes available.

CHAPTER 3

Comparative Behavior

The behavior of higher animals such as birds consists of activities
that are performed to help assure the survival of the individual
itself, such as eating, drinking, and various other self-mainte-
nance activities, including "comfort" activities such as preening
(egocentric behavior); those activities that may be directed to self-
survival but that tend to bring different individuals into chance
contact with one another, such as exploratory activities, migratory
movements, or foraging searches (quasi-social behavior); and
those activities that are specifically directed toward others of the
same or different species, such as aggression, territoriality,
courtship, nesting behavior, and parental care (eusocial behavior).
A few of these activities as they relate to the comparative biology
of hawks will be discussed in this chapter; more detailed descriptions of foraging behaviors and breeding activities can be found in
the individual species accounts.

FORAGING AND PREDATORY BEHAVIOR

As predators, a central aspect of individual fitness for raptors is
their effectiveness in finding and capturing enough food to permit

not only survival but also reproduction. In the previous chapter general aspects of food habits, foraging behavior, and foraging ecology that tend to reduce interspecific and intersexual competition were discussed; specific behavioral aspects of predation by various species or groups will be briefly discussed here.

The prey of various North American raptors ranges in size from insects to relatively large mammals such as jackrabbits and even newborn antelopes. Thus of course the techniques and difficulty of prey capture varies enormously among species. One measure of the difficulty of prey capture is to estimate the hunting success rate, or the percentage of individual hunting sorties that result in the successful capture of prey. A great deal of comparative data on hunting success in falconiform birds has been assembled by Temeles (1985), who found that hunting success rates tended to be highest for hawk species that consume invertebrate prey, and lowest among such bird-hunting raptors as accipiters and the larger falcons. Thus, among six observations cited by him for American kestrels, four hunting samples mainly involved insect or invertebrate prey (including two prey samples that also contained lizards) and had success rates of 39 to 90 percent, while the two hunting samples involving mainly vertebrate prey had success rates of 23 and 33 percent. Hunting success rates for fish-eating hawks as exemplified by the osprey are apparently rather variable, ranging in 15 samples from 19 to 91 percent, with an unweighted collective average of about 39 percent. Various mammal-hunting specialists such as the golden eagle, red-tailed hawk, ferruginous hawk, and black-shouldered kite had success rates of from 14 to 89 percent, averaging about 30 percent. Among bird hunters, nine studies of the peregrine falcon indicated hunting success rates of from 7.5 to 84 percent, the unweighted collective average being 19.5 percent. Smaller falcons such as the merlin showed even smaller average success rates, as did accipiters such as the northern goshawk and the Eurasian sparrowhawk (A. nisus).

Clearly, the effective capture and killing of vertebrate prey such as fish, mammals, and birds is a highly developed skill, and brief descriptions of these highly specialized modes of hunting may be warranted. One of the most easily observed of these techniques is that of the osprey, which can often be observed foraging along coastlines and rivers or lakes throughout large portions of North America. Fish are captured at shallow depths (up to about a meter deep) after a dive that varies in distance from as few as about 5 meters to more than 70 meters. Although such dives are sometimes initiated from a perch, the birds typically actively search for prey by a mixture of flapping, gliding,

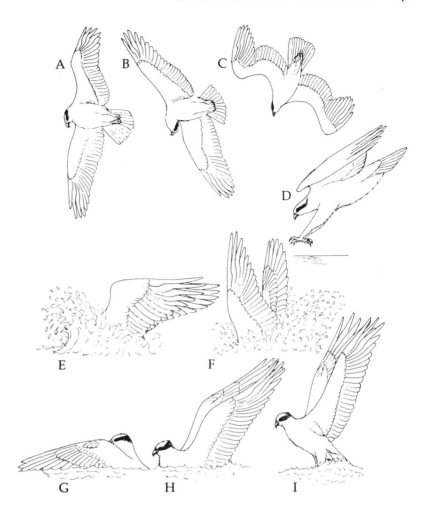

Figure 9. Stages in diving and fish catching by the osprey, based on a photographic sequence.

and hovering activities (Figure 9A,B), depending upon the speed of the wind. Often, after an initial sighting, a brief period of hovering is done, as if jockeying for position, followed by a dive (Figure 9C) that usually ranges in angle from 45 degrees to vertical. However, some dives are nearly horizontally oriented, and many are checked part-way down. The dive may thus be made in a series of steps and occasionally is abandoned altogether, probably as a result of prey movement. At the final stage of the dive the two feet are thrust forward and the wings held directly back (Figure 9D), and the bird may nearly disappear in a spray of water (Figure 9E,F). Several seconds are required for the bird to become airborne again, even if no prey is captured. This is done by a series of downward and forward thrusts of the wings, which do not however push directly against the water surface to help facilitate take-off (Figure 9G–I). Captured fish are typically carried head-forward and may be shifted to that position if necessary, but sometimes are carried as they have been captured, such as backward or even sideways (Cramp and Simmons, 1980).

Prey capture by accipiters and falcons occurs very rapidly and is far more difficult to observe in nature. However, detailed analysis of this behavior has been done by Goslow (1971) with the aid of high-speed photography, and his work offers new insights into these activities. Goslow studied the striking behavior of the Cooper's hawk, northern goshawk, red-tailed hawk, prairie falcon, peregrine falcon, and American kestrel, which thus collectively provide examples of the basic prey-catching techniques used by *Accipiter, Buteo,* and *Falco.*

The striking behavior of the northern goshawk (Figure 10A–C) consists of a flapping approach to the prey until the bird is about 9 meters away, when the wings are set into a gliding position. When about two meters away the feet are lowered, with the toes still partly flexed (Figure 10A). At the time of impact the legs are substantially extended and well out in front and the toes fully extended, while the wings and tail are suddenly placed into a maximal braking position (Figure 10C). At impact foot velocities of 2,250 centimeters per second (50 mph) were recorded. A kneading action of the talons is immediately used to kill the prey. The striking behavior of the Cooper's hawk is apparently very similar to that of the goshawk, but measured foot velocities were only about half as great.

The strikes Goslow observed by the red-tailed hawk were similar to those of the accipiters (Figure 10D–F), but occurred at considerably slower speeds. The birds set their wings at about 4 meters from their prey, and began lowering their legs when about 3 meters away (Figure 10D). The measured foot velocity on impact was only 650 centimeters per second (12 mph). Immediately after impact the toes were tightly closed, and braking was done by thrusting the wings forward and, in some cases, by dropping the heels and using the tarsometatarsus as a lever to push against (Figure 10F).

The strikes of the prairie and peregrine falcons observed by Goslow were made on flying pigeons or swinging lures. When attacking flying prey in level pursuit, the falcons continued to beat their wings until they were within a meter or so of the prey.

Figure 10. Stages in aerial predation by a northern goshawk (A–C, after Goslow, 1971), and predation on terrestrial prey by a red-tailed hawk (D–F; D and E after photos in Grossman and Hamlet, 1964; F after Goslow, 1971).

Their feet were moved forward in the last instant prior to contact, when the toes were fully extended and the wings flexed upwardly (Figure 11B). At least in one case, the strike was made with only one foot, and typically the bird would cling to the prey after contact and carry it to the ground. When attacking from a stoop (Figure 11A), the strikes were of extremely short duration, as the birds did not retain their grip on the prey. The falcon's postimpact behavior was usually a short semicircular flight back to the prey, which if sufficiently stunned fell to the ground by itself but otherwise might be attacked a second time. Upon impact the falcon's toes were immediately closed, with the hind talons optimally directed for piercing and tearing, but the contact times in two measured cases were only 30 and 100 milliseconds. The peregrine's body velocity at the time of impact was judged to be about 18 meters per second or slightly above 40 mph. This is a seemingly very slow speed considering the numerous estimates of stoop speeds of from three to five times greater than this. However, Alerstam's (1987) estimates of average (25 meters per

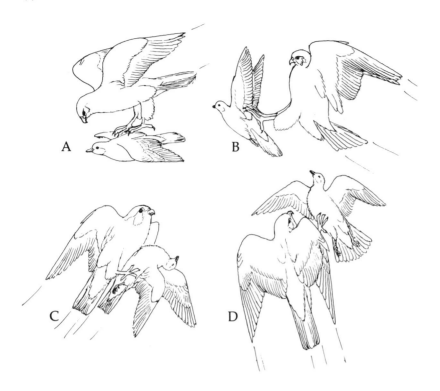

Figure 11. Methods of aerial attack by the peregrine falcon (A and B after Goslow, 1971; C and D after Glutz, Bauer, and Bezzel, 1971).

second) and maximum (31–39 meters per second) speeds attained during the peregrine's stoop suggest that these have often been greatly overestimated.

When a hawk has subdued its prey on the ground it typically "mantles" in a distinctive wing-spreading posture that is already well developed even in fledgling falcons (Figure 12C). The mantling posture probably both helps to prevent escape or further struggling by the prey, and also effectively hides it from the view of others. A very similar wing-spreading and hiding posture (Figure 13, right) is sometimes used by brooding females when disturbed at the nest. The hawk may then begin to prepare its prey for consumption by plucking out some of its larger feathers (if the prey is a bird), or may carry it to a favorite "plucking post" for this purpose (Figure 12D). When adult hawks and eagles feed young nestlings they typically tear the prey up into quite small pieces, and pass these items to the young bill to bill, typically turning the head to one side, presumably to make the food easier for the young bird to extract (Figure 13, left). Fledglings may

Figure 12. Food-related behavior of young peregrines, including food solicitation posture with calling (A), defense of food (B), mantling of prey (C), and normal eating posture (D) (after sketches by Sherrod, 1983).

continue to solicit food even after leaving the nest; young peregrine falcons do so by uttering a loud screaming or whining call. Such food solicitation is typically done while in a standing posture, with or without the back feathers variably fluffed, the wings drooped, and the tail somewhat spread (Figure 12A). Fledgling

Figure 13. Parental feeding of young by a golden eagle (left), and protective mantling of young by a brooding female northern goshawk (right) (after photos in Bailey and Niedrach, 1965).

peregrines may also perform aggressive face-offs when defending their food, flaring their body and head feathers and extending their wings and often one foot toward the opponent (Figure 12B). This posture sometimes is accompanied by hissing sounds.

MIGRATORY BEHAVIOR AND MIGRATORY MOVEMENTS

Following the breeding season, many hawk species undergo rather substantial movements, with young birds even of resident species tending to spread out and attempting to establish their own hunting territories, and both adults and young of migratory species beginning to undertake regular migrations that may range from a few hundred kilometers to as much as several thousand kilometers. Among the most spectacular migrations are those of the broad-winged hawk, of which virtually the entire North American population performs a transequatorial migration to South America. Similar transequatorial migrations are undertaken by the Swainson's hawk; one banded Swainson's hawk migrated from Saskatchewan to the Argentine pampas in no more than 126 days, a distance of some 11,500 kilometers, or about 90 kilometers per day (Houston, 1974). The arctic-breeding races of the per-egrine falcon likewise regularly winter as far south as southern South America, and one bird banded as a nestling in Canada's Northwest Territories was recovered in Argentina only 174 days later, representing a total migratory movement of about 15,000 kilometers, or about 86 kilometers per day (Kuyt, 1967). Several kites (swallow-tailed, Mississippi, and snail kites) also undertake migrations to Latin America, although these are not nearly so well understood and documented.

The degree of current interest in hawk migration is suggested by a recent book by Heintzelman (1986), which has nearly 2,000 literature citations and lists several hundred suggested locations for observing hawk migrations in the United States and Canada. In a review of hawk migration in eastern North America, Haugh (1972) classified the broad-winged hawk and osprey as long-distance migrants, the merlin and peregrine as medium- to long-distance migrants, the bald eagle, northern harrier, and sharp-shinned, Cooper's, red-shouldered, and rough-legged hawks as medium-distance migrants, and the northern goshawk, red-tailed hawk, and golden eagle as short- to medium-distance migrants. The American kestrel was considered a short- to long-distance migrant.

Haugh (1972) also provided what are perhaps the best general fall migration schedules for the commoner species of hawks in eastern North America (Figures 14, 15), as well as spring migration schedules for hawks in the same general area (Haugh and Cade, 1966). A large number of additional regional studies for both spring and fall migrations of hawks have been cited by Heintzelman (1986). These studies suggest that hawks tend to have fairly prolonged migrations, sometimes with broad or irreg-

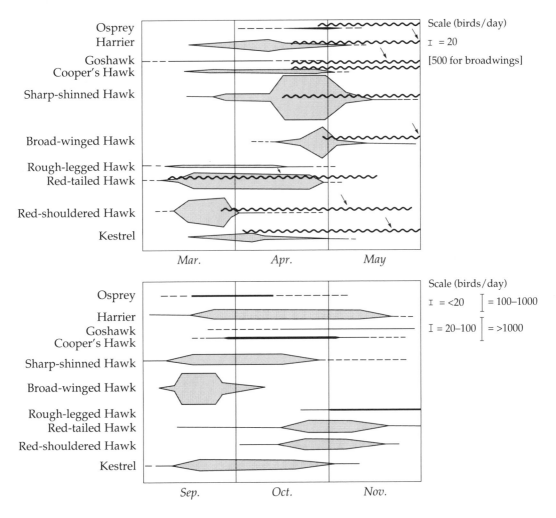

Figure 14. Spring (above, after Haugh and Cade, 1966) and fall (below, after Haugh, 1972) migration schedules observed in the vicinity of Lake Ontario, northern New York. Also shown (wavy lines) are the spread of available egg records for New York state, and (arrows) earliest reported hatching dates (Bull, 1974). The stippled area shown for each species is proportional to its observed relative migratory abundance, with the broad-winged hawk curve drawn at a scale 1/25th that of the others.

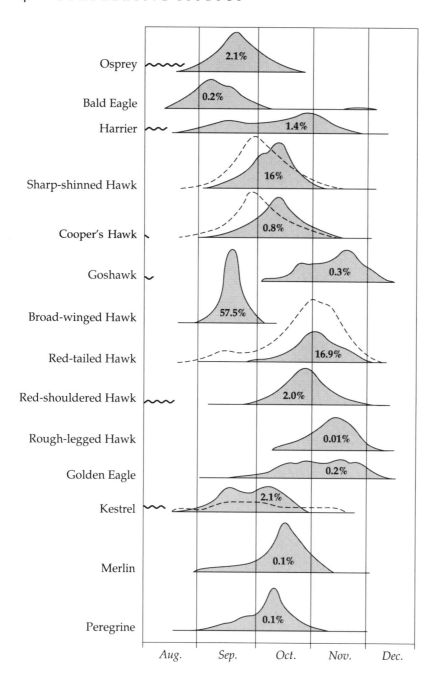

Osprey ~~~ 2.1%

Bald Eagle 0.2%

Harrier ~~~ 1.4%

Sharp-shinned Hawk 16%

Cooper's Hawk 0.8%

Goshawk 0.3%

Broad-winged Hawk 57.5%

Red-tailed Hawk 16.9%

Red-shouldered Hawk ~~~ 2.0%

Rough-legged Hawk 0.01%

Golden Eagle 0.2%

Kestrel ~~~ 2.1%

Merlin 0.1%

Peregrine 0.1%

Aug.   Sep.   Oct.   Nov.   Dec.

Figure 15. Fall migration schedules observed at Hawk Mountain, Pennsylvania (after Haugh, 1972), with the wavy lines indicating the last part of each species's breeding season (span of nestling dates) for New York state (Bull, 1974). The stippled areas represent approximate relative seasonal abundance, and the percentages shown within these areas indicate the overall fall abundance of each species relative to the total sample size for all species. The dotted lines indicate comparable smoothed seasonal abundance curves of some species observed at Point Diablo, California (Binford, 1979).

ular peak periods. The conspicuous exception to this generality is the broad-winged hawk, which, although much the most common migrant hawk in eastern North America, has perhaps the most restricted migration schedule. Like the other long-distant migrant, the osprey, the broad-winged hawk begins its migration relatively early.

Unfortunately too few broad-winged hawks have been banded to provide a clear picture of that species's entire migration pattern, but in some areas such as along the Gulf Coast of Mexico and the Panamanian isthmus it is possible to gain some idea of its movements. Thus, Thiollay (1980) estimated that over 200,000 of these hawks passed through eastern Mexico during 23 days in the spring of 1978, or a substantial part of the estimated million birds that comprise the total North American population. Smith (1985) reported annual fall counts of from 42,000 to 401,000 broad-winged hawks during nine years of study, with most of these totals being based on actual counts of individuals rather than determined from extrapolated samples. Similarly, from 77,000 to 344,000 Swainson's hawks were counted during this 9-year period, as were 2,800 to 27,400 Mississippi kites during three years of study.

Smith estimated that a typical total one-way route of a Swainson's hawk migrating between central Montana and southern Argentina would represent a flight of about 10,700 kilometers, while that of the broad-winged hawk flying from New Brunswick to northern Peru would be about 7,900 kilometers. Remarkably, these long-distance flights are apparently made without any foraging along the migratory route, which is completed during an approximate two-month period. Smith judged that an average Swainson's hawk weighing 870 grams (including about 270 grams of fat) would have no difficulty in flying from the U.S.-Mexican border to Argentina while living on stored fat alone. He estimated on the basis of personal observations that the birds typically migrate at elevations of from 270 to 2,650 meters above ground, even though the ground elevation may itself be as high as 3,500 meters above sea level. Apparently they do not fly at night, but during daylight hours the migrating birds actively seek out areas of thermal lifts or other updrafts. Localized areas of such updrafts, rather than social attraction, are probably the reason for migratory assemblages or "kettles" of hawks. By using such outside energy sources the birds maximize their own energy conservation, and perhaps flying at high altitudes also helps reduce water loss because of decreasing water loss rates with decreasing external temperatures.

In similar radar and visual observations on migrating hawks during spring in Texas, Kerlinger and Gauthreaux (1985a, 1985b) judged that these birds migrated at lower altitudes (400 to 1,100 meters) than those estimated by Smith, and suggested that flocking by migrants probably is adaptive inasmuch as it may improve their efficiency of navigation and orientation, or the ease with which birds are able to locate thermals.

TERRITORIAL AND COURTSHIP BEHAVIOR

In contrast to some other avian groups such as ducks, for example, where pair-forming or "courtship" activities are unrelated to territoriality, it is difficult if not impossible to separate these two activities in raptors such as hawks. As with many birds, pair bonding is a subtle and apparently prolonged process that is difficult if not impossible to study under natural conditions, and is probably necessarily so altered by captivity as to make its study under such circumstances of limited value. Cade (1960) believed that the normal sequence of events during pair formation in peregrines consists of mutual attraction of potential mates, mutual roosting on the nesting cliff, cooperative hunting, courtship flights, "familiarities" such as mutual preening, billing, and nibbling, courtship feeding of the female by the male, copulation, and finally nest scraping. Although the details of this sequence must of course be modified for species that construct their nests rather than use ready-made ledges or cavities, this general sequence is perhaps applicable to a variety of North American hawks. Similarly, Newton (1986) described the usual sequence in the Eurasian sparrowhawk as beginning with attraction of sexually mature birds to potential nesting places and of potential mates to one another, followed by mutual roosting, mutual calling, and aerial display, these in turn by courtship feeding, nest site inspection and stick carrying, nest building, and finally by copulation.

Probably most if not all of the species covered in this book establish breeding territories prior to the start of pair-forming behavior, and indeed it is most likely that failure to establish a territory or be accepted into one by a potential mate will normally effectively prevent an individual from further reproductive activity. Probably most raptor breeding territories are of the inclusive type that consist of relatively large areas suitable for nesting, foraging, and mating activities required to complete a breeding season. However, in some areas of restricted nest sites (such as along the Snake River Canyon or various cliff-edged arctic rivers, for example), defended territories may be rather narrowly restricted to the vicinity of the nest itself, with commonly exploited foraging areas not subjected to serious aggressive contest. Thus, perhaps the rarest resources of the territory (the nest site itself in the case of ledge- or hole-nesting species, and probably also the mate) are likely to provide the center of territorial defense, with progressively less amounts of time and energy spent in defending areas of less vital concern to the pair. This diminution of overt territorial defense with distance often makes it very difficult to differentiate between the breeding territories and home ranges of raptors.

In many North American raptors the territory is occupied continuously once it has been established, but in other migratory forms it may be abandoned for varying lengths of time. Probably in most raptors it is the male that is responsible for holding the territory between breeding seasons, or for reestablishing it each spring in the case of more mobile species. Thus, males are more likely to remain in the territorial area during winter, as in the case of gyrfalcons and peregrines, or return to it sooner in spring, as has been noted in merlins, American kestrels, and Swainson's hawks among North American species (Newton, 1979). However, among accipiters it may be the females that return first in spring, and in these species the females often take a very active role in territorial advertisement and defense. At least in the Eurasian sparrowhawk, each member of the resident territorial pair specifically challenges others of its own sex that might invade their common territory (Newton, 1986).

With the establishment or reestablishment of his territory, the male is likely to begin a period of active advertisement of territorial ownership by a variety of aerial displays as well as conspicuous perching, calling, and possibly other kinds of long-distance signaling. He may also begin nest site searches or inspections in the case of falcons, or initiate nest building or nest repairs in the case of other raptors that construct elaborate nests. The male's aerial display activities typically are centered over the nest site or potential site, and probably not only serve to keep other males away from the area but may attract the attention of unmated females.

Aerial displays may simultaneously or sequentially serve for territorial advertisement, mate attraction, and repulsion of intruders, and thus it is often difficult to interpret the meaning of particular instances of such displays that might be observed. Brown and Amadon (1968) provide an extended discussion of nuptial displays, and the accompanying descriptions and drawings (Figure 16) of aerial displays are largely based on their accounts. Prominent perching and calling is perhaps the simplest form of advertisement, and is especially common in woodland species. It is a short step to aerial display, with the bird calling as it soars above the territory. Such soaring and calling is most often done by the male, but also at times is done by the female of an established pair. In all cases, undulating flight display probably serves as much a function of aggressive territorial signaling as of sexual attraction (Harmata, 1982), and thus may be continued well beyond the period of actual pair formation.

In many species soaring and calling is variously elaborated by incorporating undulating or diving movements, perhaps best collectively termed "sky dancing." These may consist of shallow undulations (Figure 16A), deeper swooping undulations (Figure 16B), a series of interrupted shallow dives and sharp gliding rises, or "pothooks" (Figure 16C), or taking various more elaborate routes such as a pendulumlike series of horizontally oriented figure-eights or circles, as in some "booted" eagles, or even a series of repeated vertical loop-the-loop maneuvers that nearly reach the ground at their lowest points, as occur in various harriers.

At times these just-described displays may be performed by a pair in tandem, with one bird following the other (mutual sky dancing), and often the pair may specifically interact in the air with special maneuvers. A common example of this is for one of the birds (often the male) to approach the other from above and behind and dive down toward it, while the latter continues to soar in level flight. As the diving bird reaches the lower one, the lower turns over on its back and presents its talons, while the diver likewise lowers its own feet briefly before swooping up again (Figure 16D). A variant of this is "parachuting" (Figure 16E), during which the upper bird, typically a male, descends toward the other in a gentle glide while calling and holding its wings in a nearly vertical position. Again, the lower bird flips over and presents its talons momentarily to the other. The aerial displays of the falcons are basically very similar to those of the Accipitridae, with the addition of various kinds of high-speed chases between the pair, more breathtaking dives, and perhaps some unique and distinctly ritualized types of landing behavior by the male (see Figure 53).

An extension of talon presentation is talon locking (Figure 16F), in which the two birds firmly grip one another's toes as the male descends to the female from above. In species such as the *Haliaeetus* eagles (Brown, 1976c) and various falcons (Treleaven, 1980) the birds may continue to grasp each other with their

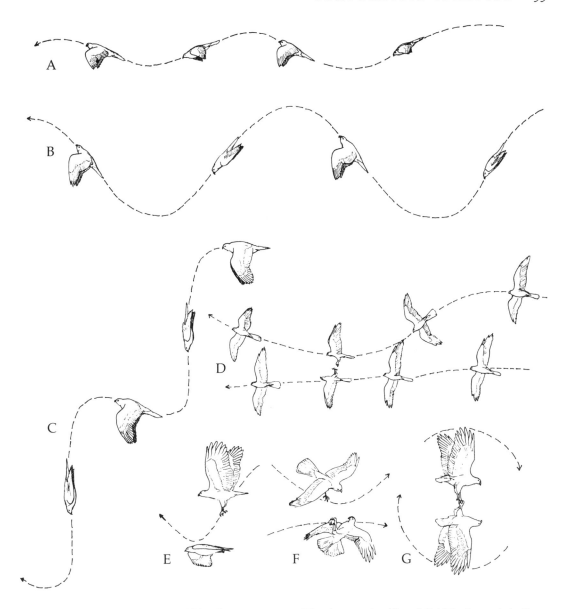

Figure 16. Aerial displays of accipitrid falconiforms, including shallow (A) and deep (B) undulating flight or "sky dancing," "pothook" dives and swoops (C), mutual soaring, diving and foot touching (D), "parachuting" (E), talon grasping (F), and "whirling" or cartwheeling (G). After Brown and Amadon (1968), except for F, which is after Cramp and Simmons (1980). See also Figure 53 for aerial displays of falcons.

talons, and the pair spectacularly cartwheels downward with wings fully outstretched, sometimes remaining thus locked together until they have nearly reached the ground (Figure 16G). This kind of activity is extremely similar to if not identical with actual aerial combat, and perhaps should simply be considered a ritualized version of it when it is performed by pairs.

　　Closely similar to talon presentation is the aerial transfer of food by paired birds. After a pair has formed, but before the female has begun intensive incubation, she typically is provided food by her mate. In some cases he may bring the food to her as she is perched in a tree or at the nesting site, but in others she

Figure 17. Courtship and copulation behavior of falconiforms, including prey transfer in Circus (A, after Cramp and Simmons, 1980), prey presentation in peregrine falcon (B, after Cramp and Simmons, 1980); copulation in Accipiter (C, after Newton, 1986), and copulation in Falco (D, after Cramp and Simmons, 1980).

may fly out to meet him and accept the food by an aerial transfer. Like talon presentation, this requires split-second timing to be achieved, and indeed talon presentation may perhaps serve as a kind of "rehearsal" activity for aerial food transfer behavior. Often the two birds pass close by in flight, the male above the female, and when directly above her he simply drops his cargo, the female neatly catching it in midair below (Figure 17A). In at least some falcons such as peregrines, prairie falcons, and American kestrels the female may also perform a food-begging flight that is very similar to the type of flight made by newly fledged birds as they fly toward their food-carrying parents (see Figures 56A, 57D) (Cade, 1960).

Aerial transfer of foods may be supplemented with or replaced by transfers between the two birds at a perch, or in the case of falcons typically at the nest site itself. These food presentation ceremonies are much more highly developed in the falcons than in the Accipitridae, and typically are stimulated by or accompanied with distinctive postures and calls (Figure 17B). The male may similarly call to attract the female's attention when he is approaching with food. Male falcons of at least the cliff-nesting species perform nest ledge displays that may involve calling, foot scraping, or prey-plucking behavior that apparently serve to advertise the locations of possible nest sites to mates or potential mates. In hole-nesting falcons such as kestrels similar male behavior occurs, during which the male attempts to attract the female to the nest site with special postures and calls, often while holding prey in his bill (see Figure 51C,D). The male may also perform stereotyped inspection behavior at the site entrance, or go inside and call softly. Either or both sexes may perform nest-scraping or nest-molding movements while inspecting the nest cavity (Cramp and Simmons, 1980).

It is generally difficult if not impossible to judge just how long may be required to establish a firm pair bond between two birds under normal circumstances, but at least in the case of mate replacement during the breeding season this sometimes can occur with astonishing rapidity. Olendorff (1971) has reviewed this general question, and found evidence in the literature for extremely rapid mate replacement in four species of falcons (prairie and peregrine falcons, American and Eurasian kestrels) but no such cases for accipitrids. Two of these cases involved replacement of mates (the male in one case, the female in the other) among kestrels that required a few hours at most, while two other cases involved mate replacement (again involving both sexes) within 40 hours.

It is also difficult to evaluate the importance of mutual preening, mutual billing, and other contact behavior between mates as a pair-forming or pair-maintaining mechanism, but it probably should not be underestimated. Similarly, duetting calls and mutual defense of the territory may help to synchronize reproductive cycles. Such mutual behavior perhaps helps to counteract aggressive tendencies on the part of either sex, but especially the larger female. Cade (1960) was of the opinion that pair bonding in the large falcons results only when the female is clearly dominant to the male, and the male in turn adjusts to his subordinate role. However, one might perhaps also argue that pair bonding will only occur when the male is "secure" or "comfortable" when near the larger female, and when she has fully inhibited any aggressive tendencies that might be present on her part. It is clearly dangerous to read too much into a bird's behavior based on human experience, and such speculation on the basis for pair bonding in any species is perhaps foolhardy in the extreme.

In any case, at some stage during pair formation copulatory activities begin. In some species, such as American kestrels, copulation may both begin well before and terminate well after the egg-laying period, suggesting that it may serve as a pair-bonding mechanism. Indeed, in species like the northern goshawk, where copulations may occur hundreds of times during a single breeding season (Moller, 1987), such a function would seem to be certain. However, in most if not all species copulation reaches a peak around the time of egg laying, and usually tapers off rather rapidly thereafter.

In the North American hawks, copulation is often achieved with little or no obvious preliminary display, and usually also with a lack of obvious ritualized activities thereafter. In many cases the female indicates her receptivity by assuming a generally horizontal position, often with the wings slightly extended and the tail occasionally raised or spread, and sometimes supplementing this posture with calling. There seems to be little difference in the solicitation postures of accipitrid hawks and falcons (Figure 18). At least in some cases the female's soliciting calls and posturing are reminiscent of those of an older nestling begging for food. Among cliff-nesting falcons copulation often occurs on the nesting ledge itself, while in hole-nesting kestrels it may occur at the nest site, on nearby branches or buildings, or rarely even on the ground. Likewise in ospreys copulation typically occurs on the nest or on nearby branches. In the accipitrids copulation is often quite varied in its location, and although in many instances it may occur at or near the nest, in some species it never occurs at

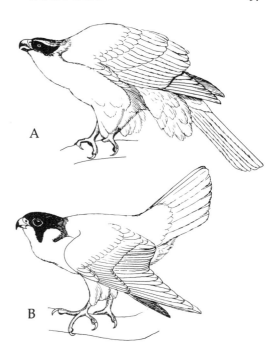

Figure 18. Copulation solicitation positions in Accipiter (A) *and* Falco (B) *(both after Glutz, Bauer, and Bezzel, 1971).*

the nest. In this group copulation often follows aerial display, and sometimes mounting occurs immediately after the male has landed after such aerial display. There are even a few reports of apparent copulation attempts by flying birds among various hawks and falcons (Cramp and Simmons, 1980).

With the successful completion of pair bonding, the pair is ready to begin or perhaps complete nest building. Alternatively, in the case of falcons, which do not need to build an elaborate nest, the pair may be ready to initiate egg laying. These topics will be considered in the following chapter, which deals specifically with reproductive biology.

CHAPTER 4

Reproductive Biology

The breeding season is here considered for the sake of convenience to begin with the choice of nest sites and onset of nest
building or nest renovation in the case of those hawks, kites, and
eagles that normally construct their own nests, or at the time of
nest ledge or nest cavity occupation in the case of falcons.
However, actual breeding activity may in effect begin much
earlier, specifically perhaps at the time that nest-related dispersion
of pairs begins, usually as a result either of territorial establishment or perhaps through environmental limitations on the numbers and distribution of suitable nest sites. Thus, nest dispersion
behavior such as territoriality, as well as actual nest site selection
and its associated effects on breeding densities, again need to be
addressed briefly.

BREEDING DISPERSION, HOME RANGES, AND
BREEDING DENSITIES

It is a well-known fact that breeding densities of raptors and other
predatory birds are extremely low as compared with nonpredatory

birds. This is mainly a result of the fact that they represent endpoints of ecological food chains. Predators necessarily must remain at population biomass levels no greater than can be supported by the biomass amounts represented in the trophic levels below them, namely their own prey and in turn the food resources of that prey. As a result, breeding territories or home ranges tend to be very large, and as a rule of thumb tend to be proportional to the species's own biomass. Small raptors, and especially those feeding on relatively abundant prey, can have smaller home ranges and thus exist in greater breeding densities than larger ones, especially those that depend on rare or relatively elusive prey. Generally, raptors can be considered as K-selected species, with life-history traits that tend to maintain their populations below environmental carrying capacities, although a few species (such as some kites) can respond rapidly to ecological opportunities by increased productivity, and thereby approach the r-selected type of life-history adaptations.

Because of this general phenomenon, the breeding territories and generally larger and more diffuse home ranges of raptors are usually inclusive ones. That is, they typically include all the basic components (those needed for feeding, nesting, roosting, brood rearing, etc.) that may be required by the pair to complete a breeding cycle successfully. Such inclusive territories provide several subsidiary functions in addition to that of offering a nesting site and helping insure an adequate food supply for the pair or family. For example, sexual isolation (freedom from disturbances by sexual competitors and potential mate stealing or cuckoldry) may be facilitated. Additionally a general degree of population dispersion tends to occur that may not only reduce intraspecific competition but may also help prevent the spread of pathogenic or parasitic agents. Furthermore, the extended use of a particular area by a pair may result in their having an improved degree of familiarity with it, so that for example the best nest sites are found, the richest hunting areas can be learned, or the most effective escape routes discovered, thereby improving a pair's reproductive and predatory efficiency (Olendorff, 1971).

Various aspects of the environment might make it suitable or unsuitable as a potential breeding territory for any particular species of raptor. Some fairly obvious components include the amount and distribution of vegetational cover (potential nest sites, perch sites, escape cover, etc.), topography (the presence of nesting cliffs, of suitable lookout points, or of areas of topographically related updrafts to facilitate takeoffs and soaring), available food supplies (quantities, qualities, accessibilities), local weather

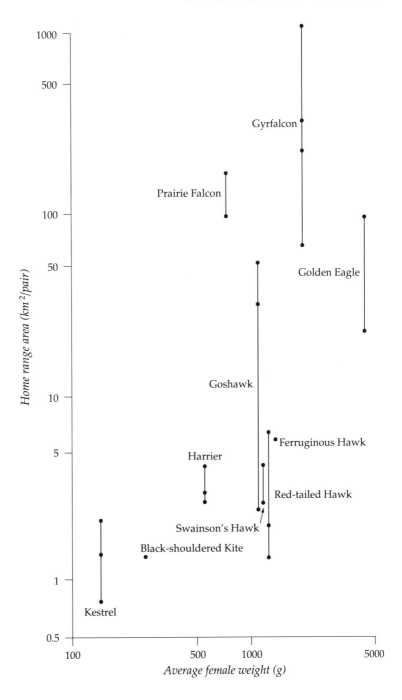

Figure 19. Home ranges of representative North American falconiforms in relation to average adult body size. Adapted from Newton (1979), with additional data from Cade (1982) for prairie falcon, and from Cramp and Simmons (1980) for gyrfalcon and northern goshawk.

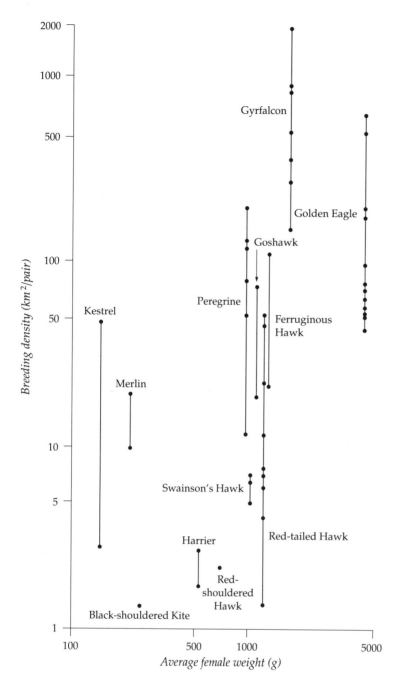

Figure 20. Breeding densities (total available habitat per breeding pair) of representative North American falconiforms in relation to average adult body size. Adapted from Newton (1979), with data added for American kestrel (Craighead and Craighead, 1956), gyrfalcon (Cade, 1982), peregrine falcon (Walter, 1979), northern goshawk (Glutz, Bauer, and Bezzel, 1971), golden eagle (Glutz, Bauer, and Bezzel, 1971), ferruginous hawk (Olendorff, 1972), red-tailed hawk (Kirley and Springer, 1980), and Swainson's hawk (Olendorff, 1972).

variables (winds, precipitation, etc.), and the possible presence of other competing and/or dominating species.

In contrast to typical territoriality and dispersed breeding are the cases of seemingly colonial breeding hawks; among North American species might be mentioned the American swallow-tailed, Mississippi, black-shouldered, and snail kites, most of which are either largely insectivorous (the black-shouldered kite the exception) or feed on localized and relatively abundant invertebrate prey, as in the snail kite. It is questionable whether any of these can truly be called colonial in the sense of being socially attracted to one another for nesting. For example, Parker (1975) determined that in the case of the Mississippi kite the birds are apparently attracted to a common suitable nesting area but tend to exhibit overdispersion tendencies for nest placement within such areas. The osprey is also seemingly colonial in its nesting in areas of high density, but again this is certainly a reflection of limited and localized available nesting sites rather than the result of social attraction.

As examples of the kinds of general dispersion and home ranges typical of North American falconiform birds, data on these species that was initially assembled by Newton (1979), redrawn and supplemented by additional information, is presented in Figures 19 and 20. In both cases it may be seen that home range sizes tend to increase and breeding densities diminish with increasing body size. Additionally, the role of foraging ecology is evident from the fact that bird-eating raptors such as the larger falcons have appreciably larger home ranges and lower densities than raptors of comparable size that feed on more accessible or less elusive prey.

NEST SITE SELECTION AND NEST BUILDING

Although it is often very difficult to determine when birds are actually seeking out a nest site, this is more obvious in the case of species with highly specific nest requirements, such as cavity nesters. Thus, in both the American and Eurasian kestrels both sexes are known to participate in this behavior for periods of as much as three weeks prior to the start of egg laying. In other species a variety of nest sites may also be systematically investigated by the pair. In some cases, as in kestrels, the male evidently takes the initiative in such inspection tours and may indeed have begun inspecting potential nest sites before a female has been established on the territory, but of course the female has the ultimate choice as it is she who finally lays the eggs (Nethersole-Thompson and Nethersole-Thompson, 1944). There is certainly a high level of site fidelity in some species, with the same breeding territory or even the same nesting site being used repeatedly by the same species or perhaps the same pair for many years if not decades. Bent (1937, 1938) has listed a number of cases of long-term occupancy for red-shouldered hawks, peregrines, and other species. Some of the most remarkable instances of this can be found for the peregrine falcon, where single females are believed to have occupied breeding territories for periods of up to 30 years, and where individual eyries have been occupied by peregrines for periods of from more than 50 years to more than 350 years (Olendorff, 1971). Such fidelity to particular sites would suggest that the birds must be responding to certain environmental features that they find strongly attractive, perhaps most often a highly favorable nest site situated near a reliable and rich food supply.

It is not uncommon for a pair to maintain two or more nest sites within a territory, and to alternate between or among them in successive years. Indeed two nests may even be constructed during a single year by a pair, with a choice finally being made between them. Brown (1976a) stated that golden eagles may have up to 14 nests on their territories in Scotland, but average about 2.3–2.6 nests. He noted that tropical species usually have fewer such extra nests than temperate-zone ones, perhaps because of the more rapid rotting and collapse of unused nests in the tropics. Such additional nests allow for the abandonment of one should it be disturbed in early stages prior to egg laying, or the first clutch lost prior to hatching, and the prompt shifting to another site. Or the use of different nests in successive years may prevent or retard the build-up of nest parasites.

Typically both sexes gather materials and help construct the nest; the male takes the major initiative in some species such as sea eagles and fishing eagles, but in most species the female apparently does most of the work. In polygynous species of harriers the female is said to do most of the nest building, with the male bringing some of the materials to her (Brown, 1976a).

The duration of nest building varies greatly according to the size of the nest, the duration of territorial residency, and perhaps the rate of nest decay or deterioration in the case of nests that may persist for years and perhaps be used for many breeding seasons. For pairs that remain on their breeding territory throughout the year, occasional branches or sticks may be added to the nest at any time. Typically a fresh green lining of nesting material is added to the nest cup shortly before the laying of the first egg, and such sprigs may continue to be brought to the nest well into the breeding period. Ideas on the functions of such materials vary from possible significance in courtship rituals to more obviously functional nest sanitation. Lining of the nest with greenery is typical of many if not most of the nest-building accipitrids and osprey, but not of the falcons (which do not construct nests) nor apparently of the crested caracara (which constructs a poorly lined stick nest). Olendorff (1971) provided a list of 15 North American accipitrids known to add green materials as nest lining, and several more might easily be added to the list (see Palmer, 1988). Among the latter are the bald eagle, which often lines its nest with soft grasses, needles, feathers, and the like. The list also lacked any species of kites, although swallow-tailed and Mississippi kite nests often have a thick lining of green leaves, and snail kites sometimes add a few leaves to their nests during incubation (Palmer, 1988).

EGG LAYING AND INCUBATION

The onset of egg laying by the female provides a convenient starting point for the breeding season, inasmuch as nest building is often lacking or at least its beginnings are frequently impossible to judge in raptors. Most raptors have relatively prolonged nesting seasons in North America, with egg laying frequently beginning very early in the year (indeed, sometimes starting before the start of the calendar year in Florida and Texas) and occasional eggs present until well on into late summer. This long period is a reflection of the fact that second or even third nesting attempts are sometimes made after the failure of the first, and rarely there may even be an attempt to raise a second brood of young after successfully fledging the first. Such double-brooding has been well documented for the black-shouldered kite in various parts of its range including North America, and for the Harris' hawk in Arizona (Mader, 1975a). It has also been found to occur in captivity and apparently also occurs occasionally under natural conditions in the American kestrel (Stahlecker and Griese, 1977; Toland, 1985c).

As an indication of the general spread of the nesting season, the spread of available egg records as originally assembled by Bent (1937, 1938) and summarized by Snyder and Wiley (1976) is shown in Figure 21. Included in this figure is an indication of each species's foraging ecology; Snyder and Wiley have pointed out that raptors feeding mostly on birds have a peak egg-laying

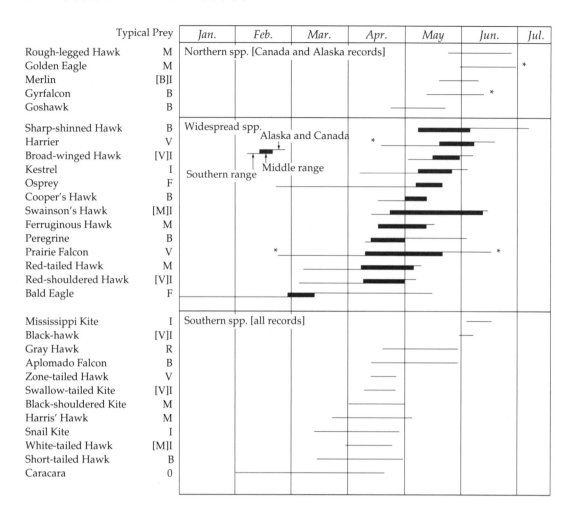

Figure 21. Temporal sequence of breeding in North American falconiforms, based on Bent's egg record data as summarized by Snyder and Wiley (1976). "Typical prey" classification is based on Snyder and Wiley's relative numerical prey frequency tabulation rather than on relative average biomass of prey; alternative prey categories are shown parenthetically where prey biomass differences would suggest a major dietary importance not apparent from prey frequency data alone. (M indicates mammals as typical prey, B indicates birds, I invertebrates, V generalized vertebrates, F fish, R reptiles, and O an omnivorous diet, including carrion.) Asterisks indicate species for which Bent's total span of egg records was used; for larger samples only the median 50 percent seasonal span is shown.

period occurring near the start of the prey population peak, while those feeding on mammals or insects have prey populations that gradually increase from spring to fall, and can afford to extend their egg-laying period later into the summer even though they may begin breeding as early as or even earlier than those feeding mainly on birds. This trait, plus a rather short fledging period, probably accounts for the ability of black-shouldered kites and Harris' hawks to raise two broods per year. Perhaps bird-eating raptors begin nesting as early as they are able to find sufficient food to form eggs (Snyder and Wiley, 1976), but even so they are constrained by their food base to only a single brood per year. Certainly at least some species such as the American kestrel begin breeding as a direct result of photoperiodic stimulation (Willoughby and Cade, 1964), and probably much the same is true of a number of the more northerly nesting species.

The incidence of clutch replacement following egg loss provides some evidence on this general problem; Morrison and Walton (1980) have summarized such data for North American falconiform and strigiform birds. They reported replacement clutches documented in 20 of the species represented in this book, including seven buteonine hawks, four falcons, all three accipiters, two kites, both eagles, the crested caracara, and the osprey. In at least ten of these species there was no apparent change in clutch size with the replacement clutch, in three cases it was smaller than the initial clutch, and in two cases it was larger. The egg development stage at the time of initial clutch loss varied from fresh to having had as much as 32 days of incubation (in the prairie falcon), and in most cases either the same nest or a nearby one was used for laying the replacement clutch. In several species (crested caracara, common black-hawk, Harris' hawk, peregrine falcon, and probably the Swainson's hawk) cases of two relayings have been reported within a single season, and in the sharp-shinned hawk at least three relayings have been found. In the Eurasian sparrowhawk the mean clutch size of 52 replacement clutches was 4.19 eggs, as compared with 5.25 for the original clutches. Such replacement clutches were usually started 12–15 days after the loss of the original clutch, if such losses occurred during the laying period. A few cases of two replacement clutches are known for this species (Newton, 1986).

The general reproductive parameters of nesting, including each species's usual clutch size, its egg-laying interval in days, its usual incubation and fledging periods, and its normal period of postfledging dependency have been summarized by Newton (1979, Table 18) for most of the North American falconiform birds. I have supplemented and modified his information with additional information (Table 4), the basis for which can be found in the individual species accounts. What is perhaps remarkable here is the general degree of similarity among nearly all these species; the only surprises come in such things as usual clutch sizes. Why, for example, do the accipiters and bird-eating falcons have larger clutch sizes on average than mammal-dependent buteos, when

Table 4.

Reproductive Parameters of North American Falconiforms[1]

Species	Usual size of clutch	Egg interval (days)	Incubation period (days)	Fledging period (days)	Postfledging period (weeks)	Adult plumage acquired (years)
Osprey	2–3	1–3	37–38	44–59	6–7	2
Hook-billed Kite	2	?	?	?	?	?
Swallow-tailed Kite	2	?	28	36–42	2+	2?
Black-shouldered Kite	4–5	2	30–32	32–40	4–10	1
Snail Kite	2–3	2–3	28–30	23–28	6–7	2+
Mississippi Kite	1–2	2–3	30	34	9+	2
Bald Eagle	2–3	2–4	34–37	70–77	8–12	4–5
Northern Harrier	4–5	2–3	32–38	42	3–3.5	2(f),3(m)
Sharp-shinned Hawk	4–5	2	30–34	24–27	2.5–3.5	2
Cooper's Hawk	4–5	2	35–36	30–34	5.5–6	2
Northern Goshawk	3–4	2–3	35–38	35–42	4.5–7.5	2
Common Black-hawk	1–2	2–3?	28–34	43–50	6–8	2
Harris' Hawk	2–4	?	33–36	40	8–12+	2
Gray Hawk	1–3	?	32	42	?	2?
Red-shouldered Hawk	2–4	2–3	33	39–45	8–10	2
Broad-winged Hawk	2–3	2–4	ca. 31	35–42	3–4	2
Short-tailed Hawk	2	?	34	?	?	2
Swainson's Hawk	2–3	2	33–36	36–46	3–4.5	2
White-tailed Hawk	2	2	ca. 31	47–57	4	2
Zone-tailed Hawk	2	?	35	35–42	?	2?
Red-tailed Hawk	2–4	2–3	32–35	43–48	5–10	2
Ferruginous Hawk	3–5	2	32	38–50	1.5–6	2
Rough-legged Hawk	3–6	2	31	31–40	3–6	2
Golden Eagle	2–3	3–4	43–45	65–77	8	ca. 5
Crested Caracara	2–3	2–3	30–32	42–56	?	2
American Kestrel	4–6	2	27–31	28–30	3	1
Merlin	3–5	2	28–32	26–32	2–3	2
Aplomado Falcon	2–3	?	31–32	28–35	?	2?
Peregrine Falcon	3–4	2	32–34	35–40	5–6	2
Gyrfalcon	3–5	2–3	34–35	46–49	4	2
Prairie Falcon	4–5	1–2?	29–33	36–41	?	2

[1]*In part after Newton (1979), but with additions and modifications to conform with present text information.*

their foods are generally rarer or more difficult to obtain? Is it perhaps that they must be greater anticipatory "gamblers," taking maximum advantage of those years or situations when a large family can be raised by having a larger number of available hatched young than can normally be reared? Or why does the largely insectivorous Mississippi kite have such a small clutch size, while the mammal-eating black-shouldered kite has one averaging at least twice as large?

Additional questions of clutch size may be raised that are not treated in Table 4. For example, are there year-to-year changes in clutch size that reflect differences in food supplies or weather? Cade (1982) believes that food-related variations in clutch size may occur in the gyrfalcon, and it seems to be generally true that buteos specializing in rodents show the biggest annual variations in clutch size, correlated with the annual fluctuations in prey abundance (Newton, 1979). Or, are there individual variations in clutch size that result from differences in age, health, or general vigor of the females? Certainly these are variables that quite probably influence clutch size, but they are generally considerably more difficult to prove without having large samples of birds of known ages. However, Newton (1986) found that the average clutch size of known-aged females of the Eurasian sparrowhawk increased progressively for the first three years, reaching a peak at 4.5 years, after which it began to decline in a nearly linear manner. Interestingly, however, the average number of young successfully raised per clutch continued to increase at least to the female's seventh year of life, suggesting the probably important role of experience in successful breeding. His studies indicated that the ages of each sex were important in breeding success, and in both sexes an adult partner contributed more to reproductive success than did a yearling, perhaps because of age-related differences in foraging abilities or in acquiring good territories.

There also appear to be regional differences in clutch sizes among various species that occur over broad geographic areas. However, there are disagreements over whether clutch sizes in raptors tend to increase or decrease toward the arctic, especially in peregrine falcons. Although early studies (e.g., Cade, 1960) supported the idea of a diminished clutch size in arctic-breeding birds, more recent data support the view that arctic peregrines are at least as productive in both clutch size and brood size as more southerly populations, if not more so (Court, Gates, and Boag, 1988).

In most falconiform birds the eggs are typically laid at 48–72-hour intervals, with the majority having alternate-day 48-hour laying schedules. Longer intervals of three or four days, typical of many eagles, cause significant differences in hatching dates that may have important effects on brood survival later on (Brown, 1976c).

In most species of hawks incubation begins, at least in a limited way, with the laying of the first egg. Thus a predictable pattern of hatching is typical, with eggs hatching in the same sequence as they were laid. However, in the falcons and accipiters in particular there is a more nearly simultaneous sequence of hatching than occurs, for example, in eagles and buteos (Brown,

1976b). Falcons apparently typically begin incubation with the laying of the penultimate egg (Cade, 1982). However, in accipiters such as the Eurasian sparrowhawk incubation seems to begin either suddenly with any egg after the second, or rather gradually, so that by the laying of the last egg the incubation drive is fully developed (Newton, 1986). In the vast majority of species the female apparently incubates all night and the majority of the day, with the males often only taking over briefly as the female is feeding, taking a bath, or doing other kinds of maintenance activities during daylight hours (Brown, 1976b; Cade, 1982). In at least some kites the males regularly incubate, and in the snail kite the male may actually perform the majority of the incubation (Beissinger, 1984).

Typical falcons of all species have relatively consistent incubation periods of about 28–33 days, while accipiters typically also hatch rather consistently in about 30–35 days, with only minor differences in incubation length as related to egg size. Not surprisingly, the longest incubation periods are to be found among the eagles, especially various African snake-eagles, which may incubate for more than 50 days. All other things being equal, tropical species of falconiforms spend more time nest-building, lay smaller average clutches, and incubate for longer periods relative to their size than do temperate-zone relatives (Brown, 1976b).

Generally speaking, at least the larger falconiforms invest a fairly small amount of energy in reproduction (when expressed as a percentage of the collective clutch weight relative to the weight of the adult female). The total clutch of a golden eagle, for example, perhaps averages up to as much as 9 percent of the adult female's weight, assuming a maximum three-egg clutch, while that of the American kestrel may equal 50–60 percent of her weight, the five eggs laid over a comparable period of about ten days. Clearly, it is probably more important for a kestrel to have a reliable food supply during this period or, put differently, it is questionable whether a kestrel could produce a much larger clutch, even assuming that the eggs could be effectively incubated or the young adequately fed.

In a few species it is not unusual for a male to support two or rarely even three females within his own territory, and to help provide food for both or all of the resulting families. Newton (1979) listed 11 species of falconiforms for which such resource-based polygyny has been reported; only four of these (osprey, northern harrier, red-tailed hawk, and peregrine) occur in North America. Of these four species, polygyny is much better developed in harriers than in any others, and indeed is fairly regular in at least some populations of these. This situation is discussed in some detail in the account of the northern harrier; in the other species it is sufficiently rare as to be probably insignificant from the standpoint of the species's normal breeding biology.

Polyandry, on the other hand, is rare in birds and not to be expected in raptors. However, the Galapagos hawk (*Buteo galapagoensis*), although normally monogamous, often forms multiple male breeding groups, a breeding strategy termed cooperative polyandry (Faaborg, 1986, 1988). This species may have up to four male mates per female, and all the males copulate with the female and help rear the young. Such groups rear more young on average than do monogamous pairs, but fewer young per male. Once a group forms no new males are added, but as the excess males die monogamy eventually results. One advantage of this unusual mating system is that territorial birds survive at a much higher annual rate than do nonterritorial ones (90 percent versus 50 percent or less), so that a male's productivity over his lifetime through polyandry may outweigh any advantages of waiting to gain a monogamous breeding territory. Like tropical buteos, this species and another insular species, the Hawaiian hawk (*B. solitarius*), have relatively small clutch sizes but long incubation and fledging periods (Griffin, 1987).

HATCHING AND BROOD REARING

As in other birds, hatching of each individual is a rather prolonged process in falconiforms, with the larger species typically requiring from about 24 to 48 hours from initial pipping of the eggs to emergence of the young. The young hatch in a relatively helpless or semialtricial state. Although they are covered with a long but rather thin coat of from white- to buff-colored or grayish down, and their eyes are sometimes partially opened, they are unable to stand and are essentially helpless.

The physical growth and behavioral development of the young have been studied by a variety of investigators, including Ellis (1979) for the golden eagle, Hardaswick, Smith, and Cade (1984) for peregrine x prairie falcon hybrids, and Sherrod (1983) for the later stages of postfledging behavioral development in peregrines. Not surprisingly, these changes are of a continuous nature, and thus it is difficult to establish convenient stages for descriptive purposes. However, a few patterns are evident and probably widely applicable. Thus, the ultimately longest feathers (the flight feathers) begin to emerge first and grow the most rapidly. As this occurs, the chick (sometimes called an "eaglet" or even an "eyass," in the case of young falcons) gradually changes from an entirely down-covered bird to one that is progressively covered by more adultlike contour feathers. However, the down itself also changes in appearance, as the original coat of down is replaced by a longer and thicker generation of down that may be the same color as or somewhat lighter than the original generation. The developing contour feathers emerge from the same papillae as this longer down, gradually pushing it out until it breaks away from the tips of these feathers, causing a transformation from a furry-looking chick to a more normal-looking bird. However, wispy remnants of these down feathers normally remain on some areas of the head, shoulders, and breast until the young are nearly fledged. As this down is dropped from the chicks into the nest it may be thrown "overboard" by the female, or swallowed by her. The female also usually removes uneaten materials from the nest, as well as pellets of undigested food materials that are regurgitated by the young.

Within a very short time after hatching, perhaps only an hour or so in some species, the young are not only able to take food particles but also can move around in the nest and can orient their excretions out toward the nest rim. These are initially semisolid, and are thrown out of the nest by the female (Newton, 1986). The young soon learn to grab food from the female's bill (Figure 13), and their feces gradually becomes a whitish liquid, which is oriented over and away from the nest rim, producing a distinctive spattering of "whitewash" below the nest.

Fledging (or nestling) periods are of course highly variable between species and among individuals, depending upon feeding rates. In the case of some of the smaller falcons and accipiters the period may be as short as or even slightly shorter than the incubation period (as little as about 24 days in male sharp-shinned hawks and Eurasian sparrowhawks), but in the progressively larger species of buteos and eagles the period is considerably longer than the incubation period, up to nearly twice its length in the case of some tropical species such as the African snake-eagles (Newton, 1979). In strongly dimorphic species such as the Eurasian sparrowhawk the female nestlings gain weight faster, but the males grow feathers more rapidly and develop skills more quickly, perhaps helping them to survive better in the presence of their larger sisters (Newton, 1986).

When their wing feathers have developed adequately, the young begin to spend a considerable amount of time flapping them, progressively rising farther and farther vertically above the nest during such maneuvers. Stretching of the legs and flexing of the toes may also begin among nestlings, and soon after they are able to stand the chicks begin mantling their food when they receive it from a parent, huddling over it with the wings spread in such a way as to hide it (Figure 12C). Such behavior may even occur among nestlings reared in isolation (Hardaswick, Smith, and Cade, 1984).

Preening begins only a few days after hatching, and after the developing contour feathers ("pinfeathers") begin to emerge the chicks typically preen after every meal. Such preening consists of nibbling, digging-in, and combing-out components, and inaccessible feathers (such as those around the head) that cannot be reached for preening are scratched with the middle talon. This talon has a central ridge on the underside that is extensively used for such grooming (Hardaswick, Smith, and Cade, 1984).

During the first week or so following hatching the young may be brooded more or less continuously by the female, but the intensity of her brooding diminishes as the young gradually become better feathered and more capable of maintaining their own thermoregulation. On the other hand, antipredator behavior by the parents tends to become more vigorous as the young become older, presumably a reflection of the high amount of energy that has been invested in them by that time. Females may also bring in fresh nesting material to add to the nest cup throughout the entire nestling period, presumably for hygienic purposes, as this behavior tends to cover pellets, feces, or un-eaten food remains that have not been thrown out of the nest.

Usually males take little or no role in such brooding activities, as they are normally fully engaged in hunting at that time. However, in many cases they are more actively involved in parental care than they are during incubation, typically taking over while the female leaves the nest briefly. More rarely do they feed the young directly, and in some species such as accipiters they apparently never feed, the young birds picking up and eating food items unaided. In accipiters and some falcons, the males continue to provide all the food for the young right up to the end of the nesting cycle, although the female may increasingly begin to hunt for herself toward the end of the nestling period (Brown, 1976b; Newton, 1986).

FLEDGING AND THE POSTFLEDGING DEPENDENCY PERIOD

As the young of most raptors have staggered hatching times, and because the larger females require more time to develop their flight feathers, the times of fledging often are also separated by several days. Thus the oldest youngster may be out of the nest but still in its close vicinity for some time while the youngest are still being cared for in the nest. During this immediate postfledging period the young may make short flights to nearby trees or perching sites and return to the nest when a parent returns with food. Later they may position themselves in such a way as to see the returning parent, and thereby intercept it on its way to the nest. At such times they may begin screaming in such a loud and conspicuous way as to stimulate calling by the other unfledged young, perhaps not only attracting the attention of their parents but also that of predators, placing them at considerable risk (Newton, 1986).

Predatory behavior by the young begins soon after they gain the power of flight, and although they may profit from the experiences gained in parental interactions, the birds are apparently genetically programmed to perform such activities even in the absence of such guidance. However, young peregrines also perform "flutter gliding," an apparent food-begging behavior done as the hungry birds pursue their parents. This behavior is perhaps a preadaptation that helps deter direct threatening pursuits of their parents by these birds. Like the corresponding flight of adult females soliciting food from their mates, the tail is spread and slightly dropped, the wings flapped so that they are strongly bent and do not appear to rise above the horizontal, in a manner similar to that of some flying sandpipers (*Actitus*). A whining call normally accompanies the flight. Like the aggressive behavior shown by a courting and food-begging female toward her mate, the birds may also aggressively attack their own parents, even at times when they may not be carrying food items (Sherrod, 1983).

In the case of young Eurasian sparrowhawks, the adults may eventually release prey to fledged young only after the young are led on a rapid chase, or the parent may even release a still-living bird for the young to catch and kill. This evidently encourages the young to perform actions associated with effective hunting for themselves. At least in this species and perhaps also in the Cooper's hawk, the final "weaning" of young away from parental assistance may be achieved by a rather sudden termination of adult feeding. This causes the fledglings to hunt completely for themselves from that point onward (Newton, 1986; Snyder and Wiley, 1976).

The postfledging dependency period varies in length from as little as about two or three weeks in some kites and harriers to as long as about five months in some Old World vultures. Postfledging periods of as long as almost a year have even been reported in the crowned eagle (*Stephanoaetus coronatus*) (Brown, 1976c). The Hawaiian hawk also has a remarkably long postfledging period of about 30 weeks (Griffin, 1987). Presumably the length of the postfledging period in most cases must be adaptively related to the probable difficulties the young are likely to have in adjusting to independent life, in terms of amounts of foods available to the young or their relative ease in obtaining adequate foods.

Dispersal or migration often begins almost immediately after the end of the postfledging period, terminating parental care and familial bonding. This is the start of an extremely critical time for young raptors. They must not only forage for themselves but also avoid being the targets of larger predators, and must begin to move through unfamiliar habitats as they undertake seasonal migration or gradual dispersal. Not surprisingly mortality rates are typically very high during this period, especially perhaps in late-hatched birds or those that terminated their postfledging period in anything less than excellent physical condition.

REPRODUCTIVE SUCCESS

Not surprisingly, the rates of reproductive success are highly variable in raptors, not only between species but locally and between years. As a result, it is difficult to generalize much on these statistics, and even studies involving large sample sizes may not be very applicable to other situations. However, a sampling of reported reproductive success rates for most North American raptors has been assembled (Table 5), largely patterned after and in part based on a similar but taxonomically more inclusive table provided by Newton (1979, Table 23). In cases of species whose reproductive performance is most likely to have been influenced by pesticides (osprey, bald eagle, and peregrine falcon) the locations and years of study are indicated; in other cases such information is not believed to be directly relevant to the reported success rates.

Long-term or regional variations in reproductive success in a species, such as those suggested in the bald eagle and peregrine falcon, for example, often provide at least indirect evidence of environmental problems, such as pesticide contamination, human disturbance, or other undesirable effects on reproduction. When the peregrine populations of Great Britain and the United States began to decline in the late 1940s and later, no clear evidence as to the cause of this decline was immediately evident. It was variously suggested that an unknown disease may have entered

Table 5.

Reproductive Success Rates of Some North American Falconiforms[1]

Species	Average clutch size	Average brood size	Fledglings per nest[2]	Fledglings per breeding pair[3]	Sample size	Sources
Osprey						
Minnesota (1968–71)	—	—	1.8	0.8	270	Mathisen, 1977
Florida (1968–74)	—	—	1.2	0.7	336	Ogden, 1975
Labrador (1969–72)	—	1.6	—	0.7	216	Wetmore & Gillespie, 1977
Washington (1972)	2.5	—	2.0	1.1	151	Melquist & Johnson, 1973
North Carolina (1973–82)	2.3	1.7	1.2	1.0	332	Whittemore, 1984
Nova Scotia (1975–81)	—	—	1.2	—	237	Seymour & Bancroft, 1983
Florida (1979–83)	2.8	1.2	1.0	1.0	135	Phillips, Westall & Zajicek, 1984
Black-shouldered Kite	4.0	—	1.9	—	20	Wright, 1978
	4.1	—	3.2	—	23	Dixon, Dixon & Dixon, 1957
Snail Kite	—	—	—	1.9	183	Sykes, 1979
Mississippi Kite	1.6	1.3	1.0	0.4	330	Parker, 1974
Bald Eagle						
Florida (1938–46)	—	—	1.7	1.0	317	Broley, 1947
Alaska (1963–70)	—	—	1.6	1.0	312	Sprunt et al., 1973
Wisconsin (1962–70)	—	—	1.55	1.0	492	Sprunt et al., 1973
Florida (1962–72)	—	—	1.45	0.7	592	Sprunt et al., 1973
Michigan (1961–70)	—	—	1.4	0.5	243	Sprunt et al., 1973
Maine (1962–70)	—	—	1.3	0.35	241	Sprunt et al., 1973
Great Lakes (1961–70)	—	—	1.3	0.3	156	Sprunt et al., 1973
Saskatchewan (1969)	—	—	1.9	1.3	51	Whitfield et al., 1974
Washington (1975–80)	—	—	0.9	0.8	275	Grubb et al., 1983
Arizona (1977–80)	—	—	—	0.7	35	Haywood & Ohmart, 1983
U.S.A. (1985)	—	—	1.55	1.0	1,482	Green, 1985
Northern Harrier	—	4.0	—	1.1	15	Craighead & Craighead, 1956
	—	4.6	—	1.9	80	Hamerstrom, 1969
Cooper's Hawk	4.2	—	—	2.3	17	Craighead & Craighead, 1956
	—	—	3.3	2.8	23	Schriver, 1969
	4.2	—	3.5	—	36	Henny & Wight, 1972
	3.3	2.9	2.65	2.2	48	Millsap, 1981
Northern Goshawk	3.2	2.9	2.7	—	40	McGowan, 1975
Common Black-hawk	1.9	1.5	1.3	1.3	21	Millsap, 1981
Harris' Hawk	3.0	—	2.3	—	50	Mader, 1975a
	2.9	—	1.9	0.8	18	Griffin, 1976

(continued)

Table 5. (Continued)

Species	Average clutch size	Average brood size	Fledglings per nest[2]	Fledglings per breeding pair[3]	Sample size	Sources
	2.3	—	—	2.1	24	Brannion, 1980
	3.0	2.3	1.6	—	71	Whaley, 1986
Red-shouldered Hawk	3.5	—	—	1.8	39	Craighead & Craighead, 1956
	—	—	2.3	1.6	74	Henny et al., 1973
	—	—	1.1	—	287	Jacobs, Jacobs & Erdman, 1988
	—	2.7	2.1	—	29	Wiley, 1975
Swainson's Hawk	2.34	1.7	1.2	1.2	95	Olendorff, 1973
	2.5	2.2	2.1	1.2	31	Dunkle, 1977
	2.2	2.0	1.8	1.2	39	Fitzner, 1978
	—	—	2.0	1.5	439	Schmutz, 1985b
Zone-tailed Hawk	2.1	1.9	1.85	1.85	22	Millsap, 1981
Red-tailed Hawk	2.2	—	—	1.2	16	Craighead & Craighead, 1956
	—	—	1.9	1.2	60	Orians & Kuhlman, 1956
	2.0	1.9	1.7	1.2	57	Luttich, Keith & Stephenson, 1971
	2.9	—	1.7	—	54	Seidensticker & Reynolds, 1971
	2.9	2.3	1.7	1.0	20	Smith & Murphy, 1973
	2.1	2.0	—	1.5	119	McInvaille & Keith, 1974
	2.6	2.2	2.2	—	51	Wiley, 1975
	3.0	2.2	1.6	—	38	Smith & Murphy, 1979
	—	—	1.8	1.3	71	Petersen, 1979
	2.8	2.4	2.3	2.2	108	Millsap, 1981
	2.79	—	—	1.47	476	Mader, 1982
Ferruginous Hawk	3.2	2.3	2.0	1.4	36	Smith & Murphy, 1973
	2.9	—	2.3	—	19	Olendorff, 1973
	4.3	—	2.9	1.5	27	Lokemoen & Duebbert, 1976
	3.5	2.2	1.9	—	88	Smith & Murphy, 1979
Golden Eagle	2.1	1.6	—	1.4	20	McGahan, 1968
	1.6	—	1.2	—	22	Olendorff, 1973
	2.1	1.8	1.6	1.1	146	Beecham & Kochert, 1975
	2.0	1.1	0.9	—	44	Smith & Murphy, 1979
	1.9	1.6	1.5	1.25	17	Millsap, 1981

(continued)

Table 5. (Continued)

Species	Average clutch size	Average brood size	Fledglings per nest[2]	Fledglings per breeding pair[3]	Sample size	Sources
American Kestrel	4.2	—	4.2	—	13	Heintzelman & Nagy, 1968
	4.7	3.1	2.3	—	22	Smith, Wilson & Frost, 1972
	5.2	4.7	2.8	1.9	9	Smith & Murphy, 1973
Merlin	—	—	4.2	3.7	82	Oliphant & Haug, 1985
	4.3	—	3.7	3.3	48	Becker & Sieg, 1985
Prairie Falcon	4.5	1.9	—	1.2	55	Enderson, 1964
	4.4	3.9	3.4	—	24	Olendorff, 1973
	3.85	3.4	3.2	2.8	26	Millsap, 1981
	4.4	3.5	3.1	—	68	Ogden & Hornacker, 1977
Peregrine Falcon						
East (1931–49)	—	—	2.4	0.9	62	Herbert & Herbert, 1965
Pennsylvania (1939–46)	—	—	2.3	1.3	44	Rice, 1969
New York (1939–40)	—	—	2.5	1.1	29	Hickey, 1942
Alaska (1950–58)	—	—	2.3	1.4	25	Cade, 1960
Alaska (1950–58)	—	—	1.9	0.8	20	Cade, 1960
Alaska (1966)	—	—	2.3	1.8	17	Cade, White & Haugh, 1968
Alaska (1970–72)	—	—	2.7	1.8	38+	White, 1975
Greenland (1984)	—	—	2.3–3.2	2.1–2.3	42	Moore, 1987

[1]In part after Newton (1979), but with additions and modifications.
[2]Based on numbers of young fledged per successful nest.
[3]Based on counts of active nests or territorial pairs present.

the species's gene pool, that food resources had declined, and that chemical pollution had perhaps somehow begun to affect the species.

The first clues to the real cause came in Britain, where in the 1950s broken and missing eggs began to be noticed in peregrine eyries under study. In the early 1960s the contents of an addled egg from Scotland showed that certain synthetic chemicals (organochlorides) that are parts of various modern insecticides such as DDT, aldrin, and dieldrin were present in quantities sufficiently high as to cause death in birds such as quails. Soon similar pesticide residues were found in the livers or fatty tissue of adult birds, in some cases sufficient to be a direct cause of death. However, the most insidious danger of pesticides such as DDT was eventually found to be its effects on the reproductive physiology of female birds, especially predators, that ingest it with their foods. When organochlorides such as DDT are ingested, the liver is stimulated to secrete certain enzymes that metabolize these poisons so that they can be broken down into products that can be excreted from the body. However, these same enzymes also destroy other physiologically important compounds, such as the hormone estrogen, that are needed by the bird for normal reproduction. Further, DDT is soon broken down into a more stable compound (DDE) that remains indefinitely stored in liver tissue and causes long-term stimulation of these estrogen-destroying enzymes. The most direct effect of this hormonal disruption in breeding females is an inability to deposit within their skeletons the calcium-rich medullary bone that is normally the source of calcium used for the formation of eggshells during the egg-laying period. Lacking this source of available calcium, the female can lay only thin-shelled eggs that are likely to crack under the bird's weight during incubation. Or the eggshell may have an insufficient amount of calcium present to allow for normal bone deposition in the developing embryo, which itself depends on the eggshell for obtaining its own source of calcium. Eggshell thinning may thus destroy a population's reproductive potential even when the degree of pesticide contamination is too low to cause mortality to adults directly (Hickey, 1969; Ratcliffe, 1980).

Such was the circuitous and unpredictable route of pesticide pollution that brought about the near demise of the peregrine, bald eagle, osprey, and various other birds over much of North America after World War II, triggered the "silent spring" crisis, and eventually initiated a general public awareness of the breadth and depth of the postwar environmental crisis. In the next chapter is a brief history of how a combination of such uncontrolled pesticide contamination and widespread habitat destruction affected three of our now most-endangered North American raptors.

CHAPTER 5

Population Biology and Conservation

The population of any species, over time, must be able to reproduce itself at an annual recruitment rate that at least balances its annual losses, or mortality rate. If one considers the species as a whole these are the only two truly relevant statistics associated with its long-term survival. However, when considering regional populations such as those of the North American falconiform species, immigrational movements of a subpopulation into an area or region must also be taken into account, as must emigrational movements out of the region. Of the 31 species considered in this book, only two have breeding ranges that are wholly encompassed within its defined regional limits, namely North America north of Mexico. These are the Mississippi kite and ferruginous hawk. The bald eagle, Cooper's, broad-winged, Swainson's, and red-shouldered hawks, and prairie falcon also nearly qualify, except for having breeding ranges that variably extend into Baja California, the mainland of northern Mexico, or the West Indies. Collectively these eight species may be considered true Nearctic endemics. For these species protection of adequate breeding habitats is critical to their long-term survival, and the same is true of protecting wintering habitats for those species that additionally remain in North America during the nonbreeding portion of the year. Ironically, three of these true North American endemics (the Mississippi kite, broad-winged hawk, and Swainson's hawk) are also among the most migratory of all our raptors and indeed spend most of the calendar year outside the limits of this book's coverage, in central to southern South America.

MORTALITY RATES AND RECRUITMENT

Because of the usually uncertain population component of falconiforms made up of nonbreeding immatures, it is not possible to calculate recruitment easily based on reproductive success rates. Instead, statistics such as the average number of young reared per breeding or territorial pair are often used (Table 5). Ideally, actual net recruitment to populations would be best estimated by comparing total population size at the start of the year with population a year later. For many raptors having distinctive first-year plumages such recruitment rates might be estimated by determining the proportion of juvenile birds in migrating fall and winter populations; in a stable population the incidence of such birds should approximate the annual mortality rate. Few such data seem to be available. However, Henny (1972) has calculated "recruitment standards" for various species of raptors, representing the average number of fledged young that he estimated must be produced by a pair annually, based on a knowledge of annual mortality rates and the age of initial breeding. He thus estimated that 2.12 young must be fledged per breeding pair of red-shouldered hawks (assuming all birds at least two years old breed) to maintain a stable population, and that each breeding American kestrel pair (assuming that 82 percent of yearlings and all older age classes nest) must produce 2.88 fledged young annually. Similarly, ospreys must produce 0.95–1.3 young per pair, assuming all birds at least three years old breed, and various populations of red-tailed hawks must produce 1.33–1.89 young per breeding pair annually. As Newton (1979) has pointed out, such figures are valid only if the mortality of adults can be accurately estimated and remains constant through time; additionally the incidence of nonbreeding in adults may vary with time, as indeed may the incidence of breeding in first-year or otherwise subadult birds. Newton (1986) calculated in the Eurasian sparrowhawk that 18 percent of the females probably breed in their first year, 49 percent the second year, 83 percent the third year, and nearly all by the fourth year. Additionally, based on estimated mortality rates, some 51 percent of the birds that survive their first year die before they can breed later in life. Such data emphasize the potential value of first-year breeding, even at a somewhat lower level of productivity than might be typical of adults. It also makes it apparent that simple but realistic hawk productivity models or recruitment standards are impossible to construct.

Newton (1979) has summarized in tabular fashion (his Table 49) the annual mortality rates of 13 species of falconiforms based on banding records, in which first-year mortality ranged from as little as 51 percent to as much as 83 percent. Annual mortality in birds at least three years old varied from as little as 18 percent to as much as 51 percent. I have similarly summarized some previously published mortality estimates for ten species of North American falconiforms (Table 6). These have first-year mortality rates ranging from 31 to 83 percent, and second-year or later mortality rates of from roughly 20 percent to 60 percent. Dif-

Table 6.
Estimated Annual Mortality Rates of North American Falconiforms

Species	No. of records	MORTALITY RATE (%)					Source
		1st year	2nd year	3rd year	4th year	5th year	
Osprey	670	30.9	22.1	30.2	26.6	30.3	Brown and Amadon, 1968
	330	53.3	19.6 (all adult age classes)				Henny and Wight, 1969
Bald Eagle	107	78.5	56.5	60.0	25.0	33.3	Brown and Amadon, 1968
	196	27–37	26–30	9 (all older ages)			McCollough, 1986[1]
Cooper's Hawk	129	77.5	48.3	20.0	33.3	62.5	Brown and Amadon, 1968
	135	79–83	37–44 (all adult age classes)				Henny and Wight, 1972
Northern Goshawk	—	63	33	19	17	11	Haukioja and Haukioja, 1970
Northern Harrier	242	67.7	38.5	47.9	56.0	26.4	Brown and Amadon, 1968
females	—	91 (0–2 yr)	10 (2–6 years)				Picozzi, 1984a[1]
males	—	80 (0–2 yr)	28 (2–6 years)				Picozzi, 1984a[1]
Red-shouldered Hawk	120	57.5	31.3 (all adult age classes)				Henny, 1972
Red-tailed Hawk	361	73.4	46.9	36.3	34.3	13.0	Brown and Amadon, 1968
	117	62.4	20.6 (all adult age classes)				Henny and Wight, 1972
American Kestrel	205	68.8	64.1	47.7	50.0	50.0	Brown and Amadon, 1968
	71	69	46.8 (all adult age classes)				Henny, 1972
Peregrine Falcon	108	57.4	32.6	25.8	26.1	29.4	Brown and Amadon, 1968
	65	70	25 (all adult age classes)				Enderson, 1969
Prairie Falcon	81	75	26 (all adult age classes)				Enderson, 1969

[1]*Based on resightings of individually marked birds rather than band recoveries.*

ferences in techniques and assumptions lead to some difficulties in comparing these calculations, but it is probably safe to say that the first year of a hawk's life is much the most risky for it, and that following first-year survival it is likely to have an expectation of additional life of up to about four years or more. This is not a long time, considering the low average productivity rates of most hawk species, and most hawk populations are perhaps rather quickly put at risk by environmental factors that significantly lower their productivity for only a few years or so. Thus, the widespread crash of peregrine populations following the introduction of hard pesticides is perhaps more understandable, even though the level of pesticide poisoning may not have been sufficiently severe as to cause direct mortality in adults.

POPULATION TRENDS AMONG NORTH AMERICAN FALCONIFORMS

Henny (1972) has reviewed the impact of modern pesticides on the population dynamics of a variety of bird species, including the osprey, Cooper's hawk, red-tailed hawk, red-shouldered hawk, and American kestrel. He judged that the mortality rates of these and 11 other species studied did not measurably increase following the 1945 advent of the hard pesticide era, and thus the severe population declines occurring in several of these species must have been the result of lowered reproductive rates. These were mainly species feeding on fish, amphibians, reptiles, and birds, whereas those foraging principally on mammals, such as most buteo hawks, had no apparent changes in their reproductive rates during this period.

As has been well documented, this phenomenon of reduced reproductive capacity in predators largely has to do with the biological magnification of toxic substances in predatory species representing the ends of ecological food chains, especially those predators feeding primarily on fish and birds (which themselves are largely predators), whereas raptors feeding on herbivorous mammals are much less likely to be affected. The physiological basis of this reduced reproductive performance in fish- and bird-eating raptors such as the osprey, peregrine falcon, and accipiters, and its consequent effects on breeding raptor populations, have been extensively documented elsewhere (e.g., Hickey, 1969; Ratcliffe, 1980) and need not be reviewed here. Suffice it to say that several of the North American falconiforms began a period of population declines following World War II that, together with habitat losses, resulted in their elimination as breeding species from many areas, and caused the near extinction of some species.

Perhaps the only good thing to come out of the postwar pesticide era was a gradually increasing awareness of the dangers that pesticides pose not only to other species of wildlife but to humans as well. Such concerns took concrete form with the passage of the federal Endangered Species Preservation Act in 1966. The resulting first list of endangered species was published in 1967, numbering 72 entries. This legislation was amended in 1969 to restrict the importation of foreign endangered species, and in 1973 the much more comprehensive Endangered Species Act was passed that included all plant and animal groups and recognized a "threatened" as well as endangered category. It also implemented the Convention on International Trade in Endangered Species of Wild Fauna and Flora (CITES), and greatly increased federal funding for endangered species preservation. By the late 1970s more than 200 taxa of birds had been identified as endangered or threatened. Critical habitat evaluations had also been completed for seven of these species, recovery plans had been approved for 28 species, and recovery teams responsible for drawing up recovery plans had been established for many more. By the early 1980s more than 1.4 million dollars had been budgeted by federal agencies for raptor programs, and more than 1.0 million by state agencies (LeFranc and Millsap, 1984).

Forms recognized as endangered in the first listing of the Department of the Interior were the southern bald eagle (*H. l. leucocephalus*), the Florida race of the snail kite (*R. s. plumbeus*), and the American and tundra races of the peregrine falcon (*F. p. anatum* and *F. p. tundrius*). Additional species such as the aplomado falcon have more recently been proposed for addition to the federal list, as have a number of variously less rare but still "threatened" forms. Canadian authorities (Committee on the Status of Endangered Wildlife in Canada) have prepared a comparable list of Canada's threatened and endangered forms, and the U.S. Forest Service has also established certain management categories for rare or ecologically sensitive species occurring on federal lands. Additionally many states and provinces have established their own categories of unique protective status; thus although the osprey has not been provided with special federal status beyond that of "special emphasis," it has received special conservation status in 15 states (Henny, 1986). Indeed, as of 1972 all of the North American falconiform birds had been fully protected by one or more federal laws, as well as by various state, provincial, and federal Canadian laws. These laws finally brought to an end a long period of apathy, if not hostility, with respect to the conservation of raptors.

On a worldwide basis, the International Union for the Conservation of Nature and Natural Resources (IUCN), in cooperation with the International Council for Bird Preservation (ICBP), prepared a "Red Data Book" listing bird species believed by authorities to be endangered, vulnerable, or rare throughout the world. This list was originally published in 1966, was revised in the late 1970s, and was finally published in book form in 1981 (King, 1981). The most recent edition includes several North American falconiform taxa, including the Florida race of the snail kite (listed as endangered), the southern race of the bald eagle (listed as rare), and the American and tundra races of the peregrine falcon (both listed as endangered). The species *F. peregrinus* was collectively considered vulnerable on a worldwide basis.

In 1972 the Environmental Protection Agency of the U.S. government finally banned the sale of DDT for further domestic use, and that year can perhaps be used as a landmark indicating the all-time low of raptor populations in North America. Since then there has been a generally favorable change in raptor populations. For example, a study of the bald eagle, four hawks, and three falcons was recently undertaken by Cornell University's Laboratory of Ornithology to compare U.S. and southern Canadian winter populations in the early 1980s to those of the early 1970s. The results of this study indicated that seven of these species have shown percentage increases of from 19 to 145 percent, and only one (Harris' hawk) has shown an apparent recent decline (Anonymous, 1986). Especially encouraging were population increases shown by such pesticide-sensitive species as the bald eagle (92 percent) and peregrine falcon (19 percent).

In a review of state and federal agency programs directed toward raptor management, LeFranc and Millsap (1984) estimated "vulnerability" indexes for all the North American raptors based on individual state assessments of their population status. Among the species covered in this book (the hook-billed kite was not included in their analysis), the ten species they considered most vulnerable to extirpation or extinction, ranked from highest to lowest risk, are aplomado falcon, snail kite, gray hawk, peregrine falcon, American swallow-tailed kite, bald eagle, common blackhawk, white-tailed hawk, crested caracara, and osprey. However, the largest number of active projects (127) were at the time of this analysis being directed toward the bald eagle, followed by the peregrine falcon (52) and the osprey (28). Only four projects were concerned with the three species they ranked most vulnerable, and three of the ten most vulnerable hawks (gray hawk, white-tailed hawk, and crested caracara) were receiving no special attention at all. Seemingly, such large and attractive species as the bald eagle, golden eagle, and osprey, which collectively account for over half of all funded raptor projects (representing 42 species of hawks, owls, and vultures), have been receiving far more attention than might seem warranted on the basis of their actual conservation needs.

For that reason a general summary may be warranted of total North American population estimates for all of the 31 falconiform species included in this book (Table 7). For many species these

Table 7.

Estimated North American Populations of Falconiform Birds

Species	Estimated population	Basis
Osprey	17,000–20,000	Poole, 1989
Hook-billed Kite	10–20 pairs	W. S. Clark (in corresp.)
American Swallow-tailed Kite	2,500–3,000	Millsap, 1987
Black-shouldered Kite	14,120	Table 3
Snail Kite	500	Rodgers, 1989
Mississippi Kite	27,400	Smith, 1985
Bald Eagle	70,000–100,000	Tables 3 and 8
Northern Harrier	111,000[1]	Table 3
Sharp-shinned Hawk	over 30,000[1]	Table 3
Cooper's Hawk	ca. 20,000[1]	Table 3
Northern Goshawk	8,500	Anonymous, 1986
Common Black-hawk	ca. 200 pairs	Palmer, 1988
Harris' Hawk	5,500	Anonymous, 1986
Gray Hawk	ca. 50 pairs	Palmer, 1988
Red-shouldered Hawk	34,250[1]	Table 3
Broad-winged Hawk	ca. 500,000	W. S. Clark (in corresp.)
Short-tailed Hawk	ca. 500	Palmer, 1988
Swainson's Hawk	ca. 500,000	W. S. Clark (in corresp.)
White-tailed Hawk	Few hundred pairs	Kopeny, 1988
Zone-tailed Hawk	Few thousand	Author's estimate
Red-tailed Hawk	350,000[1]	Anonymous, 1986
Ferruginous Hawk	5,480[1]	Table 3
Rough-legged Hawk	49,600[2]	Table 3
Golden Eagle	ca. 50,000	Palmer, 1988
Crested Caracara	2,280	Table 3
American Kestrel	over 2.4 million	Cade, 1982
Merlin	11,080[1]	Table 3
Aplomado Falcon	Few or none	Hector, 1987
Peregrine Falcon	2,800–3,800 pairs	Cade, 1982
Gyrfalcon	Few thousand pairs	Cade, 1982
Prairie Falcon	7,800[1]	Table 3

[1]Excluding unknown component wintering in Latin America (the estimates for the accipiters are extremely conservative).

[2]Excluding unknown (small?) component wintering in arctic areas.

estimates are obviously based on incompletely sampled populations, such as my winter estimates (Table 3) for those species that
have significant parts of their wintering ranges occurring south of
the U.S.-Mexican border. Such wintering population estimates
presumably provide reasonably good total North American population estimates for the largely nonmigratory falconiforms, but
offer no help in estimating populations for those populations that
are largely to entirely migratory. In a few other instances the
estimates are based on actual migration counts or extrapolations
from migration count samples reported by others. Such was the
case for using Smith's (1985) maximum estimates of Mississippi
kites counted as they were passing through the isthmus of
Panama. My assumption was that even the largest of these counts
must have missed some of the species' overall North American
migrating population. Probably only for a few species such as the
snail kite and perhaps also the hook-billed kite are the total North
American populations so small and localized as possibly to allow
for a complete census of all individuals. Indeed, in the case of the
endangered snail kite's Florida population such comprehensive
censuses have been attempted regularly.

MANAGEMENT AND CURRENT STATUS OF ENDANGERED SPECIES

Although the status of each species dealt with in this book will be
discussed in the individual species accounts, a brief review of the
three federally designated endangered species is worth making
here, as it allows for a somewhat more detailed description and
provides a greater historical perspective than is practical in the
species accounts.

Florida Snail Kite

The Florida population of snail kite has been near the brink of
extinction for some time; as early as 1950 an estimated 60–100
birds were believed to be present in the state, and in 1956 only
some 20 birds were known to be present on Lake Okeechobee. By
1964 the U.S. Fish and Wildlife Service estimated the total Florida
population at no more than 15 birds, presumably representing the
very ebb of the snail kite's status in the state. By 1969 the
estimated population was at 98, and from that point until 1975 the
population ranged from a low of 65 to a high of 120. However, in
the latter part of the 1970s it began to increase and by 1979 was

estimated at 267 birds. The increases since 1974 have been strongly correlated with years of high water levels and associated improved food supplies, while during drought years little reproduction has occurred and the birds have tended to disperse widely over the Florida peninsula (Sykes, 1979).

Probably the loss of suitable fresh-water marsh habitat has been the most important single factor affecting the decline of the snail kite in Florida; by 1980 about half of the species's original habitat had been lost, and of the remaining areas only 18 percent was actually used by the birds between 1967 and 1980. Severe droughts occurred in 1971, 1981, and 1982, causing kite dispersal and the reduction or elimination of kite habitat from large areas. Such periodic droughts are probably a natural phenomenon, but their effects have recently been made more severe as a result of drainage and heavy use of fresh water by humans. The lack of adequate water to flood the snail kite's marsh habitats during drought years is the most critical factor influencing this species's survival in Florida, and it will be necessary to manage some areas specifically for snail kites in order to save it. The establishment of such management areas would certainly also benefit other marsh-dependent wildlife (Sykes, 1983a).

Annual snail kite population surveys by the Florida Game and Fresh Water Fish Commission have been performed since 1974. In 1984 a total of 668 kites were counted, the highest number ever found. However, most birds were concentrated in a single area (Conservation Area 3A), with Lake Okeechobee of secondary importance. The previous year's total was only 231 birds, indicating the high population volatility typical of this species. The 1986 count of 564 birds was slightly higher than the previous five-year average (Rodgers, Schwikert, and Wenner, 1988), but the 1987 and 1988 surveys have again shown declines from the numbers seen in 1986 (Rodgers, 1988, 1989).

The snail kite's vulnerability index is second only to the aplomado falcon's among all the raptor species included in this book; the two were exceeded only by the now-extirpated California condor (LeFranc and Millsap, 1984). Although the aplomado falcon was not originally included on the federal list of endangered species, it has recently (1985) been recommended for such status. A brief discussion of its population trends can be found in its species account.

Southern Bald Eagle

This subspecies is easily the most famous of all of the endangered populations of North American falconiform birds, and has had the greatest amount of research (probably over a third of the total funded projects in the mid-1980s) directed toward it, judging from the vast amount of relevant technical and semipopular literature, including several books (e.g., Stalmaster, 1987). Complicating the true picture of the form's relative rarity is the virtual impossibility of separating the southern bald eagle from the northern bald eagle, which is in no danger whatsoever, with a population perhaps approaching 100,000 birds. With the advent of the Endangered Species Preservation Act of 1966, the 40th parallel of north latitude was chosen arbitrarily as a point of jurisdictional separation between the two subspecies, although little biological basis for such a choice could be advanced. In any case, this line roughly separates the Alaskan and Canadian populations from those of the 48 contiguous states, which for convenience might thus be regarded as mostly belonging to the southern population. A 1974 survey suggested that about 700 pairs existed in the 48 states, with 318 in the Great Lakes area, 150 in Florida, 101 in the Pacific northwest, 56 in the Chesapeake Bay area, 33 in Maine, and about 50 elsewhere. Counts of total pairs do not offer a fully satisfactory method of estimating population declines during that period, since many of these pairs continued to nest but produced few or no young. Thus, along Florida's west coast the number of pairs and their offspring dropped between 1946 and 1957 from 73 pairs with 103 young to 43 pairs with only 8 young. Similar declines in productivity were common through the Atlantic coast region north to Maine. During the same general postwar period the percentage of immature birds among wintering eagle populations ranged from about 20 to 25 percent, or substantially less than sample pre–World War II counts, when about 36 percent of the birds were immatures (Green, 1985).

Like other raptors, bald eagles began a slow recovery following the banning of DDT in 1972, which was supplemented by later bannings of the pesticides dieldrin (1974) and endrin (1984). Concomitantly, there was a steady reduction in mortality resulting from illegal shooting, presumably as a result of better federal protection and improved information campaigns. Not only is the destruction or disturbance of bald eagles (as well as golden eagles) a federal offense, but even the sale or possession of eagle feathers is currently prohibited by at least three federal laws. More recently, the gradual elimination of lead from shotgun shells, which began in the mid-1980s, has probably helped to eliminate lead poisoning as a significant mortality factor as well. (Eagles suffered lead poisoning from eating waterfowl that had ingested spent lead shot or been wounded by hunters.) Potentially serious problems still existing are the effects of purposeful

Table 8.

Bald Eagle Nesting and Winter Population Estimates for North America

State or province	Nests (1982)[1]	Winter Population (1980)[2]	Major areas of winter eagle concentrations
	U.S.A.		
Alabama	—	649	
Alaska	7,500	68,137	Southeastern coast
Arizona	10	399	Salt, Verde, and Gila reservoirs
Arkansas	1	902	Arkansas River; Dardanelle Lake
California	52	1,427	Klamath Basin refuges
Colorado	6	1,973	Major rivers; San Luis valley
Connecticut	—	11	
Delaware	4	5	
Florida	340	2,255	Kissimmee valley; Gainesville area
Georgia	3	71	
Idaho	15	2,128	Snake River and impoundments
Illinois	2	3,410	Illinois and Mississippi rivers
Indiana	—	59	
Iowa	1	4,127	Missouri and Mississippi rivers
Kansas	—	2,777	Quivira and Kirwin National Wildlife Refuges
Kentucky	—	357	Lake Cumberland area
Louisiana	18	94	
Maine	72	269	Penobscot and Kennebec rivers
Maryland/D.C.	58	189	Chesapeake Bay
Massachusetts	—	10	
Michigan	96	49	
Minnesota	207	409	Mississippi River
Mississippi	—	287	
Missouri	2	4,034	Squaw Creek, Swan Lake, and Mingo National Wildlife Refuges
Montana	37	4,955	Glacier National Park
Nebraska	—	1,530	Platte and Missouri rivers
Nevada	—	104	
New Hampshire	—	5	
New Jersey	1	17	
New Mexico	1	918	Wheeler Lake; Mora River
New York	2	10	
North Carolina	—	—	
North Dakota	—	178	
Ohio	7	37	
Oklahoma	2	1,297	Sequoyah National Wildlife Refuge; most reservoirs
Oregon	93	1,666	Klamath Basin refuges
Pennsylvania	4	38	
Rhode Island	—	—	
South Carolina	21	117	
South Dakota	—	2,362	Lake Andes and Karl Mundt refuges
Tennessee	—	1,567	Reelfoot National Wildlife Refuge; TVA reservoirs
Texas	14	2,716	Reservoirs in east and north
Utah	—	2,547	Northeastern rivers and lakes
Vermont	—	9	
Virginia	48	258	Chesapeake Bay
Washington	135	2,573	Skagit River; San Juan Islands

(continued)

Table 8. (Continued)

State or province	Nests (1982)[1]	Winter Population (1980)[2]	Major areas of winter eagle concentrations
	U.S.A.		
West Virginia	1	—	
Wisconsin	207	710	Upper Mississippi National Wildlife Refuge
Wyoming	22	4,617	Yellowstone and Grand Teton parks
	Canada		
Alberta	?	5,636	Waterton Lakes National Park
British Columbia	4,500	43,198	Squamish and Harrison Bay; south coast
Manitoba	1,400	—	
New Brunswick	15	158	
Newfoundland/Labrador	78	769	
Nova Scotia	83	382	
Ontario	?	140	
Prince Edward Island	2	—	
Quebec	?	140	
Saskatchewan	3,500	202	

[1]*Based on data of Green (1985) and Stalmaster (1987) for U.S.A. excluding Alaska; for Canadian breeding estimates see references in text. There have been recent increases in nesting populations in Arizona (Glinski, 1988), the southeast (Simons et al., 1988), and in the western Great Lakes region (Henderson, 1988). Schempf (1988) estimated an Alaskan population of 22,000 adults and about 11,000 immatures. This total population of about 30,000–35,000 birds is about half the winter population indicated here, and seems highly conservative.*
[2]*Based on Christmas Bird Count data presented by Gerrard (1983), who also provided alternative estimates based on the National Wildlife Federation Midwinter Surveys.*

harassment or accidental disturbance by humans, such as those caused by logging, mining, construction, or recreation (Green, 1985). More easily measurable are various kinds of habitat losses in various breeding and wintering areas, although to some extent the formation of new reservoirs in some areas has substantially improved local winter eagle populations.

The most recent comprehensive survey of nesting birds is that of Green (1985), whose 1982 data for the 48 contiguous states are shown in Table 8. Corresponding winter populations for these states and the Canadian provinces as estimated by Gerrard (1983) are also provided in that table. Productivity for the 48 contiguous states during 1982 averaged 1.0 young per nest, and 64 percent of the occupied nests were successful in rearing at least one young. According to Green (1985), productivity in the Chesapeake Bay area prior to World War II had averaged about 1.6 young per active nest, whereas in 1962 it had declined to about 0.2 young per active nest. By 1984 the estimated productivity was 1.1 young per active nest, or about the national average. In recent years the incidence of immature birds during winter counts has averaged about 35 percent.

Peregrine Falcon

The decline of the peregrine has perhaps been documented to a better degree than that of any other species of North American bird. Besides hundreds of technical papers appearing in journals, two books based on symposia proceedings (Hickey, 1969; Cade et al., 1988) provide an abundance of information dealing with the effects of hard pesticides such as DDT on this species, perhaps the finest and most spectacular of all the North American raptors.

Although many North American species of raptors were seriously affected by DDT, none was more devastated than the peregrine, perhaps in part because its original breeding range encompassed much of eastern North America where great quantities of DDT were being used on agricultural lands. Thus, in contrast to such remotely breeding boreal or arctic bird-eating forms as the gyrfalcon and merlin, the peregrine became immediately exposed to the effects of the pesticide (as it preyed on such birds as thrushes and other passerines that were themselves often being directly poisoned through its effects). Thus, of the three races of the peregrine, the most southerly race *anatum* was the first and far the most seriously affected. The arctic-breeding *tundrius* became exposed at lower levels, these effects probably mainly occurring while the birds were on migration and while in Central and South American wintering areas. The third race, the relatively sedentary *pealei* of the Pacific northwest coast, was least affected by pesticide use and has never been placed on any lists of threatened or endangered forms.

The decline of the peregrine apparently began about 1950 both in North America and in Europe, where DDT and other chlorinated hydrocarbon pesticides such as dieldrin were by then also being used in great quantities, although it was not until more than a decade later that the broad-scale declines of peregrines were causally related to this factor. A critical discovery was that during the enzymatic breakdown and detoxification of DDT certain intermediary products such as DDE occurred that had serious effects on the normal production of reproductive hormones. The consequent reduced capabilities of affected females to produce thick-shelled eggs, with resulting reproductive failures, provided an explanation for the peregrine population crash. It also provided a new example of the many unexpected and occasionally disastrous effects of introducing artificial chemicals such as DDT into natural ecosystems. By 1964, there was apparently not a single breeding pair or lone adult east of the Mississippi, in an area where nearly 200 eyries had been present during the 1940s, and the California population was reduced to about 10 percent of its historic size (Cade, 1982). By 1970, there was sufficient concern over the ecological effects of DDT to initiate legal restrictions on its use in many of the industrialized nations, especially in Britain and Europe (Ratcliffe, 1980). Canada followed with the banning of DDT in 1971, and the U.S.A. in 1972. By that time, even the boreal and arctic-breeding populations of peregrines had been seriously affected, with these populations reduced to about half of their original numbers.

The decade of the 1970s, when most raptor populations hit their lowest ebb, was one marked by numerous activities of citizen, state, and federal agencies to try to reverse the downward trend. For example, The Peregrine Fund was established with a goal of reestablishing peregrines in their lost range by building up captive-bred flocks that could provide seed stock for controlled release back into the wild, or "hacking" projects. Simultaneously, the Raptor Research Foundation became the major scientific clearinghouse for information on raptor biology, and citizen-based raptor rehabilitation or propagation groups were formed across the country.

Apparently the decline of the peregrine in North America occurred in the boreal regions in the mid-1970s, and by 1980 at least some of these were showing increases in numbers. An even earlier recovery began in Britain, and by 1981 the British populations had increased to a point that was better than at any time during the present century. However, on the European continent, where the population was more severely affected, the improvement has been much less gratifying. Also by 1980 the California breeding population was nearly double that of the early 1970s, and in that same year at least four pairs bred east of the Mississippi River for the first time in more than two decades. At least three of these pairs involved captive-raised birds that had been released (Cade, 1982).

At least at present, the future of the peregrine falcon appears more favorable than it has since the end of World War II. For this one must thank a veritable army of people devoted to wildlife conservation who were willing to make the financial, political, and personal sacrifices that were necessary to change the very philosophical face of North America, when we as a society were forced to choose between the powerful lobbying interests of agrochemical corporations and our own environmental security. The sight of a wild peregrine, which once seemed likely to become as completely a thing of the past as a flock of passenger pigeons, is now not only a possibility but a probability in many areas, a testament not only to the resilience of nature but to the power of concerted human endeavor. Likewise our national symbol, the bald eagle, is becoming increasingly abundant throughout much of North America; like the Native American's mythical thunderbird, it offers symbolic as well as real hope for the conservation ethic.

> *With the far darkness made of the rain and the mist on the ends of*
>> *your wings, come to us soaring.*
> *With the zig-zag lightning, with the rainbow hanging high on the*
>> *ends of your wings, come to us soaring.*
> *With the near darkness made of the dark cloud of the rain and the*
>> *mist, come to us,*
> *With the darkness on the earth, come to us.*

NAVAHO NIGHT CHANT (MATHEWS, 1902)

PART TWO

Natural Histories of Individual Species

Family Accipitridae
(Kites, Hawks, and Eagles)
Subfamily Pandioninae
(Osprey)

Osprey

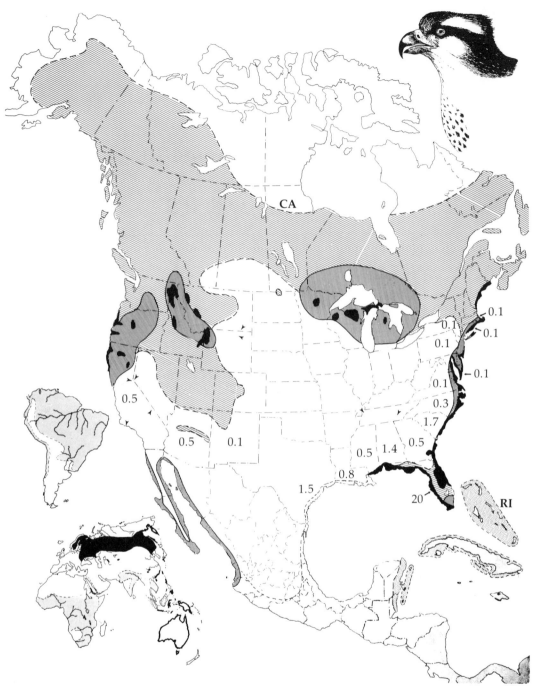

Figure 22. North American distribution of the osprey, showing wintering range (shading), and areas of dense (inked), moderate (narrow hatching), and sparse (wide hatching) breeding populations of carolinensis (**CA**), plus residential range (stippled) of ridgwayi (**RI**). Areas of local breeding outside the indicated breeding area are shown by pointers. Relative winter state or provincial density indices (average number seen per Audubon Christmas Count, 1986) are shown for major wintering areas. The breeding (inked) and nonbreeding (shading) ranges in South America and the Old World are shown on inset maps. North American distribution information mainly after Henny (1977; 1983)

Other Vernacular Names

Fish hawk; balbuzzard (French); gavilan pescador (Spanish)

Distribution

Breeds in North America from northwestern Alaska, northern Yukon, western and southern Mackenzie, northern Saskatchewan, northern Manitoba, northern Ontario, central Quebec, central Labrador, and Newfoundland south locally to Baja California, the Tres Marias Islands, Sinaloa, central Arizona, northwestern and central New Mexico, southern Texas, the Gulf coast and southern Florida, the Bahamas, Cuba, and along the east coast of the Yucatan Peninsula; also widely distributed in the Old World including Australia.

Winters in the Americas from central California, southern Texas, the Gulf coast, Florida, and Bermuda south through the West Indies, Central America, and South America, especially north of the equator; also winters widely in the Old World. First-year birds remain in the wintering range through the summer following hatching. (See Figure 22.)

North American and West Indian Subspecies

P. h. carolinensis (Gmelin): Breeds in North America as indicated above, except for the area occupied by *ridgwayi*.

P. h. ridgwayi (Maynard): Resident of the Bahamas, Cuba, and the keys off Belize and Quintana Roo; perhaps also on Hispaniola (Prevost, 1983), but the race may now be endangered (Poole, 1989).

Description (of *carolinensis*)

Adult (sexes nearly alike, but the female more noticeably banded at the breast). Forehead, crown, occiput, and nape white, washed to a varying degree with pale cinnamon-buff, the latter color often present as definite flecks on the hind crown, occiput, and nape; the anterior portion of the crown and the middle of the nape heavily marked with fuscous-black (this usually lacking in *ridgwayi*); lores bare, blackish; most upperparts and remiges varying from dark sepia to brown to fuscous-black, the lesser upper wing coverts and the interscapulars and anterior scapulars generally slightly darker than the rest of the upperparts, all the feathers narrowly and faintly margined with fulvous; the long tips of the five outermost primaries generally darker than the rest of the remiges; the primaries internally broadly edged with pale fulvous-white, incompletely barred and mottled with dark sepia; rectrices brown to sepia, all but the median pair whitish on the inner webs except for a broad brown tip and all the rectrices crossed by five to seven indistinct bands of slightly darker brown, these bands becoming distinct on the whitish inner webs of most rectrices; a broad fuscous-black postocular band extends from the rear of the eye to the side of the neck, where it connects with the transverse arm of the similar color from the nape and extends down to the sides of the breast; underparts white, sometimes tinged with pale buff; the feathers of the lower throat and breast occasionally with narrow dusky tawny to fuscous shaft streaks in males; in females the breast feathers usually showing heavier shaft streaks of brown to fuscous, the streaks generally somewhat spatulate terminally and sometimes forming a distinct breast band; under wing coverts whitish. Iris lemon-yellow; cere and base of lower mandible bluish, rest of bill black; tarsi and toes pearl-gray to grayish blue; claws black.

Subadult (sexes alike). After the wearing away of paler flight feather tips (by about 18 months), immatures are indistinguishable from adults. Two-year-olds are adultlike but nonbreeders; breeding commonly occurs initially at three years, but some individuals may not breed until their fifth year or even later.

Juvenile (sexes alike). Similar to adults, but the pale areas of crown streaked with blackish brown and the dorsal feathers and tail tipped conspicuously with white; breast band less sharply defined; primaries tipped with buffy edges; tail perhaps more heavily barred than in adults, and the under wing coverts barred. Breast banding is not diagnostic of sex in this age group (Prevost, 1983). Iris ranging from red (as nestling) through orange-yellow to brownish yellow; bill and cere black, becoming pale blue at the base of the mandible; tarsi and toes pale pearl-blue to greenish white.

Measurements (of *carolinensis,* in millimeters)

Wing, males 462–498 (ave. of 15, 477.4), females 488–512 (ave. of 11, 503.7); tail, males 199–220 (ave. of 15, 208.8), females 212–240 (ave. of 11, 225.4) (Friedmann, 1950). Average egg size 61 x 45.6 (Bent, 1937).

Weights (in grams)

Males 1220–1600 (ave. of 10, 1403), females 1250–1900 (ave. of 14, 1568) (Brown and Amadon, 1968). Fifteen adult and juvenile males of nominate *haliaetus* from Europe averaged 1428 (range 1120–1740), 14 females 1627 (range 1208–2050) (Cramp and Simmons, 1980). Estimated egg weight 70, or 4.5% of female.

Identification

In the hand. The sharp spicules on the undersides of the toes, and the talons that are deeply rounded in cross section, serve to identify this species. Additionally, the nostrils are unusually long, slitlike, and uniquely closable.

In the field. Perched ospreys appear dark brown above and white below, with a blackish stripe passing back through the eye and extending to the dark hindneck. The tips of the wings extend beyond the tail. Usually seen perched near water, often at a tall lookout perch from which prey might be seen. Females typically have better developed breast banding than males.

Flying ospreys are best identified by their long and "crooked" wings, which are mostly white below, but with contrasting black wrist marks (as in rough-legged hawks) and black barring on the undersides of flight feathers and tail feathers. Immatures have a "scaly" appearance on the back and upper wing surface because of their paler feather edges, the white under wing coverts are barred, and the tail has more distinct barring and a wider white terminal band. The usual alarm call is a series of loud *kip* notes that become more drawn out. When disturbed near the nest repeated *pee-ee* or *tchip* notes are often uttered. The repeated calls often gradually rise in pitch, and up to about 20 calls may be uttered in 6 seconds.

Habitats and Ecology

The basic habitat needs of ospreys might be simply summarized as an adequate source of fish that can be captured near the surface in water clear enough for them to be seen, and an elevated nesting site that is at least as close to the food supply as a few kilometers. Nest sites preferably are dead or open-topped live trees located beside or in the water, but in some locations rock outcrops or cliffs are also used, or even columnar cacti in treeless desert areas. In the absence of such natural nest sites or as an alternative to them, artificial structures such as utility poles, channel markers, old duck blinds, and similar structures may be utilized. Probably excessive human disturbance or destruction and declining fishery stocks as well as their contamination by persistent pesticides, especially DDT, have been major causes of population declines (Evans, 1982).

Beebe (1974) suggested that breeding habitat needs include only three components, which are fish that move slowly near the water surface, an ice-free season long enough to permit reproduction, and elevated or inaccessible nest sites, or at least freedom from molestation during the breeding season. Such conditions occur widely around the globe, and as a result the osprey like the peregrine falcon has one of the broadest geographic distributions of all birds. Both species also capture their prey in an efficient and highly specialized manner, which tends to reduce competition from other raptors. Indeed, the osprey is more isolated ecologically and taxonomically from other hawks than is the peregrine, which does have some competition from other species in the genus *Falco.*

Ospreys apparently do not defend a definable territory, either in their wintering quarters or during breeding; thus colonies can develop in areas rich in food (Brown, 1976a). However, the area immediately around the nest site is defended from usurpation by other ospreys. In dense nesting aggregations nests can be situated as close as 20 meters apart (Ogden, 1975), but in most areas they are much more widely scattered. Even in good populations the nests are often a kilometer or two apart, and colonylike nesting groups only occur in a few areas of the world, including eastern North America, western Europe, on the Red Sea coast, and on islands off the coast of Australia (Beebe, 1974). Thus, between Florida Bay and the Ten Thousand Islands in Everglades National Park an estimated 300 nests were present in 1974 (Kale, 1978), and the total Florida breeding population in the early 1980s was probably 1,500–2,000 pairs (Henny, 1983). In the Florida Bay area, nests are sometimes grouped with up to seven nests per island, or sometimes less than 50 meters apart (Kushlan and Bass, 1983). However, in some much less desirable breeding areas the birds may have to range up to nearly ten kilometers from their nest to a foraging area (Garber, 1972). In some areas at the northern periphery of the range, as in northern Quebec, the density may only be about a pair per 200 square kilometers (Bider and Bird, 1983).

Foods and Foraging

The foods of the osprey are almost exclusively fish; frogs, crustaceans, turtles, small mammals (mainly rodents), and birds have occasionally to rarely been observed as prey, but these are incidental to the species's primary diet of fish. Sherrod (1978) summarized ten primary or secondary sources of information on osprey foods, and in no case (out of seven so analyzed) did the proportion of prey other than fish attain two percent of the identified foods. The specific species of fish that are taken depend entirely upon what is locally available, and few generalizations seem to be possible other than that they tend to be species that are prone to bask near the surface or rise to the surface to forage themselves. Probably most fish that are taken weigh only about 200 grams or so, although ospreys are questionably reputed to be able to take fish weighing up to 1.8 kilograms (Brown, 1976a). Poole (1989) judged that fish weighing more than 400 grams (or more than about 25 percent of the adult bird's weight) might be difficult for the osprey to carry away.

Cramp and Simmons (1980) have summarized prey size data for Europe, noting that fish taken in Scandinavia average heavier than those in Germany (more than 250 grams vs. 97 grams), with a wide range in daily consumption of food but typically about 200–400 grams, or 1–4 fish, depending on their size. Success rates on dives vary greatly also, with estimates of 20–45 percent for West Africa, 41 percent for Norway, 65 percent for Scotland, and from 25 to 90 percent for a variety of North American estimates.

The capture method of ospreys (Figure 9) consists of initial visual scanning from above the water, often after a period of active hovering, flapping, or gliding, depending on the wind, and at heights ranging from about 5 to 70 meters. Most dives are made from about 20–30 meters above water, at angles ranging from 45 degrees to nearly vertical. At times the birds may also skim horizontally over the water and snatch a fish from the surface in a manner similar to that of the bald eagle. Some dives are made from perches, and fish are normally taken directly to a feeding perch where the osprey waits briefly for the fish to die, then begins to eat it, starting with the head (Cramp and Simmons, 1980).

Studies of foraging behavior on Yellowstone Lake (Swenson, 1978) revealed that cutthroat trout (*Salmo clarki*) comprised over 90 percent of the total diet, with immature fish 25–35 centimeters long being preferred. The diving success rate was similar (45 and 48 percent) in the two years of study, and the birds averaged from about 4 to 8 minutes between dives, or about 9–20 minutes between successful catches. The birds foraged mainly over deep water, where immature trout typically gather.

In a similar study done in California in a marine habitat (Ueoka and Koplin, 1973), the fish taken were mostly 18–23 centimeters long, and were primarily surfperch (Embiotocidae). The average hunt success rate was 82 percent, with most fishing efforts requiring only a single dive. The average length of time spent for each successful catch was about 12 minutes. Like the success rate, this statistic varied considerably from the preincubation to the postfledging period. On one occasion two fish were caught simultaneously, one with each foot.

In a Florida study, Szaro (1978) found that adults had a dive success rate approximately three times as high as did recently fledged young birds (19 vs. 6 percent), and achieved a catch in less than half the average time (38 vs. 77 minutes) spent in fishing. The overall hunting success rate of adults was 57.5 percent as compared with 29 percent for young birds. In another Florida study, Edwards (1987) found that young birds tended to avoid attempting to catch fish of the larger size classes. Fledged osprey siblings had similar prey preferences, and sibling interactions associated with hunting may have facilitated the development of improved foraging skills in young birds.

Swenson (1979) correlated dive success rates with the foraging ecology of the prey species, based on 13 study areas. Interestingly, bottom-feeding fish are apparently the most easily captured, and piscivorous fishes the most difficult. Swenson considered this possibly the result of bottom-feeders being oriented to looking below rather than being readily able to perceive attack from above, whereas piscivorous fish tend to be swift swimmers and thus more difficult to capture.

Dunstan (1974) observed osprey foraging behavior in Minnesota with the aid of radio transmitters. He found that the birds flew an average of 2.6 kilometers between their nests and fishing sites, and that they rarely fished on the lake simultaneously with others of their species. Ospreys sometimes fished over calm water by letting their feet drag intermittently, disturbing the fish and causing them to jump or move, at which they were immediately seized with one or both feet, or were later captured by stooping. Centrarchid fish accounted for 84 percent of the prey remains collected, and average weights of various species ranged from 37 to 624 grams. Twice ospreys were observed to pick up dead fish weighing more than a kilogram, only to drop them within ten meters.

Boshoff and Palmer (1983) reported that most fish in their South African sample of osprey foods weighed under 200 grams, and none exceeded 800 grams. They once observed two fish being caught with the same foot simultaneously, and once a fish was caught with each of the feet simultaneously. Most hunting was done in midmorning and late afternoon, the birds apparently not selecting particular hunting times according to cloud cover conditions or variations in wind speed.

Social Behavior

The osprey is normally monogamous, with a pair bond that at least lasts through a breeding season. However, since site fidelity to a particular eyrie appears to be strong, a long-term pair bond may be typical of the species (Cramp and Simmons, 1980). At least in Scotland (where ospreys are very rare), the birds return year after year to the same eyrie, the males usually returning first, to be joined a few days later by their mates. If one of the birds is killed or dies during the winter it will presumably be replaced if possible, and females tend to be replaced about every four years. When strange ospreys visit an occupied nest site, they are apparently tolerated by resident birds of the opposite sex, but are often repelled by birds of the same sex (Brown, 1976a).

In Scotland, the period of sexual display is probably only two or three weeks long. During that time the male may soar and make undulating dives above the nest site, either alone or with the female. This sequence is sometimes called the "sky-dance song-flight," and in its complete sequence often begins and ends at the eyrie. The male usually carries a fish or nesting material as he calls and gains altitude up to 300 meters or higher. He then hovers briefly with legs dangling and tail fanned ("hovering flight"), dives with flexed wings, and quickly rises to repeat the performance. During the dives the bird may descend in a series of intermediate steps, and as many as 27 undulations may occur. One apparently unique display is to flap his wings vigorously while high above the ground, at the same time dangling his feet (sometimes holding a fish) and calling in a frenzied manner. During this display he may even fly backward, and the display is not only directed toward females as a courtship display but also apparently intended to distract intruders (Brown, 1976a). The sequence may last for as long as nine minutes, and can be directed toward a female or done alone. Territorial calling by the male is sometimes also done from a perch (Figure 23F), and hovering flights are occasionally also performed by females. Sky dances seem to mark out nesting territories as well as to advertise the eyrie to females; intruding birds may be escorted away by the male, while the defense of the nest site itself is done by the larger female. Sky dancing reaches a climax with the arrival of the female, then declines rapidly (Cramp and Simmons, 1980).

As the intensity of sky dancing declines, that of courtship feeding probably correspondingly increases. Observations by Poole (1985) suggest that a key function of courtship feeding is to ensure mate fidelity, particularly since females are left alone while males forage, and females are usually fed exclusively by their mates between pair formation and egg laying. Older and more experienced pairs arrive earlier and lay eggs more quickly than younger pairs on average, and these factors evidently have a greater effect on a female's reproductive output and timing than does the rate of her food consumption during the period of courtship.

Copulation in ospreys often occurs after the male has returned with a fish (or branch for the nest), and is initiated by either sex performing the copulation solicitation display (Figure 23A–E). In this display the female lowers her wings and fans her tail while holding it depressed. The male

Figure 23. Social behavior of the osprey, including copulation sequence (A–E), based on sketches in Cramp and Simmons (1980), and territorial calling posture (F), after a photo in Brown (1979).

watches her intently from nearby while his body is usually oriented away from her and facing outward from the nest; then he may make a short circular flight or fly to her immediately. He lands on her back with flapping wings, his toes clutched into a fist, and may make a circular flight following copulation, finally returning to the female at the nest (Cramp and Simmons, 1980). Dunstan (1973) provided a similar description, noting additionally that the male uttered

high-pitched whistling notes while circling above the soliciting female and as he landed prior to copulation, and that the female uttered longer, ascending whistles at this same time.

Breeding Biology

It is generally believed that birds first return to their natal areas when they are two years old, but do not begin to breed until the following year

(Henny and Wight, 1969). However, Reese (1977) found that some birds of older ages did not lay eggs every year, and further believed that two-year-olds have not yet been proven nonbreeders. Dennis (1983) reported that in an expanding Scottish population initial breeding occurred at from 3 to 6 years, with the birds nesting from 9 to 100 kilometers away from their natal site. Poole (1989) noted that the average age of initial breeding in New England is 3.6 years, as compared with 5.7 years in Chesapeake Bay, where nest sites are highly limited.

The nest sites of ospreys are highly varied, depending upon locally available opportunities. Swenson (1981) examined 55 osprey nests in Yellowstone Park, of which 82 percent were in trees, and of the tree nests 89 percent were in trees with broken tops. Thus, the birds apparently selected more for tree morphology than for particular species. They also selected trees that were taller than nearby trees, and perhaps selected for dead rather than living trees. Mathisen (1968) similarly reported a high percentage (80 percent) of Minnesota osprey nests in dead trees, whereas he noted (1983) that bald eagles in the same area seldom used dead trees, but did favor trees with dead tops. Compared with bald eagle nests in the same area, ospreys are also more prone to nest in lowland (spruce and tamarack) rather than upland areas, to locate their nests at the top of the crown rather than below it, and to have smaller, more rounded and less cone-shaped nests than bald eagles. Smith and Ricardi (1983) analyzed nest site characteristics of ospreys in New Hampshire, and observed that dead or dying large pines are preferred for nests, and furthermore that such trees tend to be considerably taller than surrounding vegetation. Similarly, 65 percent of 294 nests observed by Reese (1977) in the Chesapeake Bay area were in standing dead snags. Likewise, 86 percent of 109 Ontario tree nests were in dead trees, predominantly of coniferous species. The heights of 137 Ontario nests ranged from 0.6 to 30 meters, with half located between 9 and 18 meters high (Peck and James, 1983).

Reuse of old nest sites is regular in ospreys, although old nests are renovated and new materials are added, most often during the period between the male's arrival and egg laying, especially during the last few days prior to the laying of the first egg. Eggs are laid at intervals of 1 to 3 days, with incubation probably beginning

with the laying of the first egg. Clutch sizes are fairly consistent throughout broad areas; the average of 172 Finnish clutches was 2.63, with 62 percent of the nests having 3 eggs. The average clutch size of 103 Scottish nests was 2.8, with 80.5 percent of the nests having 3 eggs. Apparently birds breeding for the first time normally lay two eggs, whereas older birds typically lay three (Green, 1976). In Ontario, 46 nests had from 1 to 5 eggs, averaging 2.67 (Peck and James, 1983), while 144 nests from New York ranged from 2 to 5 eggs, averaging 3.07 (Bull, 1974). A sample of 513 nests in Chesapeake Bay had an average of 2.91 eggs over a five-year period, with no apparent yearly variability (Reese, 1977). In reviewing egg thickness data, Reese found decreases of from 6 to 24 percent for various years and regions of the United States since the period prior to 1947, and in his study found that 52 percent of the eggs that he monitored failed to hatch. Much of this egg mortality seemed to result from increasing amounts of human disturbance to ospreys incubating thin-shelled eggs, producing egg damage, egg loss, abandonment, and similar causes of nest failure.

The incubation period has recently been estimated by Green (1976) as 34–40 days, averaging 37 days, a considerably longer period than earlier published estimates of 21–28 days. Although early literature also suggests that only the female incubates, it is now generally acknowledged that the male participates to a considerable degree (Dunstan, 1973). Hatching of the eggs is asynchronous, and hatching intervals of up to five days have been reported, producing substantial differences in sizes of nestlings. The average fledging period has been estimated as 53 days (Green, 1976), and 54 days (range 48–59 days) (Stotts and Henny, 1975). Stinson (1977) estimated that the family remains intact for at least 65 to 93 days following hatching, with the young dependent upon their parents for a period of about 93–103 days, and with dependency terminating somewhat prior to the start of fall migration.

An enormous amount of information on osprey productivity has accumulated in the past decade, much of which has been summarized in the volumes edited by Ogden (1977) and Bird (1983), and in the several regional symposia sponsored by the International Osprey Foundation that began in 1983. Reese (1977) summarized productivity data for the 1960s and early 1970s for various parts of the United States, and Poole (1989) has summarized more recent data. He observed that individual breeding success improved with increasing age and experience (at least to about 12 years), and that birds retaining their prior mates were more successful than those with new mates. Furthermore, birds nesting early in the season were more successful than tardy nesters (which had a high rate of nest abandonment), and pairs nesting on artificial nest platforms consistently raised more young than those on natural but often relatively unstable sites.

Henny (1977, 1983) has also summarized a great deal of information on the distribution, status, and population attributes of the species in the United States and North America. Based on available data, he judged that annual adult mortality rates are in the range of 15–20 percent, and that first-year postfledging mortality may be between 41 and 57 percent. Poole (1989) recently reported annual adult mortality rates of about 10–17 percent. Information on postfledging mortality is very limited, but apparently impact and gunshot injuries, electrocution, drowning, emaciation, and respiratory infections are significant contributing factors. A few individuals are known to have survived as long as 24–25 years (Spitzer, 1980).

Evolutionary Relationships and Status

The osprey has received highly variable taxonomic treatment in the past, and has usually been separated taxonomically from all other raptors at the family level (Pandionidae). At times it has been included first in taxonomic sequence because of various seemingly primitive traits such as the similarity of its pterylosis (feather tracts) to those of the New World vultures, while other taxonomists have stressed certain highly specialized traits related to its foraging adaptations (see Cramp and Simmons, 1980, for some of these). Its karyotype has been recently described (Ryttman et al., 1987) as being most like certain accipitrids (especially *Circaetus* and *Pithecophaga*), but sufficiently distinctive as probably to warrant familial distinction. However, the AOU has retained a subfamilial separation in its most recent (1983) check-list.

The osprey was included on the Audubon Society's Blue List of apparently declining species

from 1972 to 1981, and considered a species of "special concern" in 1981 and of "local concern" in 1986. It was classified as "status undetermined" by the U. S. Department of the Interior in 1973, and is now listed as a species of "special emphasis." The species was also listed as ecologically "sensitive" by the U. S. Forest Service at about the same time, and as of the mid-1980s had special conservation status in 15 states (Henny, 1986). The best numerical estimates of the U.S. osprey population are those of Henny (1983, 1986), who judged that about 8,000 pairs existed in the contiguous United States in the early 1980s, with Florida supporting the largest numbers (1,500–2,000 pairs), followed by the Chesapeake Bay area (1,569 pairs) and Maine (ca. 1,000 pairs). Additionally, more than 800 pairs were nesting in Baja California and the Gulf of California during the late 1970s (Henny and Anderson, 1979). A still largely undocumented population breeds in Canada and Alaska, which Poole (1989) optimistically estimated at 10,000–12,000 pairs. He also estimated the endemic West Indian population at only 100–150 pairs. Based on the 1986 Christmas Count, I have estimated the U.S. winter population at 7,080 individuals, with Florida supporting the largest single component of about 4,300 birds. This would suggest that well over half of the total North American population winters in Latin America and the West Indies, where pesticide usage is less controlled than in the U.S. and Canada. Among this group, probably all yearling birds and possibly some two-year-olds also remain south of the U.S. border during the breeding season as well. Although historical data are often not available, some areas are still greatly depleted in osprey populations, in part at least from pesticide effects. However, the development of large reservoirs in the western states has produced some scattered nestings there.

No complete bibliography of the osprey has yet been published, but the falconiform bibliography by Olendorff and Olendorff (1968–70) included almost 400 references relevant to the osprey. Selective bibliographies have also been provided by Zarn (1974) and Evans (1982), the latter including nearly 300 citations. Palmer (1988) cited 164 papers in his recent compilation. Poole (1989) has provided a recent bibliography of nearly 300 references.

Subfamily Accipitrinae

(Kites, Typical Hawks, and Eagles)

PERNINE KITES

Figure 24. North and Central American residential range of the hook-billed kite, including forms aquilonis *(AQ),* uncinatus *(UN), and* wilsonii *(WI). The South American residential range is shown on the inset map.*

The Cuban endemic form wilsonii *is considered by some to be a separate species. Indicated range limits of contiguous races should not be considered authoritative.*

Hook-billed Kite *Chondrohierax uncinatus* (Temminck) 1822

Other Vernacular Names

Mountain hawk (Grenada), Wilson's kite (*wilsonii*); gavilan pico ganchudo (Spanish)

Distribution

Resident (*uncinatus* group) in forested areas from Tamaulipas and recently extreme southern Texas south through Mexico and Central America to southern South America; also in Grenada and (*wilsonii* group) in eastern Cuba. (See Figure 24.)

North American and West Indian Subspecies

C. u. aquilonis Friedmann: Local resident in eastern Mexico (Tamaulipas, Veracruz south to Chiapas); recently also the Rio Grande valley of Texas. Recognized by Stresemann and Amadon (1979), but included in *uncinatus* by Brown and Amadon (1968). Smith and Temple (1982b) also found no significant differences between Mexican birds and those from other continental regions.

C. u. uncinatus (Temminck): Breeds from northwestern Mexico (Sinaloa) south locally to Argentina.

C. (u.) wilsonii (Cassin): Rare resident of eastern Cuba (often considered a full species).

Description (of *uncinatus*)

Adult male. Typical (gray-bellied) morph: Above dark lead-black, becoming more fuscous in worn plumage; the occiput with much basal white and the upper tail coverts tipped and banded with white; sides of face, ear coverts, and chin deep to dark plumbeous; under tail coverts white to ochraceous-buff; remainder of underparts deep lead-gray, unmarked or variably barred with narrow lighter bands of white to cinnamon-brown, which are variable in width and are usually narrowly bordered by fuscous or fuscous-black (the underparts rarely barred with rufous as in the female); axillaries and under wing coverts uniform deep lead-gray, barred with white or buff; primaries banded with white and lead-black below, dark lead-gray and lead-black above, the outer webs often uniform lead-black; secondaries uniform dark lead-gray; tail mostly lead-black to black, white basally, narrowly tipped with white or deep gray, and crossed by two white or

pinkish buff bands. Mandible pale yellow-green, with dusky edges, maxilla black (the bill entirely yellowish in Cuban race); iris white to bluish gray; cere, gape, and skin around mouth edges greenish yellow; eye ring and lower lores pea-green; a flap of skin above lores and in front of eyes bright yellow to orange; tarsi and toes orange-yellow, claws black.

Adult female. Typical (brown-bellied) morph: Forehead, auriculars, and sometimes the chin deep gray or dark lead-gray; crown and occiput fuscous to fuscous-black, with concealed white bases; a broad, continuous nuchal collar of ochraceous-buff, tawny, or amber-brown, remainder of upperparts fuscous to fuscous-black, darker anteriorly, often with slightly paler edges to the feathers; upper tail coverts tipped and barred with white or pale gray; entire underparts, including under wing coverts, white or ochraceous-white, with broad, nearly equal transverse rufous-brown bars; outer primaries pale fuscous above, white or pale mouse-gray below, cream or pinkish buff toward the bases of the inner webs, and distally banded with fuscous or fuscous-black; inner primaries chestnut, or russet shading to creamy or pinkish buff toward the bases of the inner webs; tail fuscous-black to black, white basally, narrowly tipped with white or pale brown, and crossed by two brown or gray bands, shading to white or pinkish buff on the inner webs; underside of primaries gray to rufous-gray, barred with white. Soft parts similar to those of male, the facial skin possibly paler.

Dark morph (both sexes). Entire plumage from dark gray to deep fuscous-black, with a slight bronzy purple-green gloss; the occiput feathers with much basal white; tail black, narrowly tipped with white and crossed by a single broad white band. Softparts as in typical morph. (This rare variant has not yet been reported from the U.S.)

Juvenile. Typical morph: Dark brown above, the feathers edged with rufous or whitish; creamy white below with variable brown barring (the bars usually narrower than adult female's, and sometimes lacking), and with a white nuchal collar and three (normally) or four light brownish gray bands on the otherwise blackish tail, which is also gray-tipped. In females the tail bars are somewhat broader and more brownish than in males. Iris dull yellow to brown; bill dusky gray, becoming dull yellowish green at base, on cere, and in anterior lores; tarsi and toes dull yellow.

Dark morph: Nearly all brownish black, the crown with much concealed white; gray-tipped tail with two (normally) or three white-bordered gray bands; the primaries barred sparingly with whitish below. Soft parts as in typical morph.

Measurements (of *uncinatus,* in millimeters)

Wing, males 265–301 (ave. of 26, 285.8), females 268–321 (ave. of 31, 289.4); tail, males 173–210 (ave. of 26, 191.1), females 191–229 (ave. of 31, 202.8) (Friedmann, 1950). Average egg size 53.6 x 40.5 (Schonwetter, 1961).

Weights (in grams)

Males 251–257 (ave. of 3, 253), females 240–300 (ave. of 3, 255) (Haverschmidt, 1962). Six males 247–274, 7 females 235–360 (Brown and Amadon, 1968). Average of 4 males, 265, of 3 females, 296 (Hartman, 1961). Average of 10 (unsexed), 270 (Willis, 1980); range of weights ca. 233–297 (sex and sample size unspecified) (Oberholser, 1974); range 215–353, ave. 277 (sex and sample size unspecified) (Clark and Wheeler, 1987). Average egg weight 33.5 (Haverschmidt, 1965), or 7.7% of female.

Identification

In the hand. The combination of the lack of a bony shield above the eye and a bill that is unusually large (culmen from cere 25–35 mm) but strongly compressed and hooked, with linear, almost slitlike nostrils, serves to identify this species.

In the field. When perched, typical (gray-bellied) males appear grayish black above and below on the wings and body, while typical (brown-bellied) females are more rufous on the underparts (including the under wing coverts when seen in flight), and have a rufous collar as well as a rufous area evident on the inner primaries (also visible only in flight). The tips of the wings reach about the midpoint of the tail. Immatures generally resemble adult females, but have a contrasting white (not chestnut) collar, are less heavily barred with brown or may be entirely white below, and have brown rather than whitish eyes.

In flight, this species's wings are distinctly rounded or paddle-shaped in outline, and the long tail is also slightly rounded, with two (in adults) or three (in immatures) lighter bands of similar width, terminated by a narrow whitish tip. The undersides of the wings are strongly barred, with the bases of the primaries somewhat lighter than the rest of the underside of the wing. Flying is done with rather slow, deep wingbeats, and gliding is done on slightly bowed wings. Calls include a musical three-noted whistle (uttered when at rest in trees) and harsh chattering and screaming when alarmed or in aggressive situations. The typical flight pattern consists of alternated rapid flapping and gliding.

Habitats and Ecology

This is a species of tropical, subtropical, and sometimes temperate forests, occurring up to about 1,800 meters elevation, perhaps most common in the tropical zone. Sometimes, however, as in Costa Rica, it is most common in the subtropical zone. Primarily a species of dense undergrowth, it also moves out into shaded coffee plantations and feeds in open marshes, where snails are abundant (Brown and Amadon, 1968). In Panama it typically occurs on wooded slopes above streams, perching in the middle tree branches or in their crowns but coming to the ground to feed on land snails. Thiollay (1985a) listed it as occurring in all five of the tropical American rain forests he surveyed, in Cuba, southern Mexico, Panama, Guiana, and eastern Ecuador. He considered it (1985b) as most abundant in virgin primary forests, and secondarily so in secondary forests and primary forest edges. Smith and Temple (1982b) reported that all of the sightings they made on Grenada (where the endemic race is nearly extirpated) occurred in xeric woodland of the southern part of that island, and that land snails were most common in that area. Six nests reported on by Smith (1982) from Tamaulipas, Mexico, were in thorn scrub woodland dominated by *Acacia* spp. Slud (1964) considered it a nonforest species inhabiting open woodlands, tree plantations, and cutover areas with low, densely foliaged trees.

Foods and Foraging

The most remarkable aspect of this species is its remarkable variation in bill size, which is not related to sex, plumage phase, or age, and is largely independent of geography. Bill polymorphism is evidently a result of disruptive natural selection; thus kites of differing bill sizes are able to feed on snails of differing sizes. The birds feed almost exclusively on tree snails of a variety of genera, which they snatch with the bill and then fly with to a tree. Typically the bird then transfers the snail to its left foot, and while the snail is braced firmly against the branch the tip of the bill is inserted into the snail's aperture. By progressively breaking each whorl of the shell, the body of the snail is gradually removed from the shell, and the extracted snail is swallowed whole (Smith and Temple, 1982a). This is a quite different procedure for snail extraction from those used by various other snail-eating birds (Snyder and Snyder, 1969).

The sizes and types of snails used by the kites vary considerably in different areas; in Tamaulipas a small form, *Rhabdotus alternatus,* is the only available tree snail, while in Colima, Mexico, a large form, *Orthalicus ponderosus,* and a smaller one, *Dryameus colimaensis,* are both used, typically by large- and small-billed birds respectively. In other areas of western Mexico where both bill-size morphs are found but where the smaller snail does not occur, selection for bill polymorphism may be maintained by the birds feeding on different size classes of snails in the genus *Orthalicus* (Smith and Temple, 1982a). In southern Texas the kites have been observed eating *Rhabdotus alternatus* (Delnicki, 1978), and in Cuba the endemic race feeds on the tree snail *Polymita picta.* Use of the aquatic apple snail *Pomacea,* on which the snail kite feeds exclusively, has been rarely reported (Brown and Amadon, 1968).

Social Behavior

According to Brown and Amadon (1968), this species is usually encountered singly or in groups of two or three, as might be expected from a highly specialized forager dependent upon a very limited food base. Smith (1982) found six nests within five kilometers of one another, suggesting moderate but not extreme social dispersion. Paulson (1983) reported seeing a soaring flock of 25 hook-billed kites, and believed that the kites may aggregate at food source areas where snails are concentrated.

The social behavior of this species is essentially unknown. Its usual call has been described as an oriole-like musical whistle, but calls also include harsher chattering and screaming notes that are uttered when chasing another hawk (Brown and Amadon, 1968). Smith (1982) said that the kite's nest defense call is very similar to that of the northern flicker *Colaptes auratus,* but louder.

Breeding Biology

Only a relatively few nests of this species have yet been described; Smith (1982) described six nests from Tamaulipas, Mexico, and noted that six had previously been described from elsewhere in the species's range. Kiff (1981) added some additional nesting records. All six of the Tamaulipas nests were in huisache (*Acacia farnesiana*) trees at heights of 5–7 meters, and all were found during June. The first nest found in Texas, at Santa Ana National Wildlife Refuge, was located in early May 1964, in a willow (*Salix nigra*) tree. The nest was located about seven meters above ground and was composed of twigs and branches (Fleetwood, 1967). A second nest was found in early May 1976, in a Texas ebony tree (*Pithecellobium flexicaule*), 6.5 meters above ground. Typically the nests of this species are small and so frail that the eggs or young may be seen from below (Smith, 1982; Orians and Paulson, 1969). Nesting also occurred in Texas during 1978, with the male observed carrying nesting material in late April (*American Birds* 32:1183). Apparently both sexes help build the nest, and both sexes evidently help with incubation (Haverschmidt, 1964; Smith and Temple, 1982a). Typically two eggs or young have been found, and in one case a second egg was found in a nest three days after the male and a single egg were collected. However, the usual egg-laying interval and incubation period remains unknown. Evidently males participate in incubation, as they have incubation patches and have been seen attending nests.

Little is known of the nestling and later stages of breeding, including the fledging period. Apparently the young are reared on snails, as the area immediately surrounding active nests becomes littered with broken shells (Smith, 1982). The fledging period is unknown, as is the incidence of renesting, but in 1985 a pair in Texas that began a nest in May began a second effort after the first was unsuccessful (*American Birds* 39:935). Intact family groups have been observed through the winter months.

Evolutionary Relationships and Status

This is a monotypic genus, which according to Brown and Amadon (1968) is related to the less specialized and mostly insect-adapted gray-headed kite (*Leptodon cayanensis*) of the New World and the cuckoo-falcons (*Aviceda*) of the Old World. The hook-billed kite is apparently not a close relative of the snail kite, in spite of similarities in bill structure and snail-feeding adaptations.

The occurrence of the hook-billed kite in the United States must be regarded as essentially accidental. Since the late 1970s the birds have reportedly nested sporadically at Santa Ana National Wildlife Refuge, at the Anzalduas unit of Rio Grande National Wildlife Refuge, at Bentsen Rio Grande State Park, and also at Rancho Santa Margita, Starr County, Texas.

The insular populations of this species are apparently in great danger, especially on Grenada, where *mirus* is on the verge of extinction, and in Cuba *wilsonii* is rare and localized in Oriente province between Moa and Baracoa (King, 1981). The species was apparently extirpated from Trinidad over a century ago (Wiley, 1985).

American Swallow-tailed Kite *Elanoides forficatus* (Linnaeus) 1758

Other Vernacular Names

Fork-tailed kite, scissor-tailed kite, snake hawk; milan à queue fourchue (French); milano tijereta (Spanish)

Distribution

Breeds locally from South Carolina south to Florida, and west to eastern Louisiana (to central Texas until about 1914); and from southeastern Mexico south through most of Central America and South America. Formerly bred north to central Oklahoma (last record 1902), Arkansas, eastern Kansas, eastern Nebraska, northwestern Minnesota, and southern Wisconsin (most of these latter records occurring prior to 1900). In the U.S. now common only in Florida, with additional local breeding in South Carolina, and perhaps locally or rarely in North Carolina, Georgia, and along the Gulf Coast to Louisiana or eastern Texas.

Winters in South America from Colombia and Venezuela southward, rarely north to Central America. Rare transient through Cuba and Jamaica. (See Figure 25.)

North American Subspecies

E. f. forficatus (Linnaeus): Breeds or has bred along the coast of the U.S. locally from South Carolina to Texas, and south locally in Mexico to Chiapas (nesting in eastern Mexico still unproven). Winters south to Brazil and Argentina.

E. f. yetapa (Vieillot): Breeds from northern Central America to Argentina, migratory in southern parts of range. Northern limits of subspecies uncertain; possibly intergrading with *forficatus* in southern Mexico, but this form is only very weakly separable from *forficatus*.

Description (of *forficatus*)

Adult (sexes alike). Tail very deeply forked, the lateral rectrices more than twice as long as the central pair; head, nape, neck, entire underparts, and a broad transverse area on the lower back white; upperparts slate to blackish slate with a slate-gray bloom on the secondaries (a patch of power down feathers is present), larger upper wing coverts, and upper tail coverts; scapulars

and interscapulars with blackish violet-gray sheen, the interscapulars like the scapulars but with relatively more bronze-green and less blackish violet-gray gloss; the anterior wing coverts and the alula slate-black and darker than the rest of the wing; secondaries white for their basal two-thirds, the terminal third slate or blackish slate; the under wing coverts and axillars white except for a few slate-black feathers at the margin of the wrist. Iris reddish brown to very dark brown; eyelids, cere, the basal half of the mandible, tarsus, and toes all pale ashy bluish, often with a yellowish or greenish tinge; tip of bill black; interior of mouth cobalt-blue; claws grayish brown.

Subadult. By the spring following hatching an adultlike plumage is present except for dark shaft streaks on the crown and chest, some white tips on the contour feathers, and a brownish cast to the dark areas. Breeding can apparently occur at one year, before the definitive plumage has been fully attained.

Juvenile (sexes alike). Similar to adults but with the head, nape, throat, and upper breast streaked with narrow fuscous shaft streaks, the streaks darker above than below, and with a rufous or buffy bloom on the breast that is lost soon after fledging; remiges and rectrices with narrow whitish edges, and the outer tail feathers considerably shorter than in adults; upperparts dull black or with a slight greenish iridescence, the upper back sometimes slightly mixed with pinkish cinnamon. Iris brown; bill black; cere, tarsi, and toes pale greenish yellow.

Measurements (of *forficatus*, in millimeters)

Wing, males 423–436 (ave. of 8, 431.2), females 436–445 (ave. of 12, 440.3); tail, males 328–343 (ave. of 8, 334.4), females 343–370 (ave. of 12, 355.6) (Friedmann, 1950). Average egg size 46.7 x 37.4 (Bent, 1937).

Weights (in grams)

Two males (of *yetapa*) 390 and 407, females 327–435 (ave. of 4, 399) (Haverschmidt, 1962). Average of 6 males 441, of 7 females 423 (various sources cited in Palmer, 1988). Average of 2 (unsexed) 400 (Willis, 1980); range of weights 325–500, ave. 430

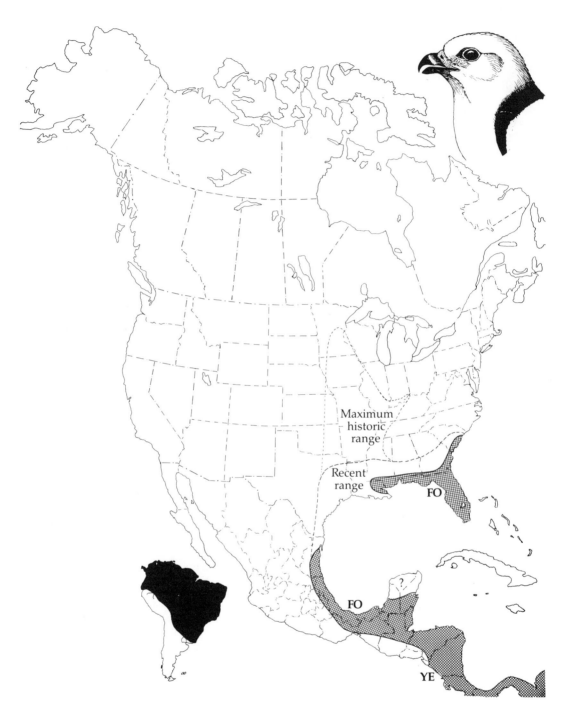

Figure 25. North and Central American breeding range of the American swallow-tailed kite, including races forficatus (**FO**) and yetapa (**YE**). Probable earlier historic U.S. breeding ranges are also indicated (dotted lines), and the South American breeding range (which includes wintering areas of northern migratory populations) is shown on the inset map. Indicated range limits of contiguous races should not be considered authoritative.

(sex and sample size unspecified) (Clark and Wheeler, 1987). Estimated egg weight 36, or about 9% of female.

Identification

In the hand. The deeply forked black tail (the lateral rectrices more than twice as long as the central ones in adults, less than twice as long in immatures) and the strongly contrasting black and white plumage pattern serve to identify this species readily.

In the field. Perched birds are easily identified by the all-white head, neck, and underparts, and blackish upperparts, including a forked tail that is nearly as long as the body itself. The long wings cross over above the tail and nearly reach its tips (adults) or beyond them (juveniles).

In flight, the white wing linings and underside of the body sharply contrast with the black flight feathers and black tail. The flight is light and swallowlike, the birds occasionally gliding on flat wings, alternating with deep and slow wingbeats. Hissing whistles and repeated, usually triple alarm *klee-klee-klee* notes are sometimes uttered; similar sharp notes are produced during courtship. Male calls are slightly lower pitched than those of the female.

Habitats and Ecology

In the United States this species's breeding habitats consist mainly of lowland cypress swamps or tall riverine woodlands (either coniferous or hardwood) adjoining open glades, prairies, marshes, or swamps. In southern Mexico it nests in pine forests at 1,600 meters or higher, but in Argentina it occurs in humid forests at much lower elevations (Brown and Amadon, 1968). Elsewhere, as in Panama and elsewhere in Central America, it is widespread in tropical and lower subtropical habitats. Since foraging is done entirely on the wing (including nest robbing and snatching lizards or snakes from branches), it is unlikely that these habitats share any special common characteristics beyond providing suitable nesting trees and an abundant food supply that can be caught in the air. Nests are typically built at the tops of fairly tall and isolated trees, in sites that are visible from above but sometimes very difficult of access from below. In Costa Rica these trees are often slender, about 30–40 meters

tall, and relatively isolated or with crowns rising above the surrounding trees (Skutch, 1965). Nests observed by Snyder (1975) in Florida were in much lower trees.

Foods and Foraging

Skutch (1965) believed that flying insects represented the mainstay of swallow-tailed kite foods in Costa Rica, but with more substantial items being taken whenever possible. These included especially the nestlings of birds that nest in exposed tree branches, plus occasional lizards. However, elsewhere such as in Florida a diversity of prey, such as frogs, tree snakes, and wasp larvae have been observed as dietary items. Birds' eggs, small mammals, and perhaps even minnows may be eaten on rather infrequent occasions, and a few isolated cases of fruit eating have also been reported. Snyder (1975) found that hylid frogs represented the bulk of foods brought to nestlings by two pairs, while anole lizards and nestling birds comprised the majority of foods observed at a third nest. Sutton (1955) also reported frogs and anole lizards as the major foods of nestlings, with insects and nestling birds next most frequent, while Wright, Green, and Reed (1971, and unpublished data cited by Snyder) found that insects and frogs were the most common food items brought to nestlings.

Besides their attraction to forests for nesting, the kites also are associated with prairies and other open spaces, where aerial foraging on insects can effectively be done (Cely, 1979). Basically two foraging methods are commonly used: capturing flying insects such as dragonflies in free flight with the talons, and grabbing animals from the outer leaves and branches of trees. The latter are often caught during continuous flight by the kites, either by swooping down from above or by coursing over and among the treetops. Sometimes entire nests are carried away, from which the nestlings are extracted and consumed. Where flying insects are abundant, as over marshes, foraging groups of up to 18 birds have been observed in Florida during the breeding season. No defense of feeding territories is evident (Snyder, 1975).

Social Behavior

Because of their long migration route, it is unlikely that kites nesting in North America are able

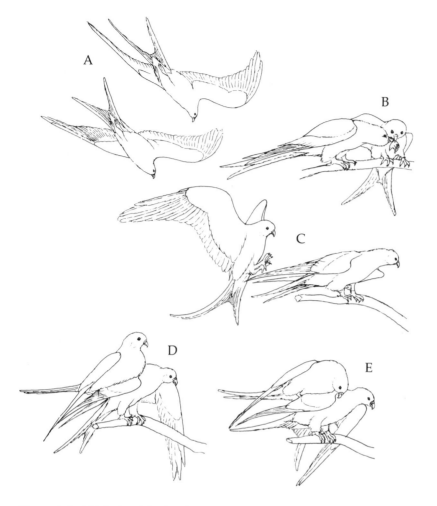

Figure 26. Social behavior of the American swallow-tailed kite, including diving swoop by a pair (A), and a copulation sequence (B–E). Based on photos in Snyder (1975).

to maintain their pair bonds throughout the year. However, at least some kites appear to arrive on their Florida nesting areas already paired. After arrival the birds spend much time soaring above their nesting areas, the female often flying above the male, and the higher bird often swooping down to the lower one. At times this may be done by a female toward her mate, and at other time a male may swoop down on birds other than his mate. These swoops (Figure 26A) are often accompanied by *klee-klee-klee* or sometimes *kees-a-wee* vocalizations (Snyder, 1975).

Kilham (1980) observed an unpaired male interfere repeatedly with the copulation of another pair, trying to copulate with and courtship-feed the already mated female. He also interfered with nest-building activities of the pair. Of 84 prey transfers during courtship feeding observed by Kilham, at least 83 involved anole lizards, leading him to speculate that perhaps females always "demand" anoles in this situation. In one case a frog was apparently rejected.

Copulatory behavior (Figure 26B–E) apparently begins soon after arrival in late February or early March (in Florida), often starting before sunrise and continuing with diminishing frequency through the day. Snyder observed six or possibly seven such copulations by one pair in a single day during early March. The behavior is mostly confined to the nest-building period, and ends at about the start of incubation. Females typically perch facing into the wind on an exposed horizontal tree limb. As the male flies over to her, the female bends forward with wings somewhat extended. The male lands on her back, and droops his wings over her as he adjusts his position. After this is completed the male's tail is lowered under the female's, the bill is brought down on her back, and mating occurs. Thereafter the male flies forward over her head as both birds call, the male uttering *klee-klee-klee* and the female *kees-a-wee* notes (Snyder, 1975).

Both sexes also usually call during nest exchanges, uttering either *eeep* or *kees-a-wee* notes; such notes were heard in 31 of 43 observed nest exchanges. These notes are uttered by both sexes, although those of males average lower in pitch. The *klee* call is usually given in groups of three syllables, but this is somewhat variable. It occurs in several situations, including interactions with potential predators, during display swoops, following copulation or food transfer, and while soaring overhead in groups (Snyder, 1975).

Breeding Biology

Both sexes actively participate in nest building, the twigs being broken from exposed branches while the birds are in full flight. The twigs are then carried back to the nest site balanced in the talons, the two sexes apparently equally sharing in nest construction (Skutch, 1965). However, gathering of nest material during the incubation period was done mainly by males in three nests studied by Snyder (1975). Lichens are later added to the nest for lining.

Of 65 clutch records in the Western Foundation of Vertebrate Zoology files, nest heights ranged from 10.6 to 30.5 meters, and 48 of the total were between 18 and 25 meters. Of the total, 51 nests were in pines, 6 in cypresses, 3 each in cottonwoods and mangroves, and 1 each in two other tree types. Of 25 Florida nests, 15 were located in pines (Meyer and Collopy, 1988). Nests are typically placed near the tops of the tallest and largest available trees, and are usually situated in open stands fairly near the forest edge. Ten Texas records were from April 20 to

June 7, and 66 Florida records were from March 23 to May 10, with 34 of the latter occurring between April 10 and 25. Bent (1938) also reported a peak between April 7 and 26 for 81 Texas and Florida nests. Of 80 clutches in the Western Foundation of Vertebrate Zoology collection, 58 were of two eggs, 19 were of three, and 3 were single-egg clutches, for an average of 2.2 eggs.

Snyder (1975) estimated the incubation period as approximately 28 days. The nestling period at one nest was between 36 and 39 days, and at a second nest was 39 days for one chick and 41 for the other. Other estimates of the nestling period are 36–40 days (Sutton, 1955), 37–42 days (Wright, Green, and Reed, 1970) and six weeks (Meyer and Collopy, 1988). Males feed their mates at the nest during incubation, but females also forage on their own while males take over incubation. Incubation is done mainly by females, judging from Snyder's observations. Similarly, females do most of the brooding of the nestlings, relying on the male for foraging. Toward the middle of the nestling period females begin to spend more time away from the nest, helping capture prey, and near the end of the period both sexes bring in prey and feed them directly to the young. In some cases extra birds have been observed at nests participating in nest building, nest defense, and other breeding activities; the identity of these birds is still unknown, but quite possibly many of these extra birds are nonbreeding adults rather than immatures (Meyer and Collopy, 1988). Following fledging the young birds apparently continue to be dependent on their parents for food for some time.

Meyer and Collopy (1988) found that of 23 Florida nests studied by them, only 9 (39 percent) were successful, and 11 young were fledged from these nests, an average of 0.48 young per nest attempt (1.2 per successful attempt). There was only one known renesting effort. They stated that this level of nesting success was well below that found by J. E. Cely in South Carolina, where there was a 73 percent nesting success in 22 nests, and an average of 1.1 young reared per nesting attempt. Losses from wind and predation are apparently the principal causes of nestling mortality. Limited data from the Florida studies suggest that the highest nest success rates are associated with nests in cypress trees, the lowest with casuarina trees, and intermediate rates are associated with pines.

Following fledging, there seems to be a concentration of birds prior to fall departure in southern Florida. Millsap (1987) reported a concentration of nearly 700 kites in late July 1986 on the western side of Lake Okeechobee, which gradually diminished to 30 by early September. He believed this to represent an annual concentration, associated with a communal nocturnal roosting area. This may represent a good part of the entire Florida population, which may number 250–400 pairs. It may also include some or all of the South Carolina population of 50–60 pairs, which nest in the Francis Marion National Forest (Cely, 1979). There are also a few nesting pairs in the Okefenokee Swamp, Georgia, some in northwestern Florida, some in the Mobile River delta, Alabama, and also some in the Atchafalaya River basin, Louisiana (Parker, 1985). All told, the late summer population in the southeastern United States might be as large as 2,600 individuals, of which perhaps a quarter may thus stage in the Lake Okeechobee area prior to migration.

Evolutionary Relationships and Status

This is a unique species of kite, which Brown and Amadon (1968) believed to be a rather isolated form, but sharing some characters with the cosmopolitan *Elanus* group of white-tailed kites and some with the Old World *Pernis* group of "honey-buzzards." Shufeldt (1891) observed that osteologically *Elanoides* is quite distinctive from the other kites he studied (but shared some similarities with *Pandion*), and judged that it belongs in a subfamily separate from *Elanus* and also from *Ictinia*, the latter of which osteologically resembled in Shufeldt's opinion "a miniature *Buteo*."

The North American population status of the swallow-tailed kites is not very favorable; as noted in the prior section it may currently (late 1980s) number only about 2,000–3,000 birds, which are mostly limited to southern Florida. It also breeds locally in northwestern Florida, especially in the Apalachicola and Ochlockonee River drainages. There is a much smaller South Carolina population; there it seems to be locally increasing along the Savannah River in Hampton County (Cely, 1987). It is probably currently a rare summer resident of the coastal plain of Alabama, and possibly breeds in North Carolina's Mattamuskeet National Wildlife Refuge. Immature kites were seen in the Atchafalaya basin of Louisiana and the Austin, Texas, region during the 1970s (Cely, 1979).

At one time the species had a much broader United States breeding range, but by 1900 this was already contracting rapidly, apparently because of habitat changes such as prairie cultivation that destroyed foraging areas and logging of streambottom woodlands used for nesting, plus indiscriminate shooting (Cely, 1979). It is thus unlikely that pesticides or other recent environmental changes can be implicated in bringing about the current situation. The species has recently become (as of 1985) a candidate for federal listing as an endangered species by the Department of the Interior.

ELANINE AND MILVINE KITES

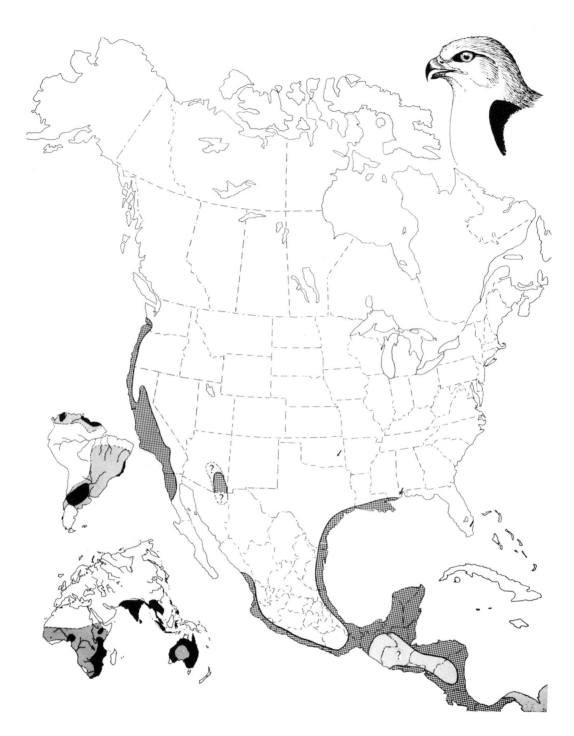

Figure 27. North and Central American residential range of the black-shouldered kite. Several recent U.S. nesting records beyond the indicated range are shown with arrows. Because of the expanding Mexican and Central American ranges, present limits there are very uncertain, *and possible recently colonized areas are shown by light shading. The breeding (inked) and nomadic or irruptive (shading) South American and Old World ranges are shown on inset maps.*

Black-shouldered Kite *Elanus caeruleus* (Desfontaines) 1789

Other Vernacular Names

Australian black-shouldered kite (*notatus* group), black-winged kite (*caeruleus* group), northern white-tailed kite (*majusculus*), white-tailed kite (*leucurus* group), white hawk; elanion blanc (French); milano maromero (Spanish)

Distribution

Resident locally from southwestern Washington (since 1982) and western Oregon (since 1977) south to northwestern Baja California; also in southeastern Arizona (locally since 1983), and from coastal Texas and southern Louisiana (since 1983) south through eastern Mexico and Central America (both slopes) to Argentina and Chile (*leucurus* group); recently also bred in peninsular Florida (nesting in 1986 was the first reported since 1910). Scattered breedings have recently occurred in Oklahoma (1981), Mississippi (1983), and possibly also New Mexico and North Dakota. Vagrants have been reported east to the Carolinas and north to Minnesota and Massachusetts. The breeding range has greatly expanded since 1960, both in Central America and North America (Eisenmann, 1971; Pruett-Jones et al., 1980). Also widespread in Australia (*notatus* group), and in southern Eurasia and Africa (*caeruleus* group), these forms often being considered allospecies. (See Figure 27.)

North American Subspecies

E. c. majusculus Bangs and Penard: Resident from southern and western U.S. south through Mexico to Nicaragua and Costa Rica, possibly to eastern Panama (where intergradation with or replacement by *leucurus* may occur; the origins of the expanding population of this region are uncertain).

Description (of *majusculus*)

Adult (sexes alike). Upper surface light to deep gray, becoming gradually white on the anterior crown and forehead; sides of head white, a semicircle of black feathers immediately anterior to the eye and continuing to the angle of the mouth and as a narrow black rim around the eyes; chin, throat, and rest of underparts pure white, the axillaries and lesser under wing coverts white, the outer median wing coverts black (this black patch is lacking in nominate *caeruleus*), and the greater wing coverts light gray; wings dark gray above, the lesser secondary wing coverts black, forming a distinct "shoulder" patch; middle pair of rectrices light gray; the rest of the tail white and more or less washed with light gray. Iris orange-rufous, chrome-yellow, or dull buffy yellow; cere chamois-colored to cream-buff; bill black; tarsi and toes dull buffy yellow; claws black.

Subadult. An adult body plumage is attained a few months after fledging, the birds retaining only their juvenal remiges and rectrices until the following summer. Initial breeding probably occurs within a year of fledging, or while the birds are still presumably carrying juvenal tail and wing feathers (Brown, Urban, and Newman, 1982).

Juvenile (sexes alike). Back of crown and nape heavily streaked with brown and white; back and scapulars brownish, the feathers narrowly edged and tipped with white, as are the inner flight feathers; throat and breast streaked or washed with cinnamon. Rectrices with a dusky subterminal band. Iris initially olive-gray, later becoming yellow-brown to orange; tarsi and toes pale yellow.

Measurements (of *majusculus*, in millimeters)

Wing, both sexes 302–328 (ave. of 14, 314); tail, both sexes 174–186 (ave. of 14, 181.6) (Friedmann, 1950). Measurements of the two sexes are not significantly different in adults, at least in the nominate race (Cramp and Simmons, 1980). Average egg size 42.5 x 32.8 (Bent, 1937).

Weights (in grams)

Two males 311 and 322, females 332–357 (ave. of 5, 350) (Dunning, 1984). Males 250–297 (ave. of 3, 273), one female 307 (Haverschmidt, 1962). Range of weights 305–361, ave. 330 (sex and sample size unspecified) (Clark and Wheeler, 1987). Six unsexed birds, 315–330, ave. 322 (P. Bloom, in Palmer, 1988). Birds of the nominate taxon *caeruleus* (often considered a separate species) are distinctly smaller; 28 males and 27 females averaged 237 and 258 gm respectively (Brown, Urban, and Newman, 1982). Estimated egg weight 25.3, or 7.2% of female.

Identification

In the hand. The combination of blackish shoulder feathers and talons that are flattened or slightly convex on their undersides serves to identify this species.

In the field. Perched adults appear white below and medium gray above, with a conspicuous black forewing patch and a white tail. Perching may be done on telephone wires as well as on poles. The wingtips reach to about the end of the tail, which is often raised and lowered in a nervous fashion. Juveniles are brownish rather than grayish above, but the blackish forewing patch is still fairly conspicuous. Juvenile Mississippi kites are somewhat similar, but are heavily streaked below, and are generally darker above, without a contrasting dark forewing area.

In flight, the black forewing patch is apparent from above, but equally evident is a smaller black carpal or wrist patch on the underside of each wing, which contrasts with the otherwise whitish underwing linings. The tail is fairly long, and in flight appears nearly square-tipped (actually the outermost pair of feathers is distinctly shorter than the second pair, and the median pair is next shortest, producing a slightly double-rounded effect) and white, and the wings are narrow and pointed. The typical flight pattern is one of slow, rather gull-like wingbeats alternating with glides while the wings are held slightly above the horizontal. The birds frequently hover in place, the tail depressed and feet dangling, often fairly close to the ground while searching for small rodent prey. They frequently hunt around dusk. Calls include a descending whistle used in alarm, but the birds are otherwise generally rather silent and their calls are relatively weak.

Habitats and Ecology

This species is adapted to a wide range of habitats throughout its multicontinental geographic range. However, it generally occupies tropical to temperate climatic zones, and habitats ranging from semideserts or open grasslands with scattered trees to lightly wooded areas and especially habitats combining open-country foraging areas with trees having fairly dense canopies for nesting. These include riparian woodlands in open country, areas near fresh-water marshes, or irrigated agricultural lands, all of which are likely to support good populations of microtine rodents. The increase in irrigation of arid lands in the southwestern states has probably tended to increase suitable habitat for the species in recent years, and may help account for its apparent recent range expansion. In a similar manner, the partial clearing of oak woodlands in western Spain, and the subsequent increase in agricultural acreage and consequent increase in small rodent populations has brought about population increases there (Molinero and Cantisan, 1985). A similar trend is occurring in Africa as a result of tree clearing, as well as from tree planting in open grasslands (Brown, Urban, and Newman, 1982). Clear cutting of forests in Mississippi may also have allowed for the species's recent invasion into that area, although when these areas are replanted to pines the habitat soon becomes unsuitable again (Toups, Jackson, and Johnson, 1985).

Foods and Foraging

Nearly all the evidence indicates that this species primarily forages on small rodents, especially diurnal forms such as *Microtus*. Bent (1937) listed other small mammals, including shrews, ground squirrels, pocket gophers, and wood rats, plus some small birds and larger mammals, and even lizards, snakes, and insects. However, these are certainly exceptional, and in southern Africa over 90 percent of the prey is comprised of small mammals in the weight range of 40–90 grams (Mendelsohn, 1981; Brown, Urban, and Newman, 1982). Likewise in Spain only a few species of small rodents are used as primary prey, with birds, insects, and lizards becoming significant only at certain times and localities when and where they become highly available (Molinero and Cantisan, 1985). Likewise in central Chile a small endemic field mouse (*Akodon olivaceus*) averaging about 30 grams was the primary prey, according to Meserve (1977). This author summarized and compared his data with those from eight food studies from California, noting that *Microtus californicus* was the dominant prey (47–88 percent of prey items) in six, and *Mus musculus* was dominant (83–85 percent) in the other two. He suggested that a minimum prey body weight (prey species of usually more than 12 grams adult weight), diurnal prey activity, and a minimum prey abundance represent the major prerequisites for specializing on a particular prey species. Waian's (1973) study in the Santa Barbara area indicated that the kite is a "nearly obligate" predator on *Microtus californicus* there, and this was also the most common mouse captured during his small mammal trapping census activities.

In a study of hunting behavior in California, Warner and Rudd (1975) found that kites of both sexes made hunting forays averaging about six minutes, a duration that did not significantly vary with season or success rate. The birds hunted most frequently from dawn until four hours past dawn, and again during the period of 10–16 hours past dawn, regardless of season. Wintering birds spent about 5 percent of their time hunting, whereas males spent from 12 to 15 percent of their time hunting after their nesting mates had ceased their own hunting. The highest intensities of hunting by males occurred during April and May, when the males were supporting nestling young as well as their mates. The kites typically hunted over open areas, usually searching while hovering about 30 meters above ground, and then dropping to about three meters before finally striking, usually with the feet but sometimes head-first. Hunting intensity and rate of prey strikes increased toward the end of the day. More than 60 percent of the individual prey strikes were successful in obtaining prey, and there was about a 40 percent overall hunting-foray success rate (percent of hunting forays resulting in prey captures).

Similar studies done in southern Africa on nominate *caeruleus* by Tarboton (1978) indicated a lower overall average success rate for prey strikes. He noted that strikes made from a hovering start are generally more effective than those starting from a perch. However, perched hunting techniques were judged by him to be nearly seven times more energy-efficient than were hovering hunts. Furthermore, the birds were found to use the former technique during 71 percent of recorded hunting episodes, as compared with 29 percent for the hovering technique. Tarboton suggested that hunting methods and success rates are likely to vary according to prey density, cover conditions, and perch availability at various times and places. Interestingly, Stendell and Myers (1973) found that in a three-year California study the birds continued to prey on *Microtus californicus* not only during years of high prey density but also during one of low prey abundance, even though more hunting time per

prey was then required. Probably this occurred because no other small mammals were apparently locally available to replace *Microtus* as easy prey.

Social Behavior

This is a relatively social hawk, the birds typically roosting in communal roosts that are often close to communal hunting areas during the nonbreeding period (Waian, 1973). However, all nesting pairs exhibited territorial behavior in Waian's study, and hunted primarily within their territories at this time. These varied in size from 17.8 to 51 hectares, and the resident male performed most territorial display activity. In Waian's study area the pairs rarely remained on their territories throughout the year.

In a South African study, Mendelsohn (1981) found that individual kites varied greatly in their social behavior and degree of dispersion. Males typically established territories and remained residents as long as the food supply lasted. Females seldom occupied territories alone, but rather usually settled with territorial males. However, females deserted territories more readily than males, and tended to move about, apparently in search of males holding territories having the best prospects for breeding. Since many territories were held for rather short periods, there was a monthly territorial turnover rate of about 26 percent, with residents breeding repeatedly and opportunistically at any time of the year. Some males thus made as many as seven breeding attempts in 19 months, sometimes starting breeding immediately after the young of the earlier brood became independent, and thereby breeding twice a year. Females, having a shorter participation than males in breeding, could breed three times in a single year. They typically deserted their young soon after the young had fledged, and apparently began searching for new males to attempt breeding again, whereas males remained with the newly fledged young and fed them for an average of 82 days. Thus, males needed a total period of about 172 days to complete a nesting cycle (24 days prelaying, 31 days incubation, 35 days nestling period, and 82 days postnestling dependency). As many as three (unsuccessful) nesting efforts by a pair in a single season have also been observed in Spain (Molinero and Cantisan, 1985).

Mendelsohn (1983) judged that by establishing territories early, males probably gain experience as to the best hunting areas, and thereby may improve their abilities to provide females and young with adequate food. He noted that early territorial establishment by males is a common trait among raptors, the males remaining more attached to territories while not breeding and returning more frequently to the same area each breeding season. (This pattern occurs in at least 14 species of six genera, in all of which males feed their females during breeding.) He also noted that the kites used two rather different breeding behavioral strategies, one associated with "stable" behavior and rather constant prey supplies, the other associated with "unstable" behavior and erratic changes in prey availability. Different areas and habitats, associated with varying climatic and weather patterns, could therefore be exploited by the kites through adopting one or the other system of social behavior patterns. The communal roosting behavior of nonbreeding kites Mendelsohn regarded as possibly serving to provide an "information center," helping the birds move more efficiently to more profitable foraging areas.

Social displays associated with breeding are fairly simple, and consist of mutual soaring above the nest site, some aggressive chasing, and a fluttering, descending flight by the male toward or around its nesting site, or mate (variously called the butterfly flight, flitterflight, and quiverflight). In this flight the male flies around and over the nesting grove, uttering a chittering sound and holding its wings in a V-shape overhead while rapidly vibrating them (Dixon, Dixon, and Dixon, 1957). As part of the territorial and courtship display the male may slowly fly from tree to tree within his territory with a small branch in his beak, or occasionally may perform such flights without anything in his beak if another male is performing the same display nearby. Once these displays begin the females rarely hunt any more (Waian, 1973).

As a threat display, either sex may call while perched and jerking its tail up and down; this same display may be directed toward the mate as well (Brown, Urban, and Newman, 1982). An aerial cartwheeling display involving talon gripping by both birds and a tumbling descent has been observed, but it is apparently a reflection of aggression rather than serving as a courtship display (Arlott, 1984). The male typically feeds the female during courtship, often near the nest,

and copulation typically occurs on, or near, the nesting tree. The male normally flies directly to the soliciting fmale, who crouches, calls, and may quiver her wings. Copulatory behavior is most frequent near the egg-laying period (Brown, Urban, and Newman, 1982).

Breeding Biology

At about the time copulation begins, the female selects a nesting site, typically rejecting any that may have been started by the male. Normally a new nest is built each year, although an old nest is sometimes rebuilt or a nest previously used by another pair may be taken over. It is probably normal for both sexes to work on the nest, but usually the female does most of this work (Dixon, Dixon, and Dixon, 1957). At times, the female may remain on the nest, accepting twigs brought by the male, and new nests may be made in as little as ten days (Brown, Urban, and Newman, 1982). However, California nest building required 14–28 days, and in that study area a wide variety of trees were used, the only general requirement seemingly that the nesting tree be of moderate height and near a suitable food source.

Eggs are laid at approximate 2- or 3-day intervals by females; the collective period of egg dates reaches over a rather wide time span in the American southwest. A total of 120 records assembled by Bent (1938) from California to Texas ranged from February 12 to June 21, with 60 of them between April 2 and 29. Dixon, Dixon, and Dixon (1957) reported southern California dates extending from February 6 to July 10, but young about 10 days old have been seen as early as February 22, indicating January egg laying. Second nestings during the same season have been observed several times in the San Diego area of California, the female usually starting a second nest before the young have been fledged from the first. This is thus perhaps the only North American hawk that normally attempts to raise two broods during a single season.

Clutch sizes in this species are fairly large, at least as compared to the other North American kites. A sample of 124 California nests had from 3 to 6 eggs, averaging 4.12, and with 4 eggs present in 106 of the nests (Dixon, Dixon, and Dixon, 1957). Very similar average clutch sizes for California were reported by Stendell (1972) and Wright (1978). However, the mean of 75 South African and 24 East African clutches of nominate *caeruleus* was 3.6 eggs (Brown, Urban, and New-man, 1982), certainly representing a significantly smaller average than that indicated by these California samples. It has been suggested (Stendell, 1972) that clutch sizes may average larger in years of good food supplies, as is common in various owl species, but it is further probable that different populations have evolved differences in average clutch size.

Most of the incubation is undertaken by the female, but at least in the African population the male may incubate about 20 percent of the daylight hours while the female incubates through the night and the remainder of the daylight period. The incubation period is from 30 to 33 days, averaging 31 (Brown, Urban, and Newman, 1982; Hawbecker, 1942). The young hatch at staggered intervals, but by differential feeding rates the size differences in the young tend to diminish, and by the time of fledging there is typically a difference of no more than two days between the departure times of the youngest and oldest birds (Dixon, Dixon, and Dixon, 1957). By the time the young are 30 days old they may perch on branches near the nest, and their first flights occur at 33–37 days after hatching, averaging 35 days. By this time the female may have begun a second nest, or have left the male to seek out a new mate. Dependence on the male for food may continue for some time after fledging; Waian (1973) observed a young bird make a kill as early as 33 days following fledging.

Breeding success rates are no doubt highly variable in different places and different times, but Dixon, Dixon, and Dixon (1957) reported that 74 young were fledged in 23 nests (average fledged brood size 3.2 young), from a total of 94 eggs, representing a remarkably high rate of fledging success. Second nesting efforts might increase this productivity rate still further. Of 26 nests studied in California by Wright (1978), 17 fledged young, with an average of 1.9 birds fledged per nest, out of an original average clutch size of 4.0 eggs. In four of his nests the birds produced fledged young during both their first and second nesting attempts, which seemed to be correlated with locally high populations of *Microtus*. Stendell (1972) similarly reported that second nestings and large clutch sizes in this species may be associated with high *Microtus* densities, whereas Waian (1973) attributed second nestings to failures of the initial effort.

To compare these observations with the Old

World population, in Spain an average of 2.0 fledglings were produced in 41 total nests (2.48 fledglings per successful nest) (Molinero and Cantisan, 1985). South African studies suggest that a loss of about 60 percent of the eggs prior to hatching and an additional mortality of 25–30 percent of the young between hatching and fledging may be typical. Additionally, irregular breeding cycles (as well as nonbreeding by some reproductively mature but nonterritorial birds), nest failures prior to egg laying, and prebreeding mortality further tend to depress actual population productivity (Brown, Urban, and Newman, 1982).

Evolutionary Relationships and Status

The genus *Elanus* has been variably constituted as to the number of species (Parkes, 1958; Husain, 1959). Parkes recommended including all of the *Elanus* populations except *scriptus* within the species *caeruleus*, a conclusion with which Husain fully agreed. The latter author postulated a southern Asian or African origin for the genus, from where it moved to Australia and America. The fomer area was later invaded a second time, by which time the earlier gene pool (producing *scriptus*) was sufficiently distinct as to be reproductively isolated from the second-arriving population (*notatus*).

More generally, *Elanus* is almost certainly a very close relative of the African scissor-tailed kite of the genus *Chelictinia*, which is very similar to it but has a long, forked tail, a more insectivorous diet, and is more gregarious (Brown, Urban, and Newman, 1982).

The recent increase in the abundance and breeding range of this species in various areas, including both North and Central America, warrants some additional comments. After the species disappeared from the southern Great Plains near the turn of the present century it appeared to be following the swallow-tailed kite into gradual disappearance. However, after the Second World War the species began to increase in both California and Texas (Waian, 1976; Larson, 1980), and since 1960 it has not only rapidly expanded

its Central American range but has begun to reoccupy much of its former southern Great Plains range and to move farther east, reaching Mississippi as a breeding species by 1983 (Toups, Jackson, and Johnson, 1985). It also bred in Arizona initially in 1983, and has since developed a small population in the southeastern part of that state (Gatz et al., 1985). During the same general period there were nesting reports for Oregon (Henny and Annear, 1978), Louisiana (*American Birds* 37:879), and Oklahoma (*American Birds* 36:993), plus nesting in Washington since 1982. There have been scattered nonbreeding occurrences elsewhere, such as in Nevada, New Mexico, Wisconsin, and Illinois (*American Birds* 41:434, 438). In 1986 the species nested in Florida for the first documented time since 1910, with three nests being located in Broward County (King, 1987). However, the birds did not return to nest there in 1987 (*American Birds* 41:420). Breeding may have occurred along the Sabine River in western Louisiana during 1988 (*American Birds* 42:458).

Other Vernacular Names

Black kite, Everglade kite, hook-billed kite, snail hawk, sociable marsh hawk; gavilan caracolero (Spanish)

Distribution

Very local resident of fresh-water marshes of southern Florida (vicinity of Lake Okeechobee), Cuba, and the Isle of Pines; also locally in the Pacific lowlands of Oaxaca and on the Gulf-Caribbean slope from Veracruz, Campeche, and Quintana Roo south locally to Nicaragua, Costa Rica, and Panama; also south through South America to Ecuador, Uruguay, and northern Argentina. (See Figure 28.)

North American Subspecies

R. s. plumbeus Ridgway: Resident in southern Florida (where classified as endangered); also Cuba and the Isle of Pines.

R. s. major Nelson and Goldman: Resident in eastern Mexico from Veracruz southeast to Quintana Roo, Belize, and the Petan region of Guatemala, there being replaced southwardly by *sociabilis*.

Description (of *sociabilis*)

Adult male. Entire head, back, wings, breast, abdomen, sides, flanks, and thighs uniform slate or blackish slate color, becoming nearly black on the remiges and rectrices, and with a faint brownish-gray tone on the upper wing coverts; upper and under tail coverts and base of tail whitish; tips of rectrices whitish with a broad dark grayish subterminal band. Iris carmine; the lores and external rim on the eyelids flame-scarlet (breeding) to yellow (nonbreeding); cere, gape, and corners of the mouth flame-scarlet (breeding) to apricot (nonbreeding); the rest of the bill black; tarsus and toes apricot-orange; claws black.

Adult female. Entire head, back, throat, breast, abdomen, sides, wings, and tail fuscous, becoming fuscous-black and black on the nape, top and sides of head, and sides of neck, with a light buff line above the eye; streaked with light buff on the throat and breast; the feathers of the middle abdomen and upper wing coverts tipped

with dull russet, the scapulars and secondaries narrowly tipped lighter, upper and under tail coverts and rectrices as in male; under wing coverts heavily spotted with russet and fuscous, the undersides of the remiges variably barred with pale russet to whitish and fuscous; thighs bright russet barred with fuscous-black. Iris orange; facial skin, cere, tarsi, and toes yellow (nonbreeding) to orange (breeding).

Subadult. By the first or second spring following hatching the sexes begin to differentiate, the males acquiring slate-colored feathers on the upperparts and breast, but in both sexes much of the new underpart plumage is edged or notched with brown. Males gradually come to resemble adults, the definitive plumage being acquired as long as three or perhaps even four (in captivity) years after hatching (Sykes, 1979). Three-year-old males are more grayish than the darkest of the adult females; similarly older adult females tend to darken by developing some slaty black on the upperparts. However, year-old birds are apparently fully capable of breeding.

Juvenile (sexes alike). Similar to adult female, but more fuscous above and heavily barred or striped with rufous below, the intervening markings tawny or cinnamon-buff. Iris more brownish; facial skin, tarsi, and toes more yellowish.

Measurements (of *sociabilis*, in millimeters)

Wing, males 325–341 (ave. of 6, 332.5), females 338–350 (ave. of 3, 342); tail, males 164–182 (ave. of 6, 172.1), females 167.5–188 (ave. of 3, 175) (Friedmann, 1950). Average egg size 44.2 x 36.2 (Bent, 1937).

Weights (in grams)

Four males (of *sociabilis*), 304–385, 2 females 384 and 413 (Haverschmidt, 1962). Range of unsexed specimens, ca. 283–396 (Oberholser, 1974). Average of 8, both sexes, 368 (Beissinger, 1983). Range of weights 340–520, ave. 427 (sex and sample size unspecified) (Clark and Wheeler, 1987). Estimated egg weight 42, or 10.5% of female.

Identification

In the hand. The combination of a very slender and strongly hooked upper mandible (culmen

Figure 28. North and Central American residential range of the snail kite, including races major (**MA**), plumbeus (**PL**), and sociabilis (**SO**). The historic range in Florida (adapted from Kale, 1978) and the residential range in South America are shown on inset maps. Indicated range limits of contiguous races should not be considered authoritative. The Cuban population is sometimes separated racially as levis.

from cere 20–25 mm) and a relatively long tail that is white basally and has a broad blackish subterminal band serves to identify this species.

In the field. When perched, adult males appear mostly or entirely black, with the white at the base of the tail typically hidden, and the tips of the wing extending slightly beyond the tail. Females and young males are more brownish, are rather heavily barred, spotted, or streaked below, and have a conspicuous whitish face that is broken by a black streak extending backward from the eye. At close range the slender and strongly hooked bill may be visible.

Flying birds of all ages exhibit rounded or paddle-shaped wings and a rather long, nearly square-tipped tail that is white basally (including upper and lower tail coverts). Otherwise, adult males appear almost slate-colored, while females and young birds are more brownish, especially below. Flight is rather slow, with deep wingbeats and gliding on slightly cupped or drooping wings. The alarm call is a series of repeated *kak* notes, sometimes described as like the bleating of a goat, or as a very dry, creaking rattle.

Habitats and Ecology

The habitats of this endangered species in North America, as well as throughout the remainder of its fairly broad range, consist of fresh-water marshes that support adequate populations of the genus *Pomacea*, generally called "apple snails." In the Everglades area of Florida these snails are most abundant in alkaline waters having a good supply of submerged vegetation and associated adequate amounts of dissolved oxygen. At one time vast areas of such marshes occurred in peninsular Florida, but drainage has eliminated more than a half million hectares of marshland, and the kite has correspondingly suffered (Hurdle, 1974). During the 1960s the birds were essentially confined to Loxahatchee National Wildlife Refuge, Palm Beach County, where they used an area of about 4,600 hectares in the southwest part of the refuge, and Lake Okeechobee, where extensive areas of sawgrass (*Cladium jamaicensis*) slough habitat with interspersed tree islands in the southwestern part of the lake were primarily used (Stieglitz and Thompson, 1967). They have also utilized several impoundments on the headwaters of the St. Johns River and in the northern part of Ever-

glades National Park (Kale, 1978). These habitats are generally characterized by areas of permanent fresh-water marsh up to nearly a meter in depth, with perches and nest sites provided by low shrubs or trees that are flooded and usually dead, including holly (*Ilex cassine*), willow (*Salix amphibium*), wax myrtle (*Myrica cerifera*), and buttonbush (*Cephalanthus occidentalis*). On Lake Okeechobee nesting has mainly occurred in areas thickly vegetated by willows and emergent marsh vegetation, and on Loxahatchee flooded tree islands and adjacent sawgrass areas have been favored for nesting (Stieglitz and Thompson, 1967). A generally low vegetative profile, with a distant horizon, scattered trees or shrubs, and water levels that can vary greatly but not drop to a point that the surface of the substrate goes dry, seem to represent a basic habitat combination that is needed by snail kites (Kale, 1978). During a drought in the early 1980s, the kites dispersed throughout peninsular Florida, mainly into the central lakes area and the east coast corridor. The lake areas (Tohopekaliga and Kissimmee) not only provided survival habitat but also enabled some breeding to occur (Beissinger and Takekawa, 1983).

Foods and Foraging

This is one of the most specialized foragers of all hawks, as it is essentially wholly dependent upon *Pomacea* snails, at least in North America. In Colombia the kite has been observed also to prey sometimes on a smaller snail (*Marisa cornuarietis*), but it has difficulties in extracting this species from its shell (Snyder and Kale, 1983). There have been single instances in which a turtle and a small mammal have been reported eaten during drought periods (Sykes and Kale, 1974), but basically the birds may be considered dependent on large fresh-water snails, which are captured by hovering above the water, then extending the talons and grasping the snails attached to the stems of marsh vegetation. The bird then flies to a perch and, while holding the snail with one or both feet, twists the operculum covering the snail's body away from the shell, tearing it off and usually discarding it. The bird then inserts its long, pointed bill into the shell's aperture and cuts the columellar muscle attaching the snail's body to the inside of the shell. The snail is then extracted and eaten whole, or torn to pieces, with the yolk gland and sometimes also the digestive tract discarded. This is a quite different method of extracting snails from that of the hook-billed kite, or of the other south Florida birds such as limpkins (*Aramus guarauna*) and boat-tailed grackles (*Quiscalus major*) that often feed on the same kind of snails (Snyder and Snyder, 1969). However, limpkins (the other snail specialist of the area) eat snails that are significantly smaller than those taken by kites, and the two species apparently do not directly compete for prey (Bourne, 1983).

`·` Studies of snail kites in Guyana (Beissinger, 1983) indicate that most daylight hours are spent perching, while only 19 percent of the observed time span was spent in active foraging. Snails were captured in 78 percent of the observed foraging bouts, with the hunting modes employed being nearly evenly split between coursing searches and still-hunting methods. The birds captured more medium-sized snails and fewer small-sized snails than were present in the general population, and took large snails at the same general frequency as their overall availability. Although the kites often have feeding perches that are close together and often forage in close proximity, they may sometimes defend small feeding territories, at least in those situations where a high local abundance of foods allows individuals to defend such territories effectively (Snyder and Snyder, 1970).

Studies in Florida (Sykes, 1987c) indicate that the birds spend nearly all their foraging time (97 percent) over marshes with sparse emergent vegetation; they are unable to forage over dense emergent vegetation or water covered by floating mats of water hyacinth (*Eichnornia crassipes*). Capture rates of adult kites averaged about 2–3 snails per hour, with the birds hunting over common foraging areas; defense of such foraging sites was rare. Snails are probably captured in waters as deep as 16 centimeters. Typically capture is achieved without the bird getting its belly feathers wet, and most snails are taken only a few centimeters below the surface. Nonsnail prey items are apparently taken mainly by young, inexperienced birds.

Of the main habitats used by snail kites in Florida since 1967, Lake Okeechobee and State Water Conservation Area 3A (adjoining Everglades National Park) have had the greatest usage. In 1987 the majority of known nesting attempts occurred on Conservation Area 3A. An estimated 49 percent of potential kite habitat has

been lost in southern Florida, and only 18 percent of the remainder was used by kites during recent years. This habitat loss is probably the most important cause of kite decline in Florida (Sykes, 1983b).

Social Behavior

Snail kites tend to be both gregarious and somewhat nomadic (but not migratory), dispersing from areas during droughts and sometimes moving substantial distances in search of new habitats (Beissinger and Takekawa, 1983). Nonbreeding birds are especially prone to form flocks, and may move considerable distances between communal roosts and foraging areas (Haverschmidt, 1970). Roosts in Florida often also serve as nesting sites; typically they consist of stands of willow in flooded marshes. Some of the roosting areas may also be used during the day as loafing sites or feeding perches (Sykes, 1985).

Courtship behavior of the snail kite is intense and conspicuous. Males perform aerial displays to potential mates that usually consist of a series of short, swooping dives, during which they close their wings and descend 2–5 meters before opening them and rising again (Beissinger, 1988). Stick carrying may be an important part of this display, the birds flying and diving with sticks carried in their bills. Immature-plumaged as well as adult males may carry sticks, and the former sometimes construct "false" nests that may remain unoccupied (Stieglitz and Thompson, 1967).

The best description of aerial displays of snail kites is that of Sykes (1987b). He observed six types of aerial display, with one maneuver usually occurring per display sequence and most displays lasting one to five minutes. Undulating flights were the most commonly observed aerial display (64 percent of all displays counted), with a series of up to six dives and swoops performed in sequence, the amplitudes usually from 10 to 30 meters. "Slow flights" were performed at 30 meters height or less, with a slow and exaggerated downstroke characteristic. The "pendulum" was a shallow dive with some wing flapping followed by a tight turn and the process repeated. Mutual soaring consisted of the male and a prospective mate flying close together while performing dives, turns, rolls, and the like. During "tumbling" the bird would dive with wings partly folded and turn end-over-end several times before leveling off. During "grappling" flights the bird rolled or turned sideways with its feet outstretched; this could be done either by a lone male or a male with a prospective mate.

Haverschmidt (1970) observed similar display flights over a Suriname kite colony, and also observed prey presentation by males toward their mates. Males would alight on a nest rim with a snail, which was then accepted by the female and promptly swallowed. The male then sometimes hopped on the female's back and copulated. Other copulations were seen on the platforms of uncompleted nests and on nearby branches. When the male alighted the female called, threw her head up and back, and pecked a few times toward the male's head. The male would stand with its bill open, uttering a gurgling or rattling call, and hop on the female's back when she rose up (Haverschmidt, 1970). Aerial talon grappling has also been reported in snail kites (Brown and Amadon, 1968), but perhaps these reflect agonistic rather than courtship activities.

It is unlikely that pair bonds are maintained during the nonbreeding period in these mobile hawks, and indeed there is no good evidence of strong pair bonds even throughout a single breeding season. The long breeding season of 5–10 months in Florida permits individual kites opportunities for producing three or four broods if they consistently desert and remate after the nestlings have hatched, but only two broods if they remain to help care for the young (Beissinger, 1984, 1988; Beissinger and Snyder, 1987). Indeed, Beissinger found that mate desertion occurred in about three-fourths of the 36 nests he observed closely. Males and females deserted their mates with about equal frequency in all years but one, when females were the more common sex to desert. Typically desertion occurred when the young were 3–5 weeks old, when they were near fledging but still about 3–6 weeks from complete independence of their parents. In one case a deserting female immediately remated, built a new nest, and was incubating eggs two weeks before her prior mate finished caring for their first brood. In all nests Beissinger observed, the deserting member of the pair was the one that provided less food for the young, but either sex was evidently equally able to complete rearing the youngsters alone. Beissinger suggested that this "fickle" mating strategy is perhaps an adaptation to the unpredictability of the nesting environment, since temporally fluc-

tuating water levels affect food supplies and nest sites and thus impact overall annual reproductive potentials. He believed that clutch sizes have declined in the Florida population during this century, perhaps also reflecting increased environmental unpredictability. Beissinger found that although mate desertion also occurred in a colony he studied in Venezuela, there it was highly dependent upon brood size, with larger broods rarely being deserted.

Breeding Biology

Nests of the snail kite are rather flat and bulky structures that are built in emergent (usually woody) vegetation above water; although bulky, they are loosely constructed, rather insubstantial, and are often subject to destruction or egg loss. Stieglitz and Thompson (1967) stated that the nests are normally less than 2.5 meters above water, but may be as high as 9 meters. Of 34 clutch records in the Western Institute of Vertebrate Zoology, the nest heights ranged from 0.6 to 9 meters, with 19 occurring between 1.5 and 2 meters. The clutch sizes of 63 nests ranged from 2 to 5, with 45 of the nests having 3 eggs and the mean being 3.14 eggs. Beissinger (1986) reported that 91 clutch records obtained in Florida between 1880 and 1925 averaged 3.23 eggs, 57 obtained between 1925 and 1959 averaged 2.96 eggs, and 156 documented between 1979 and 1983 averaged only 2.71 eggs, suggesting selection for decreased clutch size in the increasingly unpredictable nesting environment of southern Florida. Sykes (1987b) provided an analysis of 313 clutches by decade; these collectively averaged 2.92 eggs, with larger clutches more common in the period prior to 1940. The mean clutch size (of 2.66 eggs) reported by Snyder, Beissinger, and Chandler (1989) was apparently not influenced by water conditions, seasonality, or relative coloniality, but varied significantly by geographic location (probably reflecting food supply differences) and nest substrate.

Bent (1937) reported that 68 Florida egg records extended from February 14 to July 20, with half between March 13 and April 28. The Western Institute of Vertebrate Zoology's 34 egg records (which are mostly from Florida but include some Mexican records) extend from February 4 to June 28, with a peak (14) in April. Sykes (1987b) analyzed records from 294 clutches from Florida, and noted that 67 percent of these occurred during the three months of February to April, the latter about 80 days prior to the start of the rainy season there. He also (1987a) reported an average nest height of 1.6 meters above water for 99 nests, and 2.1 meters for 99 nests above the ground surface. He found that of 594 nests, willows were used as supports for nearly half, with cattails next most common and wax myrtle third. Woody-stemmed plants accounted for 77 percent of all nest supports. Green materials are added to the nest well into the nestling period, in Sykes's view perhaps to regulate humidity, influence nest temperatures, or aid in controlling nest parasites. Rodgers (1987) reported that on Lake Okeechobee willows and cattails were the most important nesting substrate, whereas on Conservation Area 3A willows and pond apple (*Annona glabra*) were most commonly used. Apparently a variety of above-water sites are used, with few definite preferences evident.

New nests are normally constructed for each nesting attempt, but sometimes are built over the previous year's nest remains or at least in the same location. Nest sites are selected by males, and almost all construction is done by males, although females assist in later nest maintenance. Typically a nest is constructed in 4–18 days (average of 13, 11.2 days). "Loose colonial nesting" may best describe the normal nest dispersion pattern; 71 nests ranged from 6 to 287 meters from nearest neighbors, averaging 84 meters. A small nesting territory, usually extending no more than 4 meters from the nest (but up to 30 meters or more at times) is actively defended (Sykes 1987a).

Although males raised in captivity were found to require four years to attain full adult plumage, at least some individuals breed while still in subadult plumage (Sykes, 1979). Indeed, Beissinger (1986) reported that at least 10 cases of one-year-old birds (both sexes) breeding have been documented, and more recently Snyder, Beissinger, and Chandler (1989) reported that nine of ten yearling breeders were successful in fledging young.

The egg-laying pattern is quite variable, with intervals between the eggs being 2–4 days, with probably 6 days required to complete a clutch of three eggs (Chandler and Anderson, 1974). Beissinger (1986) found that structural failure was the largest single known cause of nest loss, accounting for some 26 percent of nest failures. Another major cause of nest failure is predation

(Sykes, 1987b), with this accounting for 44 percent of nest losses in Sykes's study and 14 percent in Beissinger's. The rat snake (*Elaphe obsoleta*) is apparently primarily responsible for these losses. The incidence of predation was found by Sykes to be inversely related to nest distance from upland habitats, especially for nests within about 200 meters of such habitats. Sykes (1987b) reported a 50.5 percent nesting success rate for more than 200 nests, and an average of 2.0 young fledged per successful nest. However, Beissinger (1986) found a much lower rate of success, with only 32 percent of 331 nests fledging at least one young, and the five-year annual average of nesting success was only 21 percent, with substantial year-to-year variations in this statistic. A more recent analysis (Snyder, Beissinger, and Chandler, 1989) reported an even lower overall nesting success (13 percent of 236 nests) and a relatively low rate of egg hatchability (81 percent of 497 incubated eggs). These demographic features are countered by the species's extremely long potential breeding season, its strong renesting and multiple brooding tendencies, its early attainment of sexual maturity, and the high survival rates of juveniles and adults under favorable water conditions.

The usual incubation period is 28–30 days, and the period to fledging requires another 23–28 days (Chandler and Anderson, 1974). A much longer (49-day) fledging period was estimated by Stieglitz and Thompson (1967). Sykes (1987b) reported an average incubation period of 27.4 days for 21 nests, and an average nestling period of 28.7 days. The average interval between the fledging of the first and last nestlings was 2.3 days. The young continue to be fed by one of the parents until they are 9–11 weeks old. As noted in an earlier section, a substantial degree of mate desertion often occurs during the nestling period. Nestling mortality is often quite high, and has been variously attributed to parasites, predation, weather, and accidents. Additionally, older youngsters must learn to capture snails efficiently, and often become quite wet in the process. This might render them temporarily flightless, and hence less likely to survive (Sykes, 1979). Nonetheless, annual adult survival is apparently unusually high, probably over 90 percent, and some individuals may live 13 years or more (Beissinger, 1986; Snyder, Beissinger, and Chandler, 1989). This high annual adult survival may partly reflect the highly variable annual reproductive output apparently typical of the species, requiring a long potential lifespan (Nichols, Hensler, and Sykes, 1980).

Evolutionary Relationships and Status

The snail kite is one of two species in the genus *Rostrhamus*, the other being the forest-adapted slender-billed kite (*R. hamatus*) of South America. The latter species has a more chunky build, shorter wings and tail, and lacks white on the tail coverts. Like the snail kite, it also is a snail specialist. Other more distant relationships of these kites are not evident.

A great deal has been written on the population decline and recent status of the snail kite; the most complete account is by Sykes (1983b). Sykes also (1979) reviewed the recent kite population changes in Florida, and related the population data to water level changes. Apparently the birds began to decline in Florida during the early 1900s with the advent of large drainage projects and may have reached their lowest population level in the 1950–65 period, when perhaps less than 40 birds remained. After the early to mid-1970s the population increased substantially, mainly in association with normal rainfall patterns plus water management activities that brought about relatively high water conditions in the region after 1974. If such conditions were continued a further population increase and range expansion might be anticipated, the latter of which is facilitated by the species's strongly nomadic tendencies. However, a period of drought or lowered water conditions could have an equally disastrous effect on the population, according to Sykes.

Since 1974 the Florida population has been surveyed annually by the Florida Game and Fresh Water Fish Commission. More than 600 birds were counted in 1980 (653) and in 1984 (668). However, substantial year-to-year population fluctuation seems to be typical of this population. Thus, in 1985 the number counted (452) was only about two-thirds that of the previous year, while by 1986 the survey count was up to 564 birds (Rodgers, Schwikert, and Wenner, 1988). In 1987 the count had dropped to 326 birds, with 65 percent of them on the Lake Okeechobee wetlands and Conservation Area 3A, but the latter area had suffered a late-summer drought that caused substantial postbreeding dispersal (Rodgers, 1988). In 1988 the midwinter survey revealed 500 birds, mostly on Lake Okeechobee (Rodgers, 1989). A study of the nesting ecology of snail kites in Conservation Area 3A has recently appeared (Bennetts, Collopy, and Beissinger, 1988).

The population occurring in Cuba is also of interest, in part because it has been suggested (but not proven) that some population exchange with Florida birds may occur. Most of the Cuban birds are restricted to the Zapata Swamp area of south-central Cuba, where in 1982 a minimum of 55 kites were observed (Beissinger, Sprunt, and Chandler, 1983).

A bibliography of the snail kite, with 167 references, has been published (Baird, 1970).

Figure 29. North and Central American breeding (hatching) and migratory (shading) ranges of the *Mississippi kite. The migratory and wintering ranges in South America are shown on the inset map.*

Mississippi Kite *Ictinia mississippiensis* (Wilson) 1811

Other Vernacular Names

Blue kite, gray kite, Louisiana kite, mosquito kite; gavilan del Mississippi (Spanish); milan du Mississippi (French)

Distribution

Currently breeds north to central Arizona, northern New Mexico, southeastern Colorado, north-central Kansas, central Arkansas, up the Mississippi valley through western Kentucky, western Tennessee, southeastern Missouri, and southern Illinois, and northeast to the North Carolina coast. (Formerly bred north to central Colorado, Iowa, southern Indiana, and southern Ohio.) Currently breeds south to central and southeastern New Mexico, central Texas, the Gulf coast, and north-central Florida. The breeding range has expanded considerably along its western and northern borders in recent years, and some areas of New Mexico and Arizona have been recently colonized. Replaced allopatrically in Mexico by *I. plumbea*, which is sometimes considered conspecific.

Winters in central South America; casual or occasional wintering may occur north as far as southern Texas. (See Figure 29.)

North American Subspecies

None recognized, although if considered conspecific with *plumbea* the northern form would be *I. p. mississippiensis*.

Description

Adult male. Forehead, crown, occiput, cheeks, auriculars, secondaries, and entire lower parts varying from pallid gray to pale gray, lightest on the head and secondaries, where it is less ashy; a frontal area on the forehead, the chin, and the tips and outer webs of the secondaries hoary whitish; darkest on sides, lower abdomen, thighs, and under tail coverts, where it sometimes approaches medium gray; lores and circumocular region black; scapulars, interscapulars, back, and rump deep gray to slate, darkening on upper tail coverts to blackish slate; tail uniformly blackish with slate wash; lesser upper wing coverts slate to blackish slate, rest of upper wing coverts deep gray to slate;

primaries blackish and chestnut. Iris deep burnt carmine to bright red; bill, eyelids, and interior of mouth deep gray to black; corner of mouth orange-red; tarsus and toes salmon-orange-red; the lower part of tarsus and base of toes dusky.

Adult female. Similar to adult male but head and secondaries averaging darker, pale gray to medium gray, the secondaries often narrowly tipped with whitish; the primaries averaging slightly less of the chestnut-brown color. Tail paler below than that of male, with dusky tip below (W. S. Clark, in correspondence).

Subadult (first year). Body feathers similar to the adult, but with a barred tail retained from the juvenal plumage, and the juvenal remiges and wing coverts also variably retained (the underwing lining thus brown-spotted rather than uniformly gray, and the flight feathers pale-tipped, but not with distinctly white trailing edges of the secondaries, and the upper surface of the secondaries not distinctly paler than the primaries as in adults). Scattered white spots formed by whitish bases of adult feathers and retained juvenal feathers. Iris less reddish than adult; bill black; the cere and tarsi more yellowish; interior of mouth dull white.

Juvenile (sexes alike). Generally dark brown above, with a buffy superciliary stripe, and heavily streaked with brown and buffy below; secondaries not lighter than primaries; rectrices with angular whitish spots on the inner webs, producing on under surface an effect of three transverse whitish bands. Iris brownish; bill bluish gray; cere, corners of mouth, tarsi, and toes yellowish orange to orange.

Measurements (in millimeters)

Wing, males 286–305 (ave. of 6, 295), females 300–315 (ave. of 5, 309.5); tail, males 149–166 (ave. of 6, 157.1), females 154–172 (ave. of 5, 163) (Friedmann, 1950). Average egg size 41.3 x 34 (Bent, 1937).

Weights (in grams)

Males 216–269 (ave. of 11, 245); females 278–339 (ave. of 5, 311) (Sutton, 1939). Ave. of 14 males, 248; ave. of 6 females, 314 (Snyder and Wiley, 1976). Range of weights 240–372, ave. 278 (sex and sample size unspecified) (Clark and Wheeler, 1987). Estimated egg weight 26.4, or 8.3% of female.

Identification

In the hand. The combination of relatively narrow, pointed wings (but with a short outermost primary) and a long (150–175 mm), uniformly blackish or at least black-barred (in juveniles) tail that is about half the length of the wing provides a means of identifying this species.

In the field. When perched, adults of this species appear to be mostly dark grayish, the head and underparts somewhat lighter, and with no white present on the rather squared-off or slightly notched tail. The wingtips extend slightly beyond the tip of the tail when perched. Immatures are similar, but are heavily streaked with brown and yellowish buff below. Immatures also exhibit a pale buffy stripe above the eye and somewhat resemble young black-shouldered kites, but have darker tails and are not heavily streaked below.

In flight, adults appear almost uniformly medium to dark gray both above and below, including the wings and frequently flared tail. A paler silvery gray upper surface of the secondaries is apparent from above, plus a whitish trailing edge. Immatures are heavily streaked or spotted with buff and brown below, including the wing linings, and the blackish tail is crossed by three light bands. Year-old subadults have gray underbody color like adults, but retain the tail banding and brown underwing linings of immatures. The flight is buoyant, with gliding usually done on horizontal wings. The wings are relatively pointed, but the leading primary is fairly short. The typical alarm call is a thin, usually two-noted whistle, sounding like *phee-phew!*.

Habitats and Ecology

Little can be said about the South American wintering habitats of this highly migratory hawk, but probably they are rather diverse, requiring only a good supply of aerial insect prey. In the southern Great Plains the birds seem to prefer rather barren and open habitats, especially where there is broken topography, dissected by small, steep-sided ravines, often with brush and scrubby trees on the slopes (Fitch, 1963). In Oklahoma,

scattered tree growth is typical habitat, with the trees often not more than 4.5–5.5 meters high, but with occasionally taller stands (Sutton, 1939). However, the birds survive equally well in the lower Mississippi valley, where they do their foraging above forests with rather wooded canopies (Brown and Amadon, 1968).

During the approximate five-month period (May to September) that the kites are in North America, their habitats are essentially nesting habitats. Historically, kites in the Mississippi valley and southeastern parts of the U.S. were most common in the vicinity of inland, riparian forests in the lowlands, less common along the coast, and virtually absent from hilly country above the fall line. In the Great Plains they were largely associated with areas having sizable riparian woodlands along major river systems. In recent decades, however, the development of tree plantings such as shelter belts, farm woodlots, and town plantings of trees have opened up much of the previously nonwooded portions of the plains to potential nesting. As a result, substantial expansion of the nesting range has occurred in recent decades, especially as such decimating factors as shooting, egg collecting, and foraging habitat alterations have disappeared or at least declined. In particular, the increase of mesquite associated with cattle raising and farming in Texas and Oklahoma has been an important factor there, while tree-planting activities for windbreaks (shelter belts) farther north and west to Kansas and New Mexico have no doubt been especially significant. Similarly, in heavily forested areas farther east, opening of the woodlands by lumbering may have increased foraging habitats without seriously disrupting opportunities for nesting. Seemingly, tree "islands" such as farm lots and shelter belts are more attractive to the kites than are natural riparian woodlands, and may offer nest sites closely adjacent to the grasshopper-rich areas of farmland (Parker and Ogden, 1979).

A detailed analysis of nest site needs is not available, but Parker (1974) analyzed the vegetative characteristics of 50 colonies in Kansas, Oklahoma, and Texas. About 70 percent of the nesting attempts he observed were in shelter belts (extremely linear woodlots) or windbreaks (smaller and squarer woodlots), 13–21 percent occurred during different years in cottonwood woodlots, and less than 10 percent in the other kinds of available woody vegetation (oak shinn-

ery and savanna-mesquite woodlots). These latter
vegetation types did support considerably more
single (noncolonial) nests than did the shelter
belts and woodlots. Many of the colonies were
entirely surrounded by grazing lands, while others were often surrounded by various kinds of
planted vegetation or plowed areas. The diversity
of surrounding habitats did not appear to be a
significant factor in habitat use, nor did the
variations in density of woody vegetation or
amount of dead and dying timber. Patterns of
nest placement in colonies of different vegetation
types and sizes suggested that a habitat's degree
of kite usage (colony size) might be determined
by a variable degree of attraction of kites to a
common patch of suitable vegetation, accompanied by overdispersion of nests at high density
(resulting from the opposing effects of general
aggregation but nest-spacing tendencies).

Foods and Foraging

Although often described as an insect specialist,
and indeed largely dependent on flying insects,
Mississippi kites sometimes resort to other foods,
occasionally scavenging road-killed vertebrates
and periodically killing small birds (Parker, 1974).
Sutton (1939) listed the stomach contents of 16
kites collected at Oklahoma breeding areas,
which almost totally consisted of insects (mostly
orthopterans). Fitch (1963) listed the prey remains
found under perches of fledglings and obtained
from 205 pellets, mainly of adult kites, and found
that insects comprised 100 percent of the identified materials, which were mainly coleopterans
and orthopterans. Glinski and Ohmart (1983)
tabulated some 2,636 prey deliveries to 3 kite
nests in Arizona, and found that cicadas comprised 71 percent, beetles 15 percent, orthopterans 4 percent, and other insects about 6
percent. Bats, toads, lizards, and frogs also were
represented in small to moderate quantities (totaling about 6 percent). Similarly, Parker (1975)
noted that at least 15 individual birds (including a
variety of passerines and smaller nonpasserines),
various turtles, lizards, toads, and frogs (totaling
59 individuals), and 9 individual mammals have
been reported as food sources, although perhaps
many of these represented scavenged foods.
Nevertheless, the great aerial agility of this kite
might easily allow it to capture birds in flight, as
it is evidently able to do with various bats (such
as *Pipistrellus* and *Tadarida*).

The Mississippi kite is a "classic" kite in its
foraging behavior, taking most of its food in the
air during prolonged flights, soaring, circling,
and occasionally swooping down on prey below.
Sutton (1939) notes that the birds are so agile that
they can easily catch flying grasshoppers,
cicadas, and even dragonflies. They typically
suddenly spread the tail when sighting prey,
hang momentarily in midair, then close the tail
and coast giddily downward, capturing the prey
with their talons. The prey is then transferred to
the beak as the bird regains its altitude, sometimes discarding certain inedible parts along its
route.

Skinner (1962) described this species's foraging behavior, noting that the birds often circled at
a height of about 50–100 meters, then swooped
down on dragonflies and beetles in a falconlike
manner at a rather shallow angle, occasionally
supplementing the dive with a few short wing
strokes, and terminating the attack with a quick,
short turn as the insect was taken. Grasshoppers
were captured with a swift, straight vertical drop
that ended only a meter or so above the ground,
and the bird then flew off some distance to
consume the prey at an elevation of about 30–40
meters. Soaring times between strikes averaged
3–6 minutes, and 30–70 seconds were needed to
consume each prey. There was an apparent tendency for all the kites of one area to feed during

the same time period, and for all to disappear at about the same time.

Besides catching prey while soaring, some aerial hawking from perches is also done by these kites. Typically these are short flights (less than 50 meters) out from dead branches situated 2–20 meters above the surrounding vegetation. Hawking from perches within 150 meters of nest sites is apparently an important means of capturing food during the nestling period (Glinski and Ohmart, 1983)

Social Behavior

Sutton (1939) believed that the kites might pair while on wintering areas or perhaps while migrating, as many of the returning birds that he observed appeared to be already paired upon arrival in Oklahoma. He did not observe any obvious courtship, but did see both males and females displaying by "cutting capers" in the air, plunging down and swooping up again effortlessly, and pursuing others while uttering thin squeals and chippering. He believed that pairing for life was a possibility. Fitch (1963) also believed that the birds are paired upon arrival in Kansas.

Parker (1974) noted considerable nest site tenacity from year to year, suggesting that a reestablishment of prior pair bonds might occur. However, he did not tag most of his birds and so was generally unable to confirm this, except for one nest that was used for three years by a marked bird. Previously successful nests were more often reused than others, suggesting that reestablishment of previously successful pairs might indeed occur. Since adults in his colonies experienced only about a 10 percent annual mortality rate, this adds additional credence to the possibility of renewed pair bonds between years. Of 25 nest sites studied by Glinski and Ohmart (1983), 15–17 were recurrently occupied during the four years of study, and 2 were occupied all four years. However, these birds were not permanently marked for year-to-year identification. Since yearling kites have been observed to be nest helpers at 2 nests and suspected to help at 15 additional nests (Parker and Ports, 1982), it is very likely that pair bond renewal and year-to-year reuse of the same site by a single pair does often occur.

Perhaps because of such renewed pair bonds, little courtship display is evident, and pairing seemingly occurs without overt display or vocalization (Brown and Amadon, 1968). Copulation occurs with no obvious preliminaries. The male flies to the female and alights on her back; in one case this occurred on a roadside fence post (Fitch, 1963). Nest building begins about a week after arrival on the nesting grounds, and even then is desultory initially. It is not clear whether the male undertakes most of the initial nest building or nest repair, although Sutton (1939) observed a definite female carry a small twig to a nest that a male was working on. Since nests are often fairly close to one another there must not be a very concerted defense of nest territory by either sex.

Nesting sociality, as judged by nest dispersion, was measured by Parker (1974) for eight different colony sites or years. The average densities varied from 0.71 to 8.02 hectares per nest, with one shelter belt colony of 15 nests having a maximum density of 0.3 hectares per nest, and all colonies of two or more nests having an overall density of 0.8 hectares per nest. Most colony nests were no more than 100 meters from their nearest neighbor. Glinski and Ohmart (1983) reported that nests in a sparse Arizona population were spaced 125–1,700 meters apart, with individual nesting territories that consisted of areas extending 50–100 meters from the nest site. Neither study found colonial nesting to affect reproductive success positively; indeed Parker reported that lone pairs of kites were more productive on average than were colonial nesting pairs. Additionally, large colonies suffered more predation than expected, suggesting that predators might select large patches of woody vegetation that tend to support large colonies. However, collective defense of the nesting colony from possible enemies, such as hawks and owls, does occur and might influence nesting success.

Breeding Biology

Mississippi kites are seemingly adaptable to a wide range of trees as suitable sites for nests; Fitch (1963) stated that although most nests he studied were in cottonwoods (*Populus deltoides*), the frequency of cottonwood use seemed to reflect its relative abundance rather than any clear-cut preference. Parker (1974) found that 13 species of trees were used for nesting, with cottonwood, elm (*Ulmus*), osage orange (*Maclura*), and black locust (*Robinia*) accounting for more than 80 percent of the total. Apparent discrimination in choice of nest trees did occur in particular colonies, with relative shading, visual exposure to predators, wind and precipitation, and tree heights varying considerably among the trees and possibly influencing choice of nest sites. Two nests were built near wasp nests, which were left undisturbed, suggesting a possible commensal relationship. Sutton (1939) noted that nesting trees were usually in the open or, if in a woodland, near its outer edge.

It is hard to judge the period of nest building, in part because so many birds reuse or refurbish old nests. Sutton (1939) reported that in 38 of 40 nests he studied in 1936 there were two eggs, with three in the other two, while 14 found in 1937 all had two eggs. Parker (1974) noted that two-egg clutches comprised 73 percent of 156 clutches in published sources, while 25 percent were one-egg clutches. One of 68 clutches in the collection of the Western Foundation of Vertebrate Zoology apparently had three eggs. Of 391 nests observed by Parker, only half (194) contained two eggs, while 136 had single eggs and 61 contained none. Parker believed that probably most of the one-egg clutches reflect cases where the clutch was incomplete or one of the eggs had been lost to various causes.

Both sexes incubate during the 29–31 day incubation period; Sutton (1939) observed that the bird not incubating at night stayed in a nearby tree. In one case he collected both members of an incubating pair and found that no incubation patch was present in the male. He judged that most probably this male was a new replacement for one that had been recently killed. In observation of pairs with well-grown young, Fitch (1963) observed that they were fed about once every 11 minutes on average, with feeding rates gradually increasing after sunrise and tapering off again in late afternoon. Much of the food during such deliveries was carried in the throat, but insects sometimes protruded from the bills of arriving kites. These masses of food would be regurgitated in the nest, and then bits would be picked up by the adult to be fed to individual nestlings. Later, the young are fed directly bill-to-bill. They are flying by 50 days of age, but are fed until they are at least 60 days old (J. Parker, in Palmer, 1988).

Yearling birds in subadult plumages are fairly frequently found paired to and nesting with older birds; Parker and Ports (1982) reported that this was the case in 17 percent of 209 nesting pairs, and in at least two other nests (and possibly as many as 15 more) a yearling bird helped an adult pair by incubating, brooding, or defending the nest against possible predators. In their view, the value of young birds in increasing vigilance and defending the nest against predators might be the major value of such helpers. Glinski and Ohmart (1983) also observed yearling birds engaging in defensive flocking, although they did not record any paired yearlings. In both studies few nonbreeders in adult plumage were observed.

In Glinski and Ohmart's (1983) study, most nesting failures (84 percent) occurred prior to the hatching of the nests, whether during courtship and nest building (44 percent) or during incubation (40 percent), from largely uncertain causes. Mortality to nestlings occurred in only 2 of 63 nesting attempts. Overall, 34 (54 percent) of these nesting attempts produced fledged young, and averages of from 1.14 to 1.6 fledged young per successful nest were produced during three years of study. Parker (1974) found predation to be a major mortality factor, especially during the nestling period. Of 261 nests he studied, a maximum of 79 percent successfully hatched eggs, and these successful nests produced an average of 1.02 fledged young. Minimum estimates of these same statistics were 47 percent and 0.65 fledged young per successful nest, but the maximum estimates were judged to be more nearly accurate. Evans (1981) estimated a 61 percent nesting success rate in southern Illinois, with 0.6 young fledged per nesting pair. Owls, hawks, and crows were judged by Parker to be potentially important predators of eggs and nestlings. Starvation, sibling aggression, and cannibalism (siblicide) were believed to be rare.

The fledging period is approximately 34 days. In one Georgia nest the adult female disappeared 10 days after fledging had occurred, while the male continued to feed the single young for at least 16 more days (Brown and Amadon, 1968). By mid-August in Kansas the young are learning to capture their insect prey, and only a few weeks later in early September the southward migration of the entire population begins (Fitch, 1963).

Evolutionary Relationships and Status

The genus *Ictinia* includes one other closely related species, the plumbeous kite (*I. plumbea*) of tropical America from southern Mexico to southern Argentina. Although this is very similar to and allopatric with the Mississippi kite, Brown and Amadon (1968) regarded the two species as sufficiently different to warrant specific recognition. Palmer (1988) believes that the two represent only subspecies.

The interesting changes in the Mississippi kite's historical status and range in North America have been well summarized by Parker and Ogden (1979). They noted some similarities between the earlier decline and more recent recovery of the black-shouldered kite and the corresponding trends of the Mississippi kite, and attributed both in part to ecological changes as well as to some common innate attributes. Both species are capable of breeding as yearlings, are colonial, tend to wander widely, and seem little affected by pesticides. However, the black-shouldered kite is often double-brooded or at least often attempts renesting, and it has a larger clutch size than the Mississippi kite, which is not known to be a renesting species. In the case of the Mississippi kite, the various factors bringing about increased trees to the Great Plains, and the opening up of some of the more heavily forested areas of the east have evidently favored kite populations. In Illinois and in Tennessee, where the species is considered endangered and where bottomland forests are being rapidly channelized, logged, or cleared for soybeans, the resultant degree of riverine forest habitat disturbance may represent a threat to the kite populations (Hardin, Hardin, and Klimstra, 1977; Kalla and Alsop, 1983). The loss of shelter belts in recent years may likewise have disturbed the increasing population trend in the Great Plains somewhat, but there is still no apparent reason why populations should not continue to survive and perhaps increase both in the eastern and western states (Parker and Ogden, 1979). Local populations have developed in many towns in Kansas, Oklahoma, and New Mexico, and in both New Mexico and Arizona the birds have recently colonized small areas of riparian cottonwoods and salt cedars (Glinski and Ohmart, 1983). The only potential estimate of the overall continental population to my knowledge is a count by Smith (1985) of 27,400 birds migrating along the Panama isthmus in 1981.

An annotated bibliography of the Mississippi kite has been produced by Hardin and Klimstra (1976).

SEA EAGLES AND FISH EAGLES

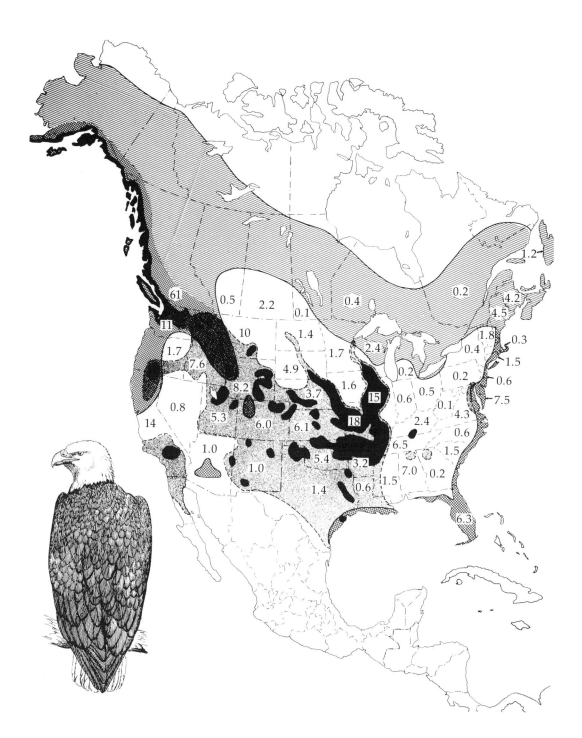

Figure 30. Breeding (hatching) and residential (cross-hatching) ranges, and primary (inked) and secondary (shading) wintering ranges of the bald eagle. Because of gradual clinal intergradation (see text) no attempt to distinguish races has been made. Relative winter state or provincial density indices (average number seen per Audubon Christmas Count, 1986) are shown for major wintering areas. Other winter distribution information adapted from Millsap (1986b).

Bald Eagle *Haliaeetus leucocephalus* (Linnaeus) 1766

Other Vernacular Names

American eagle, northern bald eagle (*alascanus*), southern bald eagle (*leucocephalus*), white-headed eagle, white-headed sea-eagle; pygargue à tête blanche (French); aguila cabez blanca, aguila calva (Spanish)

Distribution

Breeds from central Alaska, northern Yukon, northwestern and southern Mackenzie, northern Saskatchewan, northern Manitoba, central Ontario, central Quebec, Labrador, and Newfoundland south to the Great Lakes and Maine, and along the Pacific coast from the Aleutians locally to Baja California, interiorly along the Rocky Mountains south regularly to Wyoming and locally to central Arizona and southern Sonora (Yaqui River); also resident along the Gulf coast from Texas east to Florida and north along the Atlantic coast to New Jersey. Absent as a regular breeder through much of the rest of interior North America; these populations now slowly recovering from pesticide-related reductions beginning in the 1940s.

Winters generally throughout the breeding range but more frequent coastally from southern Alaska and western Canada southward along the Pacific coast to Washington (these birds mostly migrants from the nearby interior), and along the upper Mississippi River (from substantially farther north). (See Figure 30.)

North American Subspecies

H. l. leucocephalus (Linnaeus): Breeds in the southern United States and Baja California, intergrading with *alascanus* in the central and northern states (the northern limits arbitrarily set at 40 degrees north latitude).

H. l. alascanus Townsend: Breeds from Bering Island across northern North America, intergrading broadly with the preceding form, but generally assumed to breed south to Oregon, Idaho, Wyoming, Minnesota, Wisconsin, Michigan, Ohio, Pennsylvania, New Jersey, and Maryland.

Description

Adult (5–6 years). Sexes alike: entire head, nape, chin, throat, upper and under tail coverts, and tail white, often with a pale buffy or creamy tinge; elsewhere mostly dark brownish, varying from fuscous to fuscous-black on the upperparts and the remiges to dusky sepia on the underparts, the upperparts generally blacker, the underparts more rufescent and brownish; scapulars, interscapulars, under wing coverts, and feathers of the breast and abdomen narrowly edged with dirty white to pale brown. Iris pale chrome-yellow; orbital skin, cere, bill, and tarsi deep dull cadmium-yellow; claws black.

Subadult-adult transition (ca. 2–4 years). Similar to the adult, but the head (especially behind the eye) and tail feathers streaked or tipped with fuscous to fuscous-black; the breast and interscapulars blotched with creamy to white, this light predominating basally on these feathers and often showing through. The amount of white on the underparts is very variable.

Juvenile-subadult transition (ca. 1–2 years). White in the rectrices restricted to a mottled area on the inner webs of the lateral feathers and to a median streak on the central pair; head predominantly brown, the feathers tipped with tawny-fulvous; the body feathers basally whitish, this color often showing through variably, especially on the underparts; sides of the head (especially the auriculars) and neck generally brown, contrasting with the whitish interscapular area. Bill blackish gray; iris brown; tarsi and toes dull greenish yellow.

Juvenile (first-year). Entire upperparts very dark fuscous to blackish, and the underparts tawny to dark brown, the tail similar but with dirty white on some webs; under wing coverts whitish, variably streaked and tipped with dusky. Iris pale yellowish gray; cere grayish; bill brownish horn; tarsi and toes clear lemon.

Sherrod, White, and Williamson (1976) suggested that four age classes can be recognized in the field, including first-year ("brown" stage) juveniles, second- and third-year combined age classes ("mottled" stage), presumed fourth-year subadults ("eye-stripe" stage, those with some brown still present on the tail and on the head, especially behind the eyes), and adults, presumably at least five years old. These correspond approximately to the four stages listed above.

Southern (1964, 1967) reported that at least

six different age classes can be distinguished in the hand, and McCollough (1989) later provided modified descriptions of year classes based on observations of known-age wild birds:

First-year (Juvenal). Body uniformly dark brown below; no white on head; rectrices brown to sooty, often sprinkled with white toward bases (maximum up to about 60 percent on central rectrices); axillaries and most under wing coverts whitish or mostly white, with dark streaks and spots; bill and cere blackish gray, iris sepia brown, tarsi dull yellow. This plumage stage called "immature" by Clark and Wheeler (1987).

Second-year (Basic I). Upper breast forming a dark biblike band above variable (brown to whitish) underparts; interscapulars forming distinct buffy or whitish triangular pattern; some white visible in tail, which is dark-tipped; head brown, usually with a buffy superciliary line; neck often lighter than head; bill and cere mostly brownish gray, iris usually buffy brown, tarsi yellow. This plumage stage called "white-belly I" by Clark and Wheeler (1987).

Third-year (Basic II). Highly variable; throat often tan with some white, a brown breast band below, contrasting with mottled white on lower underparts; sometimes a white V or triangle present on upper back; frequently white on sides of head and throat, including a whitish superciliary line, forming ospreylike face pattern; iris light cream, cere buffy yellow with gray mottling, lower mandible with buffy yellow tip. This plumage stage called "white-belly II" by Clark and Wheeler (1987).

Fourth-year (Basic III). Crown and nape feathers often dull white, tipped with brown (often with an ospreylike dark eye stripe); white on tail more conspicuous, but with brown mottling distally; cere yellow to partly yellow, lower mandible yellow, iris pale yellow. Called "adult transition" by Clark and Wheeler (1987).

Fifth-year (Basic IV). Adultlike except for variable brown spotting or tipping on the rectrices and sometimes on head (often behind the eyes); some white on wing linings and body feathers usually still visible; bill and cere pale to bright yellow; iris pale yellow. This plumage stage called "adult" by Clark and Wheeler (1987).

Sixth-year (Basic V or Definitive). Brown flecks still sometimes present on tips of rectrices;

otherwise adultlike except for occasional white feathers still present on back or underparts; wing linings with little or no white; bill, cere, tarsi, and toes bright yellow; iris pale silver-yellow. Some birds may attain definitive plumage as early as their fifth year of life, whereas others may retain some immature plumage traits when more than eight years old.

Measurements (of *leucocephalus*, in millimeters)

Wing, males 515–545 (ave. of 16, 529.2), females 548–588 (ave. of 11, 576.5); tail, males 232–264 (ave. of 16, 248.5), females 247–286 (ave. of 11, 271.2) (Friedmann, 1950). Average egg size 70.5 x 54.2 (Bent, 1937).

Weights (in grams)

Males, ave. of 35, 4123; females, ave. of 37, 5244 (Snyder and Wiley, 1976). Range of weights 2000–6200, ave. 4300 (sex and sample size unspecified) (Clark and Wheeler, 1987). Estimated egg weight 114, or 2.2% of female.

Identification

In the hand. The absence of feathered tarsi easily separates this species from the similar-sized golden eagle, but not from the other two species of *Haliaeetus* (*albicilla* and *pelagicus*). The latter two species might occasionally occur in the Aleutian Islands or off the coast of Alaska or Greenland, but are outside the defined geographic limits of this book.

In the field. Adults can be readily identified when perched or in flight by their distinctive white heads and tails, although subadults may have these areas streaked or barred with brown. First-year birds are entirely brown-headed, but show varying amounts of white on the base of the tail. These and older age classes also have whitish axillary patches and distinctive whitish diagonal lines extending out along the under wing coverts. Older subadults ("white-belly" stage) also exhibit considerable white feathering on the belly and sometimes on the back. Curiously, juvenile birds have slightly longer wings and tails than do adults. When gliding or soaring their wings are usually held out horizontally. By comparison, golden eagles never exhibit white axillary patches or have noticeable white spotting evident on the under wing coverts, and when soaring often hold their wings at a slightly uptilted angle. The most common call is a repeated squeaky alarm note uttered near the nest. A similar note is uttered during a vertical head tossing while perching, and birds sometimes call in flight.

Habitats and Ecology

The breeding habitats of this species can perhaps be summarized simply; they require an adequate supply of moderate-sized to large fish, nearby nesting sites (usually large trees within a kilometer of water), and a reasonable degree of freedom from disturbance during the nesting period.

Because of their size, bald eagles require a substantial food base, and at least during the breeding season they are distinctly territorial. The radius of each nesting territory may be determined by the distance from the nest to perching trees up to about a half kilometer from the nest; thus 14 territories on Kodiak National Wildlife Refuge averaged 23 hectares (range 11.3–44.6), and most territories were separated by open areas (Chrest, 1965; Henzel and Troyer, 1964). Broley (1947) estimated that pairs of southern bald eagles in Florida defend an area of about 0.8 kilometers in all directions from the nest. A survey of 2,909 nests in southeastern Alaska had an average density of 0.56 nests per kilometer of shoreline (Hodges and Robards, 1982), while another nest survey in British Columbia provided a nesting density of 0.08 nests per kilometer of shoreline (Hodges, King, and Davies, 1983). A survey of the Aleutian Islands (Early, 1982) indicated 227 territories along 2,144 kilometers of coastline, or 0.1 territories per kilometer of shoreline. These are easily the densest nesting concentrations of bald eagles in North America, and provide only a crude measure of the maximum densities possible for the species in ideal habitat. Even on Kenai National Wildlife Refuge, where 23 active nests were found in 1980, only four nests occurred in randomly selected plots totaling 104 square kilometers (Bangs, Bailey, and Burns, 1982), producing a nesting density of a nest per 26 square kilometers.

Home ranges have not been studied extensively, but Frenzel (1983) estimated a mean home range of 660 hectares (range 325–1,384) for eight nesting pairs in Oregon, and a mean distance between nesting territories of 3.2 kilometers

(range 0.93–10.6). Territories had an average of about a half-kilometer of shoreline available for foraging, and the smallest body of water supporting more than one active nest was about 1,000 hectares. Haywood and Ohmart (1983) found that two breeding pairs in Arizona had variable home ranges averaging 64 square kilometers, but both pairs had rather constant river lengths of from 15 to 18 kilometers within their home ranges.

Outside of the nesting season population densities are extremely variable and probably depend entirely on the availability of an adequate supply of fish (or other foods) during that period. Again, the densest concentrations occur in coastal Alaska, where North America's largest concentration occurs in the Chilkat Valley, up to 3,500 eagles gathering during fall and winter to feed on salmon (*Oncorhynchus keta* and *O. kisutch*). Similarly, up to 1,500 eagles concentrate along the Stikine River of Alaska in April to forage on spawning eulachon (*Thaleichthys pacificus*) (Hughes, 1982), and in Glacier National Park more than 1,000 eagles now annually gather to feed on kokanee salmon (*Oncorhynchus nerka*) (McClelland et al., 1983).

Winter concentrations require the presence of suitable roosting sites as well as food supplies, and these have been studied in various areas. In Glacier National Park roosts typically consist of old-growth stands of western larch (*Larix occidentalis*), and similarly roost sites in southern Oregon and northern California typically were those with old-growth (averaging 236 years) ponderosa pines (*Pinus ponderosa*) or Douglas firs (*Pseudotsuga menziesii*) that were situated near foraging areas (Keister and Anthony, 1983). Haines (1986) found that in Virginia the birds favor trees that are easily accessible, have good visibility, are fairly large and with a strong open-branching structure, and are fairly near water and feeding areas. An analysis of roosts in Washington (Knight, Marr, and Knight, 1983) indicated that the birds select the largest and tallest trees in sites protected from prevailing winds by other trees or land-form characteristics. Conifers seemed to provide the favored sites when heavy winds occurred, as they were usually on steep leeward slopes. Apparently the birds select sites based on height, diameter, and growth form of trees for protection from inclement weather, especially wind, and those areas having a minimum of human activity. In Utah, the birds seem to favor side canyons with bowl-shaped ravines offering environmental protection, and selectively perch in large and open trees that are located near the tops of ridges, thus having easy access to valleys (Edwards, 1969).

Foods and Foraging

Although the bald eagle is an opportunistic forager, it generally favors fish (live or dead) as its primary food, and the greatest densities of birds occur only where fish are present in abundance. Stalmaster (1987) summarized 20 studies of nesting bald eagle foods having prey itemized numerically, in which fish ranged from 14 to 100 percent of the total prey items, birds from 0 to 81 percent, and mammals from 0 to 36 percent. Sherrod (1978) summarized data on 12 primary or secondary sources of bald eagle foods, eight of which provided a numerical analysis of food types. Of these, fish (or "lower vertebrates") comprised from 15 to 90 percent of food items, mammals from 1 to 36 percent, and birds from 8 to 61 percent. Mammals represented in fairly large numbers include introduced or native rabbits, hares (*Lepus*), Norway rats (*Rattus norvegicus*), and

sea otter pups (*Enhydra lutris*). Birds represented in large numbers include various smaller procellariiforms (fulmars and shearwaters), ducks, gulls, ptarmigans, and several species of alcids (murres, murrelets, auklets, and puffins). Most of these are obviously water-oriented species of at least moderate size. Varying amounts of carrion of mammalian, avian, and fish origins are also eaten at times when it is readily available. Apparently fish are preferred over birds or other prey when both are available, although in some areas waterfowl become the primary winter food of eagles (Bent, 1937), probably in the absence of an available fish supply. LeFranc and Cline (1983) found that avian prey remains comprised about 35 percent of prey remains at nests in the Chesapeake Bay area and included 45 species of birds, mainly waterfowl. The incidence of birds as prey was apparently influenced by nest location, and associated access to a prey base. Perhaps as much as 12 percent of the total food taken yearly by Alaskan bald eagles is carrion, primarily consisting of fish such as salmon (Imler and Kalmbach, 1955) but also including road-killed mammals such as rabbits (Retfalvi, 1970). Dead sheep are also important sources of carrion for eagles wintering in the Southern Gulf Islands of British Columbia and for breeding birds on San Juan Island, Washington; although afterbirths are sometimes consumed there is no evidence of mortality caused by eagles to lambs or ewes (Hancock, 1964; Retfalvi, 1965, 1970).

McEwan and Hirth (1980) compared food estimates based on numerical frequency with estimated prey biomass for Florida bald eagles, based on 788 prey remains at nests, and found that fish comprised 78 percent of the items and 70 percent of the estimated total biomass, whereas birds comprised 17 percent of the items and 26 percent of the total biomass. Mammals and reptiles were represented in very small amounts. Three species of catfish (*Ictalurus*) made up the majority of the prey items and biomass, and catfish are evidently a staple food item in several parts of the bald eagle's overall range, as for example in Minnesota (Dunstan and Harper, 1975) and in Maine (Todd et al., 1982). American coots (*Fulica americana*) were the most important bird species, representing 19 percent of the total biomass, and similarly were the most common prey at one Washington study site but not at another fairly nearby site, emphasizing the opportunistic feeding tendencies of bald eagles

(Fielder, 1982). Sherrod, White, and Williamson (1976) estimated that on Amchitka Island, Alaska, the biomass of foods brought to nests averaged 14 percent fish, 49 percent birds, 36 percent mammals, and a trace of invertebrates (octopus and amphipods). They judged that the birds may be able to carry prey as large as baby sea otters (or about 1,850 grams), at least for short distances, and once an adult emperor goose (*Chen canagica*) (which weigh an average of about 2,800 grams) was observed to be killed and carried to a sea stack. The data of these authors indicate a higher level of birds and mammals in the prey than most other data, although Murie (1959) reported an even higher percentage of bird remains at nests (86 percent) and even lower values for fish (6 percent). Both studies, however, probably greatly underestimated the percentage of fish in the diet, since their remains are much more likely to go undetected in pellet samples and probably don't persist long as prey artifacts around nests. In general, prey remains collections underestimate the intake of small, soft-bodied fish, while pellet analyses underestimate overall fish use (Mersmann, Buehler, and Fraser, 1987). Several authors have emphasized the underrepresentation of fish in their food estimates based on their prey samples (e.g., Cash et al., 1985; Todd et al., 1982). Probably the highest incidence of mammals and birds such as waterfowl occurs in winter, after lakes and reservoirs freeze over and waterfowl become increasingly vulnerable (LeFranc and Cline, 1983). Most studies of bald eagle foods tend to support the idea that the species is opportunistic rather than preferential in its food choice, taking whatever may be most available and most vulnerable at a given time and place, which produces considerable seasonal and locational differences in diets (Cash et al., 1985; Grubb and Hensel, 1978).

Haywood and Ohmart (1983) estimated that breeding birds in Arizona consumed foods consisting of about 63 percent fish, 16 percent birds, and 21 percent mammals. The fish varied in length from about 8 to 60 centimeters, and mostly consisted of rough fish such as suckers (*Catastomus*), carp (*Cyprinus*), and catfish (*Ictalurus* and *Pylodictus*). The species of fish that are taken by bald eagles are highly variable with season and locality, and it is perhaps impossible to generalize much on them. A list of fish reportedly taken in the United States (including Alaska) and Canada provided by Snow (1973b) included 27 taxa, with

a concentration on herring (Clupeidae), salmon (Salmonidae), Percidae, and various catfish (Ictaluridae). In a study not summarized by Snow, Imler and Kalmbach (1955) noted that salmonids (salmon and trout) and gadids (pollock and cod) were the most important prey on a volumetric basis in an Alaskan sample of 435 eagles, and that fish comprised 66 percent of the entire prey volume total, while birds comprised 19 percent, of which about half were waterfowl and half other birds, mostly alcids and other sea birds. Probably as much as 90 percent of the salmon eaten by eagles is carrion, whereas many of the other species are probably taken alive while feeding near the water surface or when isolated in shallow tidal pools.

In a Virginia study (Haines, 1986), where more than half of the prey were fish (mostly of bullheads *Ictalurus nebulosus*), most fishing occurred on days when the wind speeds were under 16 km/hour and the temperature was 18 to 26 degrees centigrade, although hunting success was not related to these variables or cloud cover. The four hunting techniques included swooping from flight, swooping from a perch, wading from shore and grabbing prey with bill or talons, and gliding out from ice or a low perch on piles of ice. Similarly, in an Illinois study, where most of the prey were gizzard shad (*Dorosoma cepedianum*), foraging was greatest during morning hours and the most common hunting strategy (successful about 70 percent of the time) was an aerial search, followed by swooping and capturing the prey. Adults were more successful at this technique than immatures, and most fish thus captured were no more than 15 centimeters long (Fischer, 1982).

Southern (1963) also found that wintering bald eagles used four techniques for capturing live fish. The first two, swooping from flight or from a perch, were each about 25 percent successful in obtaining fish. The most successful technique was that of wading in water and catching fish with the bill, while the last and rarest technique was to stand at the edge of ice and capture fish with the talons or bill. Russock (1979) noted that although the overall success rate on swooping dives for catching fish was 48 percent, an adult and subadult had a success rate of 62.5 percent, whereas two younger birds had a success rate of only 35 percent.

Obviously, other techniques are used for taking birds and mammals, and Edwards (1969) described some of these. Generally, bald eagles are prone to hunt cooperatively in hunting mammals such as rabbits. They use short flights back and forth above vegetation in an attempt to flush prey out, or watch from a perch and wait for prey to appear. The latter technique was used more often than extended coursing, and a variety of elevated sites were used as perching posts.

Stalmaster (1981, 1987) has described hunting behavior of adult and immature bald eagles in Washington; he calculated that a daily consumption of 500 grams of salmon, 364 grams of jackrabbit, or 296 grams of duck is required by a wild eagle to meet the daily energy requirements, which amount to some 494 kilocalories per day for a 4.5 kilogram eagle. Beebe (1974) stated that the birds are adept at capturing fish up to the size of a 3 kilogram salmon, but that the larger of these fish may be hauled ashore by alternated periods of towing or flying with the prey for short distances, interspersed with periods of resting half-submerged on the water.

Social Behavior

Bald eagles have a prolonged period of immaturity and a very long potential lifespan after attaining sexual maturity, allowing for the possibility of very extended pair bonds. Bent (1937) believed that bald eagles have lifelong pair bonds, and mentioned some examples of nests of great size and obvious old age, one of which had been used continuously for 35 years until it was blown down in a storm. Of course, several generations of birds might well have been involved in this nest's long history, and there seems to be little good evidence as to average or maximum lengths of pair bonds in bald eagles.

It is believed that bald eagles probably normally attain sexual maturity at five years, with variation around this mean, including some known four-year-olds breeding. About 30–35 percent of normally reproducing wild eagle populations probably consists of juveniles and subadults, whereas the incidence of such immature-plumaged birds during periods of reproductive decline may average less than 25 percent (Sherrod, White, and Williamson, 1976). Furthermore, even in an area of excellent eagle habitat with an abundance of nests, an apparently substantial portion of the adult population may be nonbreeders. Thus, in southeastern Alaska the incidence of nonbreeding in adults increased

from 26 to 86 percent between 1970 and 1979 (Hansen and Hodges, 1985), and it appears that this phenomenon may be associated with intense intraspecific competition and highly variable annual food supplies. Although average clutch size is apparently not related to food availability, offspring survival, the timing of early laying, and the proportion of active nests may all be influenced by food availability (Hansen, 1987).

Probably more southern populations of bald eagles are essentially sedentary, and thus pair bonds might be readily maintained through the entire year. However, more northerly ones are distinctly mobile and migratory, and it is unlikely that pairs remain together during winter periods. Beebe (1974) noted that, depending on climate and latitude, the birds may begin their reproductive activity any time between October and May, and often lay eggs before the shortest day of the year in southern parts of their range, while photoperiods are still decreasing. In the south and along the Pacific coast the birds probably become settled on nesting territories during fall, but in midcontinental areas they arrive in spring, following thawing of ice.

One aerial display involves a talon-locking and spinning descent, which may be associated with an undulating pursuit of the female by the male; this has been variously interpreted as aerial copulation, courtship display, and aerial combat (Beebe, 1974). Of these interpretations the two latter functions are most likely; and certainly at times talon locking occurs between birds that are obviously not paired, as during encounters between adult and immature birds (King, 1982), or even between immatures. Aerial displays also include pursuit flights, undulating sky dancing, and a "fly-around" display. During calling, a vertical "head toss" is frequent and may serve as an alarm signal or perhaps may attract a female's attention and help initiate a duet (Palmer, 1988).

Copulation certainly does not occur in flight, but rather on tree branches or other supports. Gerrard, Wiemeyer, and Gerrard (1979) provided one description of this behavior, which they observed 16 times in a captive pair. Nearly half of these times were in early morning, and copulation sequences always began with a period of the pair perching side by side. Then one bird (of either sex) approached the other (Figure 31A, B). When the female approached first she would lower her head and spread her wings slightly, calling softly with soft, high-pitched notes. When

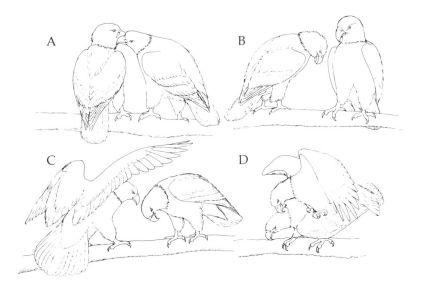

Figure 31. Copulatory behavior sequence in the bald eagle, including female (right) approaching male (A), male approaching female (B), male wing flapping and tail pumping while female bows (C), and male mounting female (D). After drawings in Gerrard, Wiemeyer, and Gerrard (1979).

the male approached he would edge close to the female, sometimes calling in a similar manner to the female. Then the male called, flapped his wings, and moved his tail up and down, often hitting the perch with his tail feathers. The female would then position herself with her feet farther apart, head lowered, and wings slightly apart (Figure 31C). The male would then step with balled talons on the female's shoulder and back, flapping his wings to maintain balance (Figure 31D). After treading the eagles remained perched together for a variable length of time, and then one or both would usually go to the nest and arrange nest materials. Mutual billing and stroking of the head, neck, shoulder, back, and breast feathers was also observed on various occasions independent of copulation behavior.

Breeding Biology

Nest site selection has been very extensively studied in bald eagles, and as summarized by Snow (1973b) several elements of preference seem to be consistently present. These include a clear flight path to a close point on a beach or river. The largest tree in a stand is normally chosen, and an open view of the surrounding area is an associated characteristic. Proximity to water is probably a normal requirement, because of the usual dependency on fish for food, and most nest trees are less than a kilometer from water. Finally, freedom from human disturbance is usually an important but highly variable factor.

Anthony (1983) has also summarized nest site characteristics of bald eagles in the western United States. There they nest in a considerable variety of coniferous forest types, as well as on cliffs and broad-leaved riparian vegetation in the Sonoran desert area of Arizona. Nesting in deciduous trees has also been reported in Oregon and the greater Yellowstone region. Apparently the birds select sites based on general tree structure (size, limb and visibility features, etc.) rather than tree species. Most nests are located within 1.6 kilometers of large bodies of water.

In a study of 3,850 nests in southeastern Alaska (Hodges and Robards, 1982), nearly all the nests were in coniferous trees, with only six percent in dead trees and no nests found on the ground. Among nests in live trees, the most common tree forms were those with bushy, live, broken, and deformed tops. Nests were usually located in old-growth forest stands, and nests placed on islets with little timber and those located in heavy old-growth timber stands had the highest activity rates. Second-growth timber supported almost no nests, and clear-cutting of old-growth stands was considered likely to destroy potential nesting for the entire expected 100-year rotation period. Nests averaged 25 meters high, and nesting trees averaged 30 meters; activity rates increased significantly with nest height. The average distance to the waterfront was 37 meters, and nearly all nests were within 183 meters. Probably the ideal nest location is on a prominent point or islet exposed to a broad channel or to a narrow passage if tidal currents are present.

In coastal Alaska and the Aleutian Islands where no trees are available, nesting commonly occurs on isolated or connected sea stacks, peninsular ridges, islets, hillsides, and the like (Sherrod, White, and Williamson, 1976), all of which variably provide height and isolation, especially from mammalian predators such as arctic foxes (*Alopex lagopus*) and red foxes (*Vulpes fulva*).

In a Minnesota study, Mathisen (1983) found that long-needled species of pines ranging in height from 15 to 37 meters are evidently the most desirable nest trees, and that although dead trees are seldom used, those with dead tops are apparently preferred. Most nests there are located in northern hardwood forest habitats, near openings and close (average 59 meters) to water. Closeness to a forest opening not only provides nest accessibility, but may favor the survival of young that have fledged and fallen to the ground. In Minnesota, the directional aspect of the nests was apparently random, although in other areas a tendency toward directional nest orientation has been noted.

In most areas the typical clutch is of two eggs, which are laid two to four days apart (Herrick, 1934). In some nests as much as two or three weeks' difference in the apparent ages of the young has been reported (Broley, 1947). Bent (1937) stated that the clutch of southern bald eagles is "almost invariably" of two eggs, with three-egg clutches rare. A sample of 43 Ontario nests had from 1 to 3 eggs, but averaged 2.05 eggs (Peck and James, 1983). In Alaska, 57 nests had from 1 to 4 eggs and averaged 2.31 (Sherrod,

White, and Williamson, 1976), suggesting that a higher average clutch size may be typical of northern populations. However, it would appear that only infrequently (up to three percent of the active nests, based on various studies) are three young ("triplets") ever fledged from nests, and four fledged young are very rare indeed. The incidence of "twins" being fledged successfully generally ranges from 3 to about 35 percent, with the highest observed rates of "twinning" being reported from Kodiak Island (Sprunt et al., 1973) and Amchitka Island (Sherrod, White, and Williamson, 1976).

The incubation period has been generally estimated at 34 to 35 days, with cases of interrupted incubation requiring a somewhat longer time of up to 45 days (Herrick, 1934; Bent, 1937). Eggs incubated artificially have generally hatched in 35 to 38 days. Both sexes participate in incubation, with the female taking the larger share during the day. Apparently the birds are able to adjust the length of time the eggs are exposed in relation to the cooling power of the air, and thus regulate egg cooling rates (Gerrard, Wiemeyer, and Gerrard, 1979).

The young remain in the nest for about 10 to 12 weeks (Herrick, 1934), during which time they are initially closely brooded and later left increasingly alone as both parents hunt to provide them with food. Not surprisingly, average rearing success is improved with increased food availability during the nesting period, even though this variable does not appear to influence initial clutch size (Hansen, 1987).

Productivity in three declining eagle populations averaged 0.34 young per occupied nest, as compared with 0.75 young in seven stable populations and 1.03 young in four increasing populations (Swenson, Alt, and Eng, 1986).

Evolutionary Relationships and Status

The bald eagle is part of a group of "fish eagles" that also includes the Asian white-tailed sea eagle (*H. albicilla*) as well as other species in Asia (*H. leucoryphus*), Africa (*H. vocifer*), Madagascar (*H. vociferoides*), southeast Asia (*H. leucogaster*), and the Solomon Islands (*H. sanfordi*). Of these, the bald eagle and white-tailed sea eagle probably represent an allopatric superspecies (Brown and Amadon, 1968).

There is now an enormous literature on bald

eagle productivity statistics, much of which has been summarized recently by Stalmaster (1987). He provided recent productivity data for 20 different areas and estimated that, considering all of these data collectively, about 43 percent of territorial pairs successfully produced young, that about 58 percent of the occupied nests were successful, that an average of 1.05 young were produced per active nest, and that 1.61 young were produced on average per successful nest.

As has been documented extensively, bald eagle productivity began to decline greatly during the 1950s, at least in the southern populations such as in Florida. Unfortunately, productivity data prior to this period are limited and thus somewhat unreliable, but a comparison of six populations during the 1960s indicated that three were then declining and three were apparently stable (Sprunt et al., 1973). In 1966 the International Union for Conservation of Nature and Natural Resources included the southern bald eagle as an endangered taxon in its Red Data Book, a classification that is still in effect (King, 1981). In 1973 the U.S. Fish and Wildlife Service listed the southern race *leucocephalus* as endangered. The entire breeding population of bald eagles in the conterminous United States was classified as endangered in 1978, except for those populations in Washington, Oregon, Minnesota, Wisconsin, and Michigan, which were listed as threatened.

King (1981) estimated that fewer than 500 pairs of the southern bald eagle existed in the late 1970s, of which two-thirds (300–325 pairs) were in Florida. In the Chesapeake Bay area 72 pairs were present, 16 pairs were in South Carolina, and 4 nests were active in Delaware. There were also about 20 pairs in California, 8 pairs each in Texas and Louisiana, 7 pairs in Arizona, and 2 pairs in Baja California. Single pairs were known for Mississippi, New Jersey, and possibly Tennessee. Most of the population reductions in Florida have been attributed to loss of habitat and the effects of DDE on reproduction (Evans, 1982), and probably the same factors have been important elsewhere in this form's range. In recent years the Arizona population has been increasing, with 26 known nesting sites in 1988. Hacking programs in Oklahoma and Alabama are also currently under way to try to reestablish nesting bald eagles in these states (Glinski, 1988). Similarly, nestling eagles hatched in Alaska have been

transported to and released in several other states (Tennessee, New York, Indiana, Missouri, and North Carolina), with more than 250 such birds having been relocated through 1988.

The northern bald eagle's overall population has never been subjected to such serious declines as southern populations, although productivity estimates in Michigan, Maine, and the Great Lakes averaged even lower than that of Florida during the 1970s (Sprunt et al., 1973). Generally, the Alaskan population has apparently never been seriously affected by human disturbance and habitat loss to the extent of more southern breeding populations, and populations nesting on interior lakes west of Lake Erie appeared to be stable or increasing slowly by the late 1970s. However, those from Lake Erie eastward, especially in northern Ohio, as well as those in Maine were still experiencing poor reproduction at that time (Evans, 1982).

By 1982, it was estimated that nearly 1,500 pairs of bald eagles were breeding in the conterminous United States, with about 65 percent of the pairs raising young successfully, and the average productivity being about 1.0 young per occupied nest (Green, 1983a, 1983b, 1985). Using Green's data, plus additional information for Canada and Alaska, Stalmaster (1987) has provided individual state and provincial estimates for breeding bald eagles (see Table 8). Gerrard (1983) estimated on the basis of Christmas Count data that about 70,000 bald eagles occurred in all of North America in the early 1980s, with 48,000 in British Columbia and Alaska and about 22,000 in the remainder of the continent. In central and western regions the numbers appear to have increased by a factor of two or three during the past 25 years, while on the west coast the changes are less clear. The east coast and Florida populations may have reached their low point about 1970 (at 25 to 33 percent of historic populations) and perhaps are now slowly increasing (Simons et al., 1988).

Complete information is not yet available for Canada, although a good deal of population status information has recently been summarized (Gerrard and Ingram, 1985). It is clear that British Columbia has the highest population of potentially breeding bald eagles, which perhaps numbered about 15,000 adults on the coast and another 6,000 in the interior during the early 1980s (Davies, 1985). Data are lacking for Alberta, but an aerial survey of Saskatchewan in 1974 resulted in an estimate of 3,900 "breeding areas" (areas with one or more active or inactive nests), 6,900 adults, 4,700 immatures, and 2,500 nestlings (Leighton et al., 1979). There may have been about 1,400 active breeding sites in Manitoba during 1984 (Koonz, 1985). Farther east, in Ontario and Quebec the populations are less well documented but generally apparently are rather small, the Maritime Provinces (Newfoundland with Labrador, Nova Scotia, New Brunswick, and Prince Edward Island) perhaps collectively supporting close to 200 nests in the early 1980s (see Table 8).

The literature of the bald eagle is extremely large, probably the largest of any raptor in North America, if not the world. An extensive bibliography, containing more than 2,000 references, is available (Lincer, Clark, and LeFranc, 1979), and additional selective but very useful bibliographies may be found in Evans (1982) and Stalmaster (1987).

TYPICAL HARRIERS

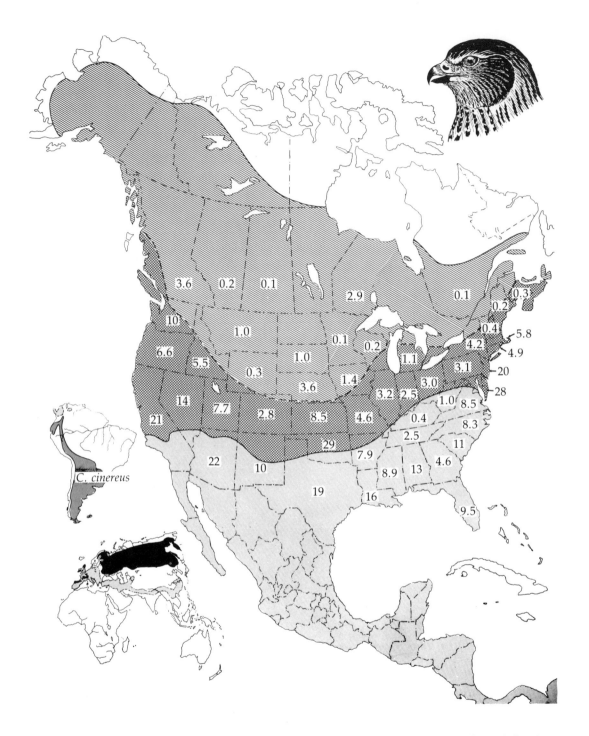

Figure 32. North American breeding (hatching), residential (cross-hatching), and wintering (shading) ranges of the northern harrier. Relative winter state or provincial density indices (average number seen per Audubon Christmas Count, 1986) are shown for major

wintering areas. The Old World breeding (inked) and wintering (shading) ranges of this species (there called the hen harrier), and the South American range of the closely related (possibly conspecific) cinereous harrier are shown on inset maps.

Northern Harrier *Circus cyaneus* (Linnaeus) 1766

Other Vernacular Names

Blue hawk (male), cinereous harrier (*cinereus* group), frog hawk, hen harrier (Britain), marsh hawk, mouse hawk, white-rumped harrier; busard Saint-Martin (French); gavilan ratonero (Spanish)

Distribution

Breeds (*hudsonius* group) in grassland or marshy habitats of North America from northern Alaska, northern Yukon, northwestern and southern Mackenzie, northern Saskatchewan, northern Manitoba, northern Ontario, central Quebec, and Newfoundland south to southern California, northern Arizona, northern New Mexico, northern Texas, central Oklahoma, southern Missouri, southern Illinois, Ohio, Pennsylvania, and northern Virginia. Formerly south to northern Baja California and Florida, the population now generally declining and the southern breeding limits retracting northward. Also widely distributed in Eurasia (*cyaneus* group); *cinereus* group is widespread in South America (usually considered an allospecies).

Winters (*hudsonius* group) in the Americas from Alaska, southern British Columbia, southern Alberta, southern Saskatchewan (rarely), Nebraska, Wisconsin, southern Michigan, southern Ontario, New York, and Massachusetts south through the United States and Central America to Panama, occasionally to Colombia and the Greater Antilles (Cuba and Hispaniola); also (*cyaneus* group) widespread in Eurasia, and (*cinereus* group) in South America. (See Figure 32.)

North American Subspecies

C. c. hudsonius (Linnaeus): Breeds in North America as indicated above; winters south to northern South America.

Description (of *hudsonius*)

Adult male. Head, neck, breast, and most upperparts excepting the tail coverts light neutral gray to deep neutral gray, washed with brownish cinnamon; the upper wing coverts neutral gray with dusky shafts; remiges similar, but the five outermost primaries fuscous to fuscous-black on the inner webs for their distal half, occasionally also on the outer web; the inner primaries and the secondaries with a broad but rather indefinite subterminal band of dark brown to light fuscous; upper tail coverts white; tail gray to deep gray, whitish basally and crossed at the tip with a fairly broad subterminal fuscous band, and anteriorly by five to seven narrow and often indistinct fuscous bands; the rectrices whitish on their inner webs; lower breast and posterior underparts white; the sides, lower breast, and upper abdomen usually with small brown to tawny-olive wedge-shaped spots; under wing coverts mostly white. Iris bright lemon-yellow; cere and edge of eyelid orange-yellow; inside of mouth purplish pink; bill mostly blackish, but abruptly bluish at base from below nostril; tarsi and toes deep cadmium-yellow; claws black.

Adult female. Head, neck, breast, and most upperparts excepting the tail coverts sepia to umber or olive-brown, the feathers of the crown, occiput, and nape edged laterally with rufous to tawny-buff; remiges (especially the primaries) becoming whitish basally on the inner web, and more or less obscurely banded with four to six narrow bars of dark sepia, the bars becoming distinct on the whitish areas of the inner webs; feathers of the rump like the back but with narrow rufescent or tawny tips and edges; upper tail coverts white, contrasting sharply with the rump and with the tail; central pair of rectrices umber to sepia, but basally whitish, tipped with tawny whitish and crossed by six or seven regular, sharply defined, but not very contrasting dark olive-brown bands; other tail feathers paler; lores, superciliaries, and suborbital areas dull buff to buffy white, leaving a dusky stripe between them; auriculars dull dark rufous to snuff-brown, indistinctly streaked with darker brown; feathers of the facial disk pale cream to pale tawny; underparts whitish and more or less washed with cinnamon to buff and usually with numerous longitudinal stripes umber to orange-cinnamon, these stripes becoming smaller posteriorly; under wing coverts whitish and more or less washed with buff or tawny streaked with dark umber to fuscous; under surface of primaries dull white and crossed at wide intervals with wavy dark sepia bands. Iris gradually changing from amber to pale yellow (by fifth year); cere light greenish; lower base of maxilla and basal half of mandible bluish, rest of bill blackish; tarsi and toes pale lemon-yellow; claws black.

Subadult. Year-old birds of both sexes resemble adult female but lack spotting and streaking on the underparts. Second-year females closely resemble adults but have fainter breast spotting, and second-year males are mostly gray above and whitish below, with some chest mottling. The iris of females changes from chocolate-brown to amber in the second year, and in males from grayish brown to pale straw the first winter. Breeding may occur at two (usually) or three years, or even occasionally at one year, especially when prey is abundant (Watson, 1977; Hamerstrom, 1986).

Juvenile. Young of both sexes resemble adult females during their first year, but are browner above and initially more tawny or cinnamon below (this color gradually fading), with little or no abdominal streaking. Iris yellowish gray or grayish brown (males) to dark brown (females); bill black, becoming lead-gray basally; cere and corner of mouth yellowish olive to greenish yellow; tarsi and toes orange-yellow to chrome-yellow.

Measurements (of *hudsonius,* in millimeters)

Wing, males 327.6–351.8 (ave. of 48, 339.6), females 335–405 (ave. of 54, 367.5); tail, males 228.6–251.5 (ave. of 48, 241.5); females 223–266 (ave. of 54, 249.5) (Friedmann, 1950). Average egg size 46.6 x 36.4 (Bent, 1937).

Weights (in grams)

Males, ave. of 90, 350; females, ave. of 97, 531 (Snyder and Wiley, 1976). Males 298–372 (ave. of 15, 364), females 473–595 (ave. of 42, 529.9) (Hamerstrom, 1986). Five adult males of the nominate race *cyaneus* averaged 346, and 8 females averaged 527; 14 males at least a year old averaged 347, and 40 females of this age group averaged 518 (Cramp and Simmons, 1980). Estimated egg weight 34.1, or 6.4% of female.

Identification

In the hand. The presence of a distinct facial ruff, and the unusually large ear openings (these hidden by the facial feathers, see Figure 68E), provide a ready means of identifying this species.

In the field. When perched (usually on the ground, rarely if ever in trees), the long tail is evident but the white rump is largely or entirely hidden by the folded wings. The wingtips of perched birds do not reach the end of the tail. Females and young males appear mostly rusty brown below and dark brown above, while adult males are rather uniformly medium gray above and whitish below.

In flight, birds of both sexes and all age groups show conspicuous white rump patches, and long, variably barred tails. Adult males show blackish wingtips that appear to have been "dipped in ink," whereas adult females and younger birds of both sexes have rather uniformly barred underwing surfaces. The birds often course close to the ground while quartering over grasslands and marshes, gliding with the wings held at a slightly uptilted angle. Vocalizations include high, chattering and repeated notes, and the female often utters a long, wailing *peee-e* during the breeding season. Males utter a series of short, dry *kek* notes during aerial display, which become higher and more rapid *ke* notes in alarm.

Habitats and Ecology

This is an essentially open-country species, breeding at generally lower altitudes (up to about 300 meters in western states) in medium to tall prairie grasslands and associated wetlands, fresh and salt-water marshes, swamps and bogs, wet meadows, hay meadows, logged-over or burned woodlands, open muskeg, and tundra. Outside of the breeding season it extends up to alpine tundra, but especially favors coastal marshes. Besides being associated with wet to mesic habitats, the birds also sometimes occur in habitats fairly far from water, a habitat usage that is especially typical of the Palearctic population (called the "hen harrier" in Britain), where other competing species of harriers are present (Brown and Amadon, 1968; Nieboer, 1973).

In a habitat usage study in southwestern Idaho, Martin (1986) found a disproportionately higher use of riparian habitats (mostly consisting of *Phragmites, Scirpus, Salix,* and *Tamarix*) and cultivated areas (mainly of alfalfa, wheat, and sugar beets), and a correspondingly lower use of shrub-steppe habitats than would have been expected on the basis of available habitat areas. As alfalfa growth approached 46 centimeters in height, males moved to open-shrub steppe habitats increasingly, with a shift from mammalian to reptilian prey, but returned to the alfalfa fields following their cutting. However, breeding females were found to use only riparian and cultivated habitats directly adjacent to their nests.

Apfelbaum and Seelbach (1983) reported that of 721 nest sites in North America, 38 percent were situated in dry-land habitats (cultivated or uncultivated fields, pastures, and grasslands), 18 percent were in fresh-water marshes, 17 percent were in marsh meadows, 7 percent were in bogs, swamps, brackish marshes, or dried wetlands, and 20 percent were in miscellaneous habitats (sage rangelands, brushlands, deciduous woods, stubble, brushy rangelands, or unknown). This kind of habitat selection tends to remove the northern harrier from competition with other North American hawks, although it does compete with short-eared owls and evidently partitions its food resources with that species (Linner, 1980).

Home ranges are quite variable in this species, probably as a reflection of differing habitat resources. Breckenridge (1935) made an early estimate of 2.5 square kilometers for harriers in Minnesota, and Craighead and Craighead (1956) estimated that breeding birds may have a home range of 0.6–6.3 square kilometers (averaging 2.1), but wintering birds were judged to have a smaller home range averaging about 1.5 square kilometers. Martin (1986) determined a marked difference in the home ranges of radio-tagged breeding males and females, with males having an average home range of 15.7 square kilometers, while females averaged only 1.13 square kilometers. A similar smaller home range for females was reported by Watson (1977) for British harriers. In favorable habitats the birds sometimes attain quite high breeding densities; Bent (1937) mentioned three nests within 366 meters of one another, and a breeding density estimate of a pair per 1.5 square kilometers in good North Dakota breeding habitat has been made (Stewart and Kantrud, 1972).

Foods and Foraging

This is a broadly adapted species of hawk, with an ability to diversify its diet according to time and place. Nevertheless it is largely specialized for feeding on small to medium-sized mammals captured in rather low vegetation while flying at low levels in open vegetation. In many ways it is like the short-eared owl in its hunting techniques,

and it is the only North American hawk that not only has evolved an owllike facial disk but also is able to locate and capture hidden prey using only acoustic clues (Rice, 1982).

Early studies (Fisher, 1893; Bent, 1937; Errington and Breckenridge, 1936) indicated that though this species consumes a wide range of vertebrate prey, it clearly specializes on meadow voles (*Microtus* spp.); Hamerstrom (1986) referred to the northern harrier as "the hawk that is ruled by a mouse," in reference to the strong influence of vole populations on the mating systems and associated productivity of harrier populations (Hamerstrom, Hamerstrom, and Burke, 1985). It should be noted, however, that many harrier populations in North America have been found to make a shift from small rodents to young passerine birds during the breeding season (Randall, 1940; Barnard, 1982; MacWhirter, 1985). Thus, Barnard et al. (1987) noted that the birds tend to switch from *Microtus* voles to preying on recently fledged grassland and marsh passerines, such as bobolinks (*Dolychonyx oryzivorus*), as soon as they become seasonally available. Earlier, Errington and Breckenridge (1936) also found a series of "waves" of new foods utilized during summer as various vulnerable prey such as young cottontails, ground squirrels, and small birds became successively available. Simmons, Barnard, and Smith (1987) summarized adult nest-provisioning rates as reported from the literature, which indicates that *Microtus* voles, juvenile birds, and juvenile lagomorphs are usually the principal prey during this period.

Sherrod (1978) summarized data from 11 earlier primary or secondary sources of information on harrier foods. These included 9 summaries having a numerical percentage analysis of prey items, which collectively indicated a range of mammalian prey comprising from 14 to 82 percent of total identified foods, and bird prey ranging from 8 to 81 percent of the total, with reptiles, amphibians, and invertebrates constituting the remainder. These quite variable figures suggest a considerable degree of prey flexibility and opportunistic hunting in the species.

Not surprisingly, given this opportunistic hunting technique, prey weights vary greatly, from about 9 to 500 grams in males and about 9 to 600 grams in females (Schipper, 1973). Sometimes, however, birds weighing more than 900 grams such as greater prairie-chickens (*Tympanuchus cupido*) are taken (Toland, 1985a;

Hamerstrom, 1986). Attacks on adult waterfowl, including the killing not only of wounded but also of healthy birds as large as American wigeon (*Anas americana*), have been documented (Godfrey and Fedynich, 1987).

The northern harrier hunts for small, ground-dwelling rodents by flying at low speed very close to the tops of low vegetation, its wings tilted upward for maximum aerodynamic stability, and systematic quartering over the landscape. Brown and Amadon (1968) suggested that over rough vegetation such as weeds and fallow land it cruises at only about 20–30 kilometers per hour, versus up to about 55 kilometers per hour over ploughed fields, adjusting its speed to allow for the relative visibility of prey in differing habitats. They estimated that the birds may be on the wing as much as 40 percent of the daylight hours, perhaps flying about 160 kilometers per day.

Schipper, Buurma, and Bossenbroek (1975) analyzed the hunting behavior of this species in Europe, finding that the birds used the benefits of surprise attack by approaching likely prey sites such as ditches transversely, by using zigzag flight routes, and by suddenly appearing around vertical "edges," such as trees. Their flight speeds during hunts ranged from about 20 to 40 kilometers per hour, with faster average speeds used by males than by females, and by both sexes when flying higher or when using tail winds. The altitudes of hunts varied with vegetation height, being lowest in hunting over crops, followed progressively by flights over reeds, willows, and tree plantings; flights were also lower against head winds. Generally, surprise tactics were more often used by the smaller and more agile males than by females, which concentrated on larger and slower prey. Not only do females hunt more slowly and tend to hunt in higher vegetation, they may also exclude males from favored hunting areas by their larger size and associated social dominance, forcing males into less desirable areas (Temeles, 1986).

When food is seen, the harrier typically holds a hovering position directly overhead, stalling and blocking every move this prey makes, until it is able to drop down and strike, either by a direct vertical fall with upraised wings or with a kind of corkscrew twisting motion (Beebe, 1974). Prey can be accurately detected acoustically at distances of up to about 3–4 meters, and successfully attacked without additional visual or olfacto-ry cues (Rice, 1982). Success rates in capturing voles are rather low, perhaps averaging around 6–8 percent of attempts (Schipper, Buurma, and Bossenbroek, 1975). Most hunting is done during morning hours, with a midday lull typical, followed again by a late afternoon period of hunting (Hamerstrom and Wilde, 1973).

Social Behavior

Perhaps the most remarkable feature of this species's social behavior is its tendency for males to adopt polygamous (bigynous or polygynous) mating under suitable ecological conditions. In this respect it is unique among North American hawks and remarkable among all birds of prey in the world (Brown, 1976b), although several owl species are now known occasionally to form bigamous pair bonds, and the Harris' hawk regularly forms nesting groups that often involve an extra female at the same nest. In the northern harrier, however, each female establishes her own nest, each of which is separately tended by the male.

Northern harriers show little tendency toward prolonged pair bonds; in 25 years of study involving color-marked mates at 70 nests, Hamerstrom (1986) only once found a female to breed again (two years later) with the same male (who that year had two other mates as well); in the intervening year she paired with a different male. There is also little tendency for harriers to locate their nest in the same place in succeeding years; nests of 20 males averaged 546 meters away from their previous year's nests, and those of 18 females averaged 1,092 meters (Burke, 1979).

The spectacular courtship flights of northern harriers have been described by a variety of authors. Hamerstrom (1986) has pointed out that they often differ greatly in details, apparently because of the differences in appearance resulting from seeing them from differing angles. When seen from the side, the "sky dance" appears to be a series of shallow or deep (to about 20 meters) steep dives and equally steep ascents in rapid succession. When seen from the front, these flights actually appear as a series of "barrel rolls," the bird turning over sideways to either the right or left as it begins a dive and then righting itself by the same kind of sideways turn as it reaches the bottom of its dive, producing a series of successive looping arcs in the sky. This is done mainly by males, and probably these flights are centered over his territory. Although aerial talon grasping has been observed a few times in northern harriers, this has occurred during situations of agonistic encounters between an adult of either sex and an intruder into the territory that may be an adult or subadult (Craig, Craig, and Marks, 1982).

Another territorial display of this species is "leg lowering," which mainly occurs as an intrasexual confrontation display performed by either sex. It involves slow, close, and parallel flights by the participants along territorial boundaries as they conspicuously lower their legs and repeatedly glide to the ground. Such encounters typically end with the departure of one of the birds (Barnard and Simmons, 1986).

Brown (1976a) has described the display flights of the European race of this species, which may begin with the pair soaring together, the male often diving at the female, who often turns over and presents her talons to his (see Figure 16F). He may sometimes display in company with two females, and the flights are accompanied by a staccato *chek-ek-ek-ek* call that later in the breeding cycle indicates to the female that the male is bringing food. In the most intense form of display the male may appear to be out of control as he dives toward the ground, only to check his fall a meter or so above the substrate and then quickly rise and repeat the performance. The sequence is rarely extended to more than a hundred times in succession; normally it is performed directly above a nest site, but the displaying bird may also range over nearly a kilometer. Females may occasionally also perform the diving display, especially where females outnumber males, possibly thereby insuring spacial separation of each nest from those of other females mated to the same male. Diving displays by the female may continue late into the fledging period, suggesting that each female maintains her own smaller territory within the male's larger home range.

Simmons (1988) reported that the intensity of the males' aerial displays provides an accurate measure of their potential to provision their mates, and that the most vigorously displaying males attract the largest harems, the males thus engaging in "honest advertising."

Copulation was observed once by Clark (1972), who saw a female alight on a post after hearing loud calls. She was followed immediately by a male, who landed on her back and remained there about four seconds, opening one wing for balance. He then flew to a fence post nearby as the female began to preen. The male soon flew up over the presumed nest site area, followed by the female. She landed on the ground, and the male made four or five passes above, performing shallow dives but not turning over. He finally flew back to the post where copulation had occurred and remained there. Observations on the European race indicate that mating can occur on the ground or on a low perch, usually near the nest site, and typically occurs after the female has received food from the male. She usually attracts the male with her soliciting call, which is a high-intensity food call, and assumes the receptive posture as the male flies to her (Cramp and Simmons, 1980).

The frequency of nonmonogamous matings probably varies with time and locality, depending upon food supplies and perhaps also the sex ratios in the adult population. Hamerstrom (1986) tallied 330 nesting pairs between 1959 and 1983 in Wisconsin, of which 252 (76 percent) were monogamous, 54 (16 percent) were bigamous, and 24 (7 percent) were trigamous. In Picozzi's (1984a) study area in the Orkney Islands, off the north coast of Scotland, the incidence of polygyny was found to be related to male age. Of 19 yearling males, only one was bigamous, whereas most older males were polygynous, with 15 two-year-olds having an average of 1.9 mates, 16 three-year-olds having an average of 2.1 mates, 10 four-year-olds having an average of 2.7 mates, and older age classes having averages of 2.3 to 4.0 mates. The oldest marked male, a 10-year-old, had three mates. As harem size increased, males had greater probabilities of rearing at least one young, but any possible advantage in polygyny for females, other than the opportunity to breed at all, was less clear. About the same percentage (26–31 percent) of both sexes bred as yearlings. Reproductive efficiency apparently increased in females up to five years of age and then declined, while that of males did not differ significantly with age.

Simmons, Barnard, and Smith (1987), comparing the social behavior of North American and European harrier populations, have concluded that only minor sexual behavioral differences occur between these populations, such as in the extent to which adult and immature males and females display. They stated that tentative pair bonds were indicated by mutual male-female territorial soaring and food transfers, while territorial behavior by females was manifested by leg lowering and talon grappling, with the most persistently aggressive females getting the most productive nesting areas. Males asserted their territoriality by leg lowering, which was directed only toward other males, and by sky dancing, mainly directed toward females. Over 90 percent of the sky-dancing sequences they observed were by males, and 96 percent of the male sky-dancing displays were performed near perched or flying females. These authors found that adult males display more often and more intensively in high than in low-vole years, while immatures display only during high-vole years. Some monogamous males rejected additional females, while some females also rejected males with low provisioning rates.

Breeding Biology

The establishment of a nest site perhaps initiates the breeding cycle for northern harriers, and this is typically done in low, grassy, weedy or shrubby vegetation, avoiding very open sites (Brown and Amadon, 1968). Hamerstrom and Kopeny (1981) analyzed the vegetation at 184 nest sites in Wisconsin, finding that certain types of vegetation, especially grasses, sedges, willows (*Salix*), and various forbs (*Solidago* and *Urtica*) or shrubs (*Spirea*), were disproportionately frequent at these sites. Somewhat different cover plants were found typical in Manitoba (Hecht, 1951), but most nest sites were in mainly dead grass (*Fluminea* and *Phragmites*) cover that produced some visual protection from the start of the breeding season. Among nearly 200 nests from Ontario a wide range of habitats and cover types were used, but open areas seemed to be preferred, and even in wet situations the nest sites were more often dry than wet (Peck and James, 1983). In a New Brunswick study (Simmons and Smith, 1985) it was found that most birds selected wet nest sites that were surrounded by cattails (*Typha*) and had high visibility features. Although moisture at the nest site was the best predictor of nest success, the female's choice of nest site was only weakly correlated with this factor. The authors suggested that nest site choice may be a compromise between having a wet site, proximity to optimum foraging areas, and access to a good mate. Nests

tend to be rather uniformly dispersed; Picozzi (1978) reported an average internest distance of 1.52 kilometers in Scotland.

Simmons, Barnard, and Clark (1987) reported that in nearly all of 19 observed cases, males initiated the nest building, building platforms ("cock nests") in the presence of both monogamous and polygynous females, which appeared to stimulate the females into completing their nests. Eight of the females later built and used alternative sites within 20 meters of the platforms, but most of the latter were adopted by the females and lined by them. Platform nests are apparently relatively rarely built by males in Scotland (Watson, 1977).

Eggs are laid at intervals of 48–72 hours, or occasionally longer, generally during morning hours, and incubation may begin with the second or any later egg. Thus there is considerable overlap between the end of the egg-laying period and the onset of incubation, which is normally done only by the female (Brown, 1976a). There is some disagreement about the possible participation of males in incubation and brooding in North America, with some reports (Bent, 1937; Sealy, 1967) indicating that it occurs, but others (Breckenridge, 1935; Hecht, 1951; Simmons, Barnard, and Smith, 1987) finding it not to occur. It is probable that male incubation occurs only after the loss of a mate, and that male incubation behavior has gradually been lost in association with the evolution of a polygynous mating system (Simmons, Barnard, and Smith, 1987).

Clutch sizes of northern harriers in North America occasionally range to 7 eggs, but usually are of 4 or 5 eggs. Peck and James (1983) found that 65 percent of 171 Ontario nests had 4 or 5 eggs, with the mean clutch size 4.58 eggs. A total of 39 North Dakota nests had 4–7 eggs, averaging 5.0 (Stewart, 1975). An earlier North Dakota sample of 60 clutches also averaged 5 eggs (Hammond and Henry, 1949). One sample of 13 clutches from Manitoba averaged 5.07 eggs (Clark, 1972), while another of 21 clutches from Saskatchewan averaged 4.18 eggs and 11 clutches from Alberta averaged 4.45 eggs (Sealy, 1967). Of 200 clutches from the Orkney Islands, the average was essentially the same, 4.46 (with annual variations of from 3.8 to 5.2), while clutches in Norway averaged larger (4.69) in vole plague years than in nonplague years (3.4) (Cramp and Simmons, 1980). Average clutch size was also found to vary annually (five-year average of 80 clutches, 4.3) with varying microtine populations in a New Brunswick population studied by Simmons et al. (1986), who also found the number of polygynous males and the number of harem females to be significantly correlated with this variable, as were measures of breeding density and age-related productivity.

The incubation period lasts from 29 to 31 days per egg, or about 29–39 days per clutch (Cramp and Simmons, 1980). Sealy (1967) estimated a somewhat longer (34.3 day) average period for Canadian birds, based on the laying and hatching of the last egg.

Hatching occurs over a highly variable period of 1 to 10 days, reflecting both the laying interval and the variable onset of incubation, but typically there is a considerable disparity in age and size of the nestlings. Females continue to incubate eggs even when they have a brood to deal with, and even in large clutches hatching success is apparently not influenced by clutch size. Repeat clutches are often laid if the first clutch is lost early in incubation; such clutches often average about three eggs. Females continue to brood for about a week after initial hatching, and thereafter may begin to take part in hunting (Brown, 1976a). At least in some areas males are able to support at least four mates and their young, which must place a tremendous burden on such birds.

Interestingly, the average number of young fledging tended to increase with increasing clutch size in Picozzi's (1984a) study, and in general males with two or more mates tend to fledge more young (although not at a statistically significant level). Of 562 Orkney nests, 231 (41 percent) failed during laying, incubation, or the nestling period. In 455 nests the hatching success was 53.5 percent and the rearing success was 67 percent, with an average of 2.4 young reared per successful nest, or about 1.6 young per breeding pair (Brown, 1976a). Hamerstrom (1986) reported that of 252 monogamous nests studied in Wisconsin, 76 percent were successful, as compared with 61 percent of 54 bigamous nests, and 71 percent of 24 trigamous nests. In all, 726 young were fledged from 330 nests, or 2.2 young per nest and 3.0 per successful nest. Nonmonogamous males fledged more young on average (3.7–4.7) than did monogamous males (2.3), but nonmonogamous females fledged fewer young on average (1.6–1.9) than monogamous ones (2.3), thus producing conflicting selective tendencies toward monogamy or polygamy in the two sexes.

In a similar study, Balfour and Cadbury (1979) found in Scotland that most first-year birds formed monogamous pairs, but polygyny allowed a greater number of females to breed in a particular area. Hatching and rearing success was higher in monogamous than in polygynous nests, but males having three or more female mates produced more young than did monogamous ones.

Although Picozzi (1984b) judged that the main reason for the high level of polygyny in the Orkneys was a shortage of males in the breeding population (males having an estimated 72 percent annual survival rate as compared to a 90 percent survival rate in females), Simmons, Smith, and MacWhirter (1986) suggested that at least in some populations individual differences in male quality (as nest provisioners) may contribute to the development of polygyny. They found that within harems there was a dominance hierarchy, with dominant (alpha) females receiving more prey than the others and thereby raising more young on average. Females forming polygynous attachments to already mated birds had higher nest predation rates than monogamous females or alpha females, perhaps because they had to spend more time hunting for their own young, thereby leaving the nest more vulnerable. Simmons et al. (1986) also found significant correlations between male food-provisioning rates, clutch sizes, and reproductive success rates over three years, thus providing a proximate pathway possibly leading toward polygyny.

Fledging occurs at about 35 days on average, with male young typically leaving at about 32 days, whereas the heavier females may remain in the nest until 42 days (Brown, 1976a). Their growth and development have been documented by Scharf and Balfour (1971). The young remain dependent upon their parents for some time after fledging; after about three weeks the young may begin to leave their home ranges, although some may stay near their nest site for as long as 50 days. Apparently most young birds migrate alone rather than with siblings or their parents (Beske, 1978). Based on studies on the Orkney Islands (Picozzi, 1984b), females have a significantly higher annual survival rate than males during their first two years of life (29 vs. 14 percent), as well as at least the next four years (90 vs. 72 percent). Individuals of at least 12 years old have been recorded (Watson, 1977).

Evolutionary Relationships and Status

Brown and Amadon (1968) suggested that the genus *Circus* is a rather isolated one, with no obvious close relatives. Nieboer (1973) and Watson (1977) have discussed the ecological and geographical differentiation within this genus, which consists of numerous widely spread species or superspecies. The South American cinereous harrier (*C. cinereus*) sometimes is considered conspecific with the northern harrier.

The northern harrier was placed on the Audubon Society's Blue List of potentially seriously declining species in 1972 and has remained on it through 1986, but has not yet been given special attention by the U. S. Department of the Interior or the Canadian Wildlife Service. It is probably stable or perhaps is declining slowly in population, with local habitat loss probably the species's greatest threat (Evans, 1982). My own estimate of the species's North American population during the early winter of 1986, based on the Audubon Society's Christmas Bird Counts, is 111,500 birds, with high densities in the Chesapeake Bay area and in Texas, California, and Arizona, which had estimated state populations of 20,000, 13,200, and 9,900 birds respectively. A selective, nonannotated bibliography of more than 130 references has been published (Evans, 1982).

TYPICAL ACCIPITERS

Figure 33. North American breeding (hatching), residential (cross-hatching), and wintering (shading) ranges of the velox race (**VE**) of the sharp-shinned hawk, together with residential ranges of the forms chionogaster (**CH**), fringilloides (**FR**), madrensis (**MA**), perobscura (**PE**), and suttoni (**SU**). The white-breasted form chionogaster is considered by some to be a separate species. Relative winter state or provincial density indices (average number seen per Audubon Christmas Count, 1986) are shown for major wintering areas. The South American breeding or residential range (of chionogaster) is shown on the inset map. Indicated range limits of contiguous races should not be considered authoritative.

Sharp-shinned Hawk *Accipiter striatus* Vieillot 1808

Other Vernacular Names

Bird hawk, chicken hawk, little blue darter, sharpshin, sharpy, white-chested accipiter (*chionogaster*); epervier brun (French); esmerejon coludo (Spanish)

Distribution

Breeds from western and central Alaska, northern Yukon, western and southern Mackenzie, northern Saskatchewan, central Manitoba, central Ontario, central Quebec, southern Labrador, and Newfoundland south to central California, central Arizona, southern New Mexico, southern Texas, locally in the upper Great Plains to the Black Hills, northern Nebraska, and south-central Minnesota (no recent records for Iowa), Missouri, and the northern parts of the Gulf states east to South Carolina, and south through the highlands of Central America (*chionogaster* group) to South America; also resident locally in the Greater Antilles (Cuba, Hispaniola).

Winters from southern Alaska, the southernmost portions of the Canadian provinces, and Nova Scotia south through the United States and Middle America to South America, casually to the Bahamas, Jamaica, and perhaps elsewhere; also winters in the breeding range of the Greater Antilles. (See Figure 33.)

North American and West Indian Subspecies

A. s. velox (Wilson): Breeds widely in North America from Alaska and Canada south to the southern U.S.

A. s. perobscurus Snyder: Resident in the Queen Charlotte Islands, and perhaps the adjacent British Columbia mainland.

A. s. suttoni van Rossem: Breeds from extreme southern New Mexico to central Mexico.

A. s. madrensis Storer: Endemic to Guerrero (and perhaps also western Oaxaca) in the Sierra Madre del Sur, Mexico.

A. s. fringilloides Vigors: Resident in Cuba.

A. s. striatus Vieillot: Resident in Hispaniola.

A. s. venator Wetmore: Resident in Puerto Rico.

A. (s.) chionogaster (Kaup): Resident from Chiapas to Nicaragua (often considered a separate species).

Description (of *velox*)

Adult male. Forehead blackish plumbeous (lead-colored) with pinkish cinnamon, pinkish buff, or hazel-tinted margins to the feathers; crown and occiput blackish plumbeous-black; nape blackish plumbeous with concealed white bases; remainder of upperparts shading from blackish plumbeous on the upper back to dark or deep plumbeous on the rump and upper tail coverts; tertials with concealed large white and smaller rusty spots; tail square-tipped (when closed) and gray above, pallid gray below, white toward the bases of the inner webs of the outer rectrices, narrowly tipped with white and crossed by four bands of grayish fuscous; cheeks, ear coverts, and sides of neck ochraceous-tawny pinkish buff, with fine blackish shaft lines; chin and throat white to pinkish buff, with conspicuous fine blackish shaft lines; under tail coverts pure white; remainder of underparts white, irregularly barred with brownish, usually less dense on the lower abdomen, occasionally with the breast and flanks more or less uniform tawny; under wing coverts white or pinkish buff, streaked, spotted, or barred with fuscous; wings fuscous above, shading to deep plumbeous on the inner secondaries, white toward the bases of the inner webs, pallid gray below and broadly barred with dark fuscous or fuscous-black. Iris ruby-red; bill black or blackish; cere yellow or greenish yellow; interior of mouth light cobalt-blue; tarsi and toes yellow; claws black.

Adult female. Similar to the male, but larger, more fuscous, less gray above, and on the average less brightly colored below. Softparts similar to those of the male, but slightly less bright.

Subadult. The juvenal plumage is held without much change through the first winter, except for wear and fading. Initial breeding is done in this plumage. By the second fall the plumage is nearly adult, but averaging browner above, and the breast and flank feathers have darker brown and less white present (Bent, 1938). The iris color gradually changes from yellow to intense ruby-red with increasing age.

Juvenile (sexes nearly alike). Browner than the adult (males averaging darker than females) and the upperpart feathers often edged with russet; underparts white to buffy, heavily striped with cinnamon (in females) to fuscous (in males). Iris grayish to greenish or sulfur-yellow; bill blackish, becoming lead-gray basally; cere yellowish green to greenish yellow; tarsi and toes yellowish.

Measurements (of *velox*, in millimeters)

Wing, males 162–185 (ave. of 51, 171.1); females 180–210 (ave. of 40, 200.3); tail, males 134–152.5 (ave. of 51, 140.8), females 150–179.5 (ave. of 40, 165.6) (Friedmann, 1950). Average egg size 37.5 x 30.4 (Bent, 1937).

Weights (in grams)

Males 82–125 (ave. of 435, 103); females 144–208 (ave. of 487, 174) (Dunning, 1984). Males 87–114, ave. 101, females 150–218, ave. 177 (sample sizes unspecified) (Clark and Wheeler, 1987). Estimated egg weight 19.1, or 10.9% of female.

Identification

In the hand. The combination of a small body (under 225 g) with short, rounded wings (160–210 mm) and a barred tail that is about 80 percent as long as the wings, distinctly square-tipped, and no more than 180 mm, should serve to identify this species.

In the field. When perched, this pigeon-sized species's wingtips fail to reach the midpoint of the tail, and the underparts are barred (adults) or streaked (immatures) with rufous. Besides their smaller size than the very similar Cooper's hawk, adult sharp-shinned hawks typically have a uniformly dark crown and nape (instead of a lighter nape than crown, as typical of the Cooper's hawk), and if the folded tail can be observed it appears to be square-tipped or slightly notched rather than rounded. In juvenal plumage the underpart streaking of the sharpshin is somewhat less well defined than in the Cooper's. Additionally, the smaller sharpshin has relatively larger and more conspicuous eyes than does the Cooper's. Both the Cooper's hawk and sharpshin have bright red eyes as adults, helping to separate them from falcons, which have dark brown eyes.

In flight, the species's small size and rapid wingbeat are evident, and the bird's relatively small head barely projects beyond the leading edge of its wings. Unlike the similar-sized merlin, this species has rounded rather than pointed wings. The larger Cooper's hawk has a somewhat more rounded tail (but even the sharpshin's tail appears rounded when spread) and the whitish trailing edge of the tail is somewhat broader. Both species are associated with woodlands, and have darting flights that cause them to be lost from view quite rapidly. The sharpshin has a somewhat faster wing stroke, relatively longer wings, and a relatively shorter tail having a less conspicuous white tip. Both species are fairly silent except during the breeding season, when various alarm and territorial calls are frequent, including duetting by paired birds.

Habitats and Ecology

This is a relatively migratory species of accipiter, with apparently rather generalized winter habitat requirements. These requirements probably consist of little more than woodlands, especially fairly open or young forests with a variety of plant life forms present, which in turn support a large diversity of small bird species on which the sharpshin forages. This is also true of its breeding habitats, such as the boreal forest where perhaps up to 80 percent of the total North American population breeds, and where it favors young forests with a variety of spruce thickets, meadows, or lightly treed areas of aspens or pines (Beebe, 1974). Similarly, mixed coniferous-deciduous forests are important breeding habitats; mixed woodlands (55 percent of total) and pure coniferous forests (46 percent) made up nearly all the breeding habitats used in a survey of 37 North American nesting locations (Apfelbaum and Seelbach, 1983). In northeastern Oregon sharpshins and Cooper's hawks tend to nest in younger successional stands of coniferous forests than do goshawks, and also favor sites with larger crown canopy volumes than those selected by goshawks (Moore and Henny, 1983). In the northeastern states the species breeds mainly in open mixed or coniferous woodlands, clearings, and edges, as well as in heavily wooded areas (Bull, 1974). In northern Utah a proximity to water appeared to be an important breeding habitat component, and the birds also favored the more mesic conditions associated with northern to eastern slopes (Hennessy, 1978). Somewhat similar preferences for nearby water and north-facing slopes were found in Oregon (Reynolds, 1978; Reynolds, Meslow, and Wight, 1982). With the end of the breeding season in the western states the birds often move downslope, gradually leaving the denser forests and often wintering in oak woodlands or even occasionally moving into mostly grassland areas for their foraging.

Home ranges and territories in this species are probably much the same; two such home ranges in Jackson Hole, Wyoming, were estimated as 65 and 130 hectares (Craighead and Craighead, 1956). Reynolds (1978) observed an average distance between four Oregon nests of 4.3 kilometers, the four nests being located within an area of 117 square kilometers. The estimated home range of one pair was 460 hectares. Clarke (1984) estimated a similar breeding home range of 380 hectares for an Alaskan pair but with much denser nest spacing (24 pairs per 100 square kilometers), while Mueller, Mueller, and Parker (1981) estimated a larger home range of 796.5 hectares for a breeding pair in Ontario.

Foods and Foraging

This is a highly specialized bird-catching hawk, with about 90 percent of its diet coming from this source (Fisher, 1893; Snyder and Wiley, 1976). Otherwise, small mammals, lizards, frogs, and various large insects comprise the remainder.

Storer (1966) did an extensive survey of foods taken by sharpshins and found a total of 81 genera of animal prey, nearly all of which were birds, among 833 food items. The prey species most frequently taken were song sparrows (*Melospiza melodia*) and American robins (*Turdus migratorius*), with other species of sparrows and sparrow-sized birds making up most of the commonly taken foods; the greatest number of prey species were in the weight category of 15.6–27 grams (see Figure 7). The warbler genus *Dendroica* was especially highly represented, with 153 of 833 prey items in this group. The mean weight of all prey was 17.6 grams for males and 28.4 grams for females, which represents approximately 17 percent of the average adult weight of each sex. Clarke (1984) estimated that 97 percent of prey biomass associated with 14 nests in interior Alaska consisted of small birds, with seven taxa (of thrushes, warblers, and sparrows) contributing over 70 percent of the total, and small mammals representing only about 3 percent.

Mueller and Berger (1970) analyzed the sexual differences in prey preferences of migrating sharp-shinned hawks, finding that males showed a stronger tendency to strike at smaller prey

(house sparrows *Passer domesticus*) than did females, which were more likely to attack starlings (*Sturnus vulgaris*). This was at least the case for juvenile birds, although adult birds showed no clear-cut sex differences in this regard. Older birds of both sexes were less likely to attack inappropriately large prey (such as rock doves *Columba livia*), and, because the hawks that actually struck such prey averaged lighter in bodily weight than those that simply passed close by above, it was judged that hunger probably influences a hawk's tendency to respond selectively to the size of particular potential prey.

Two basic methods of hunting are used by sharpshins: still-hunting attacks on unsuspecting prey that are initiated from inconspicuous perches, and fast bursts of speed along woodland routes, causing small birds to flush and be overtaken in flight. Prey is typically mostly plucked before being eaten, usually at a favorite "plucking post" (Evans, 1982).

Social Behavior

Although pair-forming behavior is essentially unstudied in the sharp-shinned hawk, it has been well documented (Newton, 1986) in the closely related Eurasian sparrowhawk (*Accipiter nisus*). In that species, the success of a territorial male in attracting a mate during spring centers on its ability to obtain food sufficient to allow it to remain within its nesting area and devote its time to courtship and nest building. Females have much larger and more overlapping home ranges than males, and the extent to which a female remains in one place depends at least in part on whether she can obtain enough food there, either from her individual efforts or with assistance from the male. Thus a female unable to obtain enough food in one area may move into the territory of another male, and perhaps mate with him. Through staying with a male while being fed and leaving when hungry, the female also determines the quality of both the male and his home range. Males establish their home ranges well before breeding, and often reoccupy areas they had used the previous year. When intruders are detected the resident male (or female) responds with "slow flight" (flying at a constant level above the trees with exaggerated wing-beats) or with "undulating displays" (flying along an undulating path that gradually loses height and often ends with a spectacular swoop into the trees). Such displays may also occur spontaneously, presumably to advertise the territory, and are most common as the birds prepare for breeding. Most such displays are apparently directed by one female toward another that has intruded on the former's territory, but males likewise sometimes respond thus to other intruding males. Either sex may simply soar above the canopy, alternating wing flapping with gliding, the display flight often ending with an abrupt stoop or a long glide. Females perform these flights, which probably signal occupation of a nesting area, more often than males, but sometimes both sexes do so simultaneously.

According to Newton (1986) pair formation in Eurasian sparrowhawks occurs in six phases: attraction of birds to nesting places and potential mates to one another; mutual roosting, mutual calling, and aerial display; feeding of the female by the male; nest site inspections and stick carrying; nest building; and copulation. It is likely that the same approximate sequence applies in general to North American accipiters such as the sharpshin, but few details are available. Clarke (1984) judged that birds not already paired on arrival at their Alaskan breeding grounds quickly found mates, with only a brief courtship period. On three occasions he saw a male flying in broad circles just above the treetops with the tail closed and under tail coverts flared, while uttering a repeated nasal call. He never observed copulation, but heard apparent copulation calls at two nest sites. Of fourteen pairs, six females and one male were still in immature plumage, the yearling male also being paired with a yearling female.

Breeding Biology

Following the attraction of a female to a male's territory and the onset of mutual activities including courtship feeding, the selection of a nest site and the beginning of nest building represent a critical stage in the breeding cycle. As with general habitat preferences, coniferous trees seem to be preferred over deciduous trees for nest sites across the species's North American range (Apfelbaum and Seelbach, 1983). Sites chosen may be those that tend to provide concealment overhead from larger predators such as goshawks or great horned owls (*Bubo virginianus*), and overhead vegetation may also provide shading during periods of warm temperatures (Moore and Henny, 1983). In Utah the nesting trees consistently had dense foliage, which is why conifers may be preferred (85 percent of 27 nests) over deciduous trees, although the majority of the conifers were in small groves within deciduous stands, rather than in generally coniferous woodlands. Old nests are rarely used, but groves are commonly reused (Platt, 1976a). Of six nesting areas used once in Alaska, three were occupied a second year and two others for three and four years respectively. In Alaska the birds typically nest in conifers that are within dense and young stands of mixed deciduous and coniferous trees. The latter provided a well-developed understory vegetation. The nests were usually in trees about 16 meters high, and the nest itself averaged about 6.5 meters above ground. The nest is usually well shaded from above, placed on a limb close to the trunk, and only dimly lit, providing concealment and possible protection from heat stress. Nearby plucking posts, often of bent-over willows, are usually present (Clarke, 1982, 1984).

Egg laying begins soon after the completion of the nest, with eggs being deposited on alternate days. Platt (1976a) reported an average clutch of 4.3 eggs for 34 Utah nests, and Apfelbaum and Seelbach (1983) indicated that an average of 3.9 is typical of their entire North American sample. A total of 34 New York nests ranged from 2 to 6 eggs, averaging 4.1 (Bull, 1974), and 33 Ontario nests had 3–6 eggs, averaging 4.4 (Peck and James, 1983). Eight Alaska nests had an average of 5.1 eggs (Clarke, 1984), suggesting a possible trend toward larger clutches northwardly.

Incubation is done primarily and probably normally exclusively by the female, with the male providing her with her food. Incubation requires an average of 30 days (Platt, 1976a; Clarke, 1984), with some estimates of up to 35 days perhaps atypical. For the first two weeks or so of the nestling period the male provides nearly all the food, but increasingly the female also begins to gather food during the later stages of the nestling period. In one nest a male fledged in 24 days and a female in 27 days (Platt, 1976a), while nestling periods of 21–24 days for males and 24–27 days for females have also been reported (Evans, 1982). The very closely related but slightly larger Eurasian sparrowhawk has been reliably reported to have an incubation period of 31–37 days, and average fledging periods of 26.2 and 30 days for males and females respectively (Newton, 1986).

Nesting success data are very limited, but for various samples Reynolds and Wight (1978) reported an average of 2.7 young fledged per nesting attempt for 11 nests, a 69.9 percent hatching success (based on 23 eggs), an 81.2 percent fledging success (based on 16 young), and 91.7 percent of 12 nests fledging at least one young. Infertility, deaths of embryos, and egg breakage were apparently the major causes of nest losses, the egg breakage possibly a result of high pesticide residues. Of 41 eggs laid in 8 Alaska nests, 34 (83 percent) hatched and 27 of those hatched (79 percent) were reared (Clarke, 1984).

Once the young fledge, they become very difficult to follow, but one family in Utah remained intact for at least nearly a month before disappearing as a family group (unpublished data of J. Platt, cited by Reynolds and Wight, 1978). A brood studied in Alaska dispersed at 20 to 25 days after fledging (Clarke, 1984). Data from birds banded in northern Minnesota suggest that females tend to migrate before males of corresponding age, and tend to winter substantially farther south (Evans and Rosenfield, 1985), so maintenance of pair and family bonds over winter seems quite unlikely. Although full adult plumage is not attained by yearling birds, the successful breeding by a pair of birds in immature plumage has been reported (Fischer, 1984). In the closely related Eurasian sparrowhawk yearlings comprise about 18 percent of the total breeding population in males, and about 15 percent of the females. An estimated 18 percent of first-year females breed, 49 percent of second-year females, 83 percent of third-year females, and nearly all the females of older age classes

(Newton, 1986). Probably similar trends apply to the sharp-shinned hawk.

Evolutionary Relationships and Status

Apart from the Central American form *chionogaster* that is sometimes given specific recognition, the nearest relatives of the sharp-shinned hawk perhaps are the African rufous-breasted sparrowhawk (*A. rufiventris*) and the Eurasian sparrowhawk. These two latter species comprise a fairly clear nuclear superspecies, to which several other forms including *striatus* might well be added (Stresemann and Amadon, 1979).

It is very difficult to judge the population status of this small and elusive species, but since the advent of the DDT era following World War II there have been clear declines, probably at least in part attributable to the effects of this pesticide and its breakdown products such as DDE on reproduction. Until about 1971 this population decline continued in the eastern states. In 1971 the species was included on the Audubon Society's first Blue List of apparently seriously declining species, but since then there has been a definite upward population trend. Large-scale use of hard pesticides such as DDT continues to occur in the Central and South American wintering areas of both the hawk and the prey species on which it depends, so the sharpshin is by no means out of danger from this factor. Similarly, forest management practices producing monoculture forest habitats in the western states do not tend to favor it at the present time. However, large-scale wanton shooting of the birds during migration, which once was a serious problem, has been eliminated (Evans, 1982). This is one of only three species of hawks (the red-shouldered hawk and northern harrier the others) that have been continuously on the Audubon Society's Blue List between 1972 and 1986.

There are apparently no previous estimates of continental populations, but on the basis of the Audubon Society's 1986 Christmas Count data (which certainly underreports such elusive species as the accipiters) I have very conservatively estimated a Canadian and U.S. wintering population of 30,100 birds. Highest densities occurred from Massachusetts to Virginia on the Atlantic Coast, and in California and Arizona in the west. The population wintering in Mexico and southward is unknown but may be quite substantial.

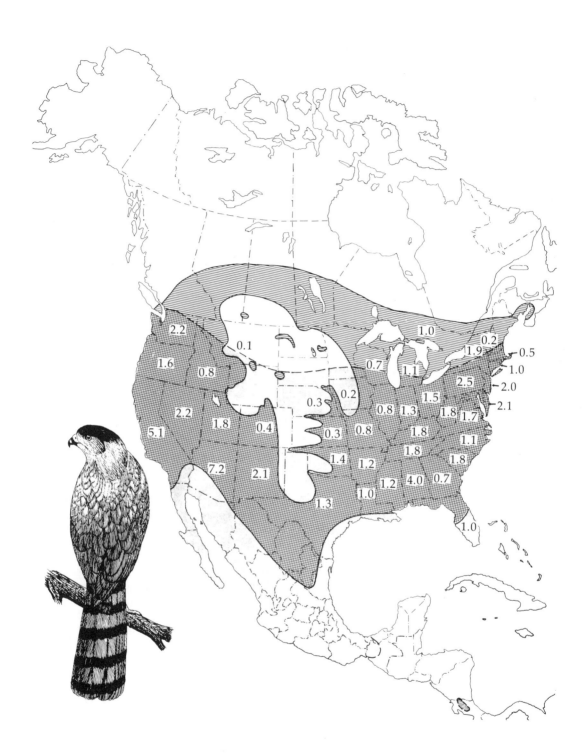

Figure 34. Breeding (hatching), residential (cross-hatching), and wintering ranges (shading) of the Cooper's hawk.

Cooper's Hawk *Accipiter cooperii* (Bonaparte) 1828

Other Vernacular Names

Big blue darter, chicken hawk, quail hawk; epervier de Cooper (French), esmerejon de Cooper (Spanish)

Distribution

Breeds from southern British Columbia, central Alberta, central Saskatchewan, central Manitoba, western and southern Ontario, southern Quebec, Maine, New Brunswick, Prince Edward Island, and Nova Scotia south to Baja California, Chihuahua, Nuevo Leon, southern Texas, Louisiana, central Mississippi, central Alabama, and central Florida.

Winters from Washington, Wyoming, South Dakota, southern Minnesota, southern Wisconsin, southern Michigan, southern Ontario, New York, and New England south through the southern United States and Mexico to Guatemala and Honduras, casually to northern South America. (See Figure 34.)

North American Subspecies

None recognized.

Description

Adult male. Forehead blackish plumbeous (lead-colored), each feather edged with cinnamon-buff, light ochraceous-buff, or vinaceous-buff; crown and occiput sooty black to plumbeous-black, becoming more fuscous-black in worn plumage, with snow-white bases to the feathers of the latter area; remainder of upperparts dark plumbeous to deep gray, becoming somewhat fuscous in worn plumage, the nape abruptly lighter than the occiput; feathers of the nape, back, scapulars, and rump with darker shafts; scapulars with concealed white spots; upper tail coverts sharply tipped with white and often with a more or less developed single white band, tail somewhat rounded and more brownish than the rump, becoming whitish toward the bases of the inner webs of the outer rectrices, pale gray below, sharply tipped with white, and crossed by three broad, sharply defined bands of sooty black to fuscous-black, the last the broadest; lores grayish;

ear coverts and sides of neck varying from white to brown or russet, finely streaked with black; throat white, occasionally pinkish buff or cinnamon-buff, with fine, hairlike shaft streaks of blackish; remainder of underparts white with fine black shaft streaks broken by detached transverse bars that are extremely variable in form and width; sides of chest usually shaded with gray; thighs regularly banded with orange-cinnamon, brown, tawny, or russet, sometimes uniform tawny or russet; abdomen less heavily marked, sometimes uniform white; under tail coverts pure white; under wing coverts and axillaries white, barred or spotted with tawny or russet; remiges fuscous above, darker on the outer webs, white toward the bases of the inner webs, silvery white to pale gray below, tipped and barred with darker fuscous or fuscous-black. Iris deep orange-red; terminal half of bill deep black, basal half pale blue; cere greenish yellow; inside of mouth dull yellow; tarsi and toes deep lemon-yellow.

Adult female. Very similar to the male, but the upperparts often lacking a bluish cast, and usually less heavily marked below. Crown and occiput fuscous to fuscous-black; remainder of upperparts brown to fuscous.

Subadult. The juvenal plumage is worn through the first winter, with only fading and wear occurring. Initial breeding occurs in this plumage, and by the second fall a nearly adult plumage is present (Bent, 1938). The iris color changes from bluish gray to increasingly yellow during the second year of life (Palmer, 1988).

Juvenile (sexes alike). Similar to the adult female but medium to pale brown above, the feathers streaked and edged or tipped with tawny, underparts creamy white, streaked with fuscous, tibia spotted or barred, and the tail with four exposed fuscous bars. Iris grayish blue; bill grayish, becoming black terminally; cere yellow; tarsi and toes greenish yellow.

Measurements (in millimeters)

Wing, males 214–238 (ave. of 34, 231), females 247–278 (ave. of 27, 259.9); tail, males 181–211 (ave. of 34, 198.8), females 215–242 (ave. of 27, 220.9) (Friedmann, 1950). Average egg size 49 x 38.5 (Bent, 1937).

Weights (in grams)

Males 297–380 (ave. of 51, 349), females 460–588 (ave. of 57, 529 (Dunning, 1984). Males 302–402, ave. 341, females 479–678, ave. 528 (sample sizes unspecified) (Clark and Wheeler, 1987). Estimated egg weight 39.5, or 7.5% of female.

Identification

In the hand. The combination of medium body size (300–700 g),with relatively short, rounded wings (214–278 mm) and a long, barred, and slightly rounded tail that is about 85 percent as long as the wing and no less than 180 mm, should identify this species.

In the field. When perched, this medium-sized (370–500 mm long) hawk has a dark brown to slate-colored back, a long, white-tipped tail that extends well beyond the tips of the folded wings, and barred rufous (adults) or streaked (immatures) brownish underparts. Adults have a dark blackish crown that is noticeably set off from a lighter nape, whereas the smaller (250–350 mm long) sharp-shinned hawk has a less distinctively delineated dark crown area and a more squared-off (when slightly fanned) or even slightly notched (when closed) tail.

In flight the Cooper's hawk exhibits a long barred tail and rather short and rounded wings. It has a somewhat more rounded and more broadly white-tipped tail than does the sharp-shinned hawk, although sometimes one must rely on relative size in distinguishing these very similar species. A swift flyer, the Cooper's hawk has a rapid wingbeat and is able to negotiate its often heavily vegetated woodland habitats very well. Except during the breeding season both sexes are quite silent, but territorial birds regularly perform duetting calls and also utter alarm calls sounding like repeated *ki* or *kak* notes. These differ but little from those of either the sharp-shinned hawk or goshawk, and are unlikely to aid in identification.

Habitats and Ecology

In contrast to the mixed and coniferous forest habitats favored by the smaller sharp-shinned hawk, the Cooper's hawk is more closely associated with deciduous and mixed forests and open woodland habitats such as woodlots, riparian woodlands, semiarid woodlands of the south-west, and other areas where the woodlands tend to occur in patches and groves or as spaced trees, rather than in dense, continuous stands (Beebe, 1974; Evans, 1982). Where the species does occur in large forests, it is more likely to be found near forest edges, along roads or clearings, or at other forest openings such as stream or lake edges. In mountain regions it tends to occupy somewhat lower and warmer climatic zones than does the sharpshin (Wattel, 1973). Nesting often occurs near man-made clearings both in the eastern and western states (Meng, 1951; Hennessy, 1978). As with the sharpshin, nests are also often located fairly close to a source of water (Hennessy, 1978; Reynolds, Meslow, and Wight, 1982). Forests used for nesting in Oregon averaged somewhat older than those used by sharpshins (50–80 vs. 40–60 years), and typically contained larger and more widely spaced trees (Reynolds, 1978). Wintering habitats are similar to breeding habitats, and the birds are less prone to migrate than are sharpshins. At that season they may occupy almost any habitat having trees or shrubs and a good population of wintering birds.

Home ranges of these hawks are relatively large, with three breeding home ranges in Michigan estimated from 105 to 404 hectares, averaging 291 hectares (Craighead and Craighead, 1956). Probably all of the home range is defended as a territory. In Wisconsin, a breeding male was found to have a seasonal home range of 784 hectares, and an average daily home range (ave. of 6 sample days) of 231 hectares (Murphy, Gratson, and Rosenfield, 1987). The average winter home range was judged to be 2.4–3.2 kilometers in diameter (Craighead and Craighead, 1956). Because of these large home ranges, the densities of hawks are usually quite small; Stewart and Robbins (1958) estimated a density of only 0.2 pairs per 40 hectares in Maryland.

Foods and Foraging

Probably about 80 percent of the prey taken in the eastern states is avian (Meng, 1959; Storer, 1966; Wattel, 1973), while in the southwest and west up to half of the food intake or more may consist of mammals and lizards (Fitch, Glading, and House, 1946; Wattel, 1973; Evans, 1982). According to Beebe (1974) prey is captured mostly by deliberate or planned surprise, rather than by opportunism, the hawk sometimes taking a circuitous sneak approach on a prey, remaining well hidden until the potential victim wanders away from protective cover and is momentarily off-guard. The prey may be taken from the ground, from branches, or in full flight. Experienced birds usually make only a single try at a prey, and then abandon the effort should that try fail.

Storer (1966) calculated that the average prey weight for Cooper's hawks was 37.6 and 50.7 grams for males and females respectively, which is approximately 12 percent of average adult weight. The most commonly occurring prey species in his sample were the house sparrow (*Passer domesticus*) and northern bobwhite (*Colinus virginianus*), and the prey genera most commonly encountered in the sample were *Melospiza, Colinus,* and *Passer.*

Studies by Toland (1985b) in Missouri on 259 prey items brought to 14 nests indicated that birds comprised 87 percent of prey numerically and 65 percent by biomass, while mammals made up 34 percent of the biomass and represented 12 percent of the prey numerically. Blue jays (*Cyanocitta cristata*) were the most frequent prey, followed by young eastern cottontails (*Sylvilagus floridanus*). In a Washington study, Kennedy and Johnson (1986) found that American robins (*Turdus migratorius*) and California quail (*Callipepla californica*) comprised 85 percent of 110 identified prey brought to the nests of five pairs. Males brought in more than 63 percent of the total prey, and there were not significant differences between the sexes in size of prey brought to the nests.

The hunting behavior of Cooper's hawks is similar to that of the sharpshin, consisting of a combination of still-hunting from perches followed by fairly short attack flights, plus extended searching flights along woodland edges or other natural routes, during which small birds may be flushed out or taken by surprise. Toland (1985b) estimated a hunting success rate of 53 percent for Cooper's hawks on mammals, compared with a success rate of 21 percent on birds.

Social Behavior

As with the sharp-shinned hawk, little has been written on courtship in this species. The general pattern of pair bonding is undoubtedly like that of the other accipiters, in which a territorial male attempts to attract a female into his territory and

keep other males out by periodic display flights above it. Apparent courtship behavior has been seen involving a male flying around a female while exposing his expanded under tail coverts to her view (Mockford, 1951). A similar aerial display was observed by Fitch (1958), in which two birds were observed in flight, the male raising his wings high above the back and flying in a wide arc with slow, rhythmic flapping somewhat like that of a nighthawk (*Chordeiles*). Typically these display flights occur on bright, sunny days in midmorning, and begin with both birds soaring high on thermals, followed by the male diving down at and behind the female. Then a very slow-speed display chase begins, the birds moving with slow and exaggerated wingbeats alternated with glides while the wings are held at a dihedral angle and the white under tail coverts conspicuously spread (Beebe, 1974). Although observations so far suggest that males are at least two years old prior to breeding, a small proportion (6–20 percent) of breeding females have been found in first-year plumage (Reynolds and Wight, 1978; Fischer, 1984).

Males show a strong attachment to traditional territories, and up to a half-dozen old nests may be located within one territory (Beebe, 1974). Reynolds and Wight (1978) reported that individual "nesting sites" (defined by them as an area of 6 hectares around a specific nest location) may be used for up to three years (compared with two for sharpshins and five for goshawks).

Breeding Biology

Nests are built, or an old one repaired, during the spring display period. The female may select the site but the male does much of the building

(Rosenfield, 1988). Nests are usually well spaced, the two nearest ones Meng (1951) found in New York being 2.4 kilometers apart. In a California study (Fitch, Glading, and House, 1946), internest distances averaged about 1.6 kilometers, while in Oregon they averaged 3.5 kilometers (or a nest per 2,200 hectares). In Oregon the birds were found to favor sites on gentle slopes with northern exposures, and the nests were usually placed on horizontal limbs against the trunk and just below the crown or in the lower crown. Typically a plucking post is located within about 50 meters of the nest (Reynolds, Meslow, and Wight, 1982). These posts are normally situated in direct view of the nest, and may consist of a log on the ground, a nearly horizontal large branch, a lodged tree trunk, or an old nest (Beebe, 1974). Like sharpshins, Cooper's hawks select nesting sites that apparently offer overhead concealment from possible predators, and seem in Oregon to favor using mistletoe (*Arcenthobium* spp.) as a nest platform (Moore and Henny, 1983). Compared with broad-winged and red-shouldered hawks, Cooper's hawks tend to nest proportionately higher in trees and to favor more mature forests with a well-developed understory and ground cover (Titus and Mosher, 1981). Nesting in isolated trees occurs only rarely (Asay, 1980, 1987).

Copulation occurs several times per day during the nest-building and egg-laying phases of reproduction. According to Meng (1951, and in Palmer, 1988), copulations often follow a period of morning duetting, during which the birds sing a great variety of notes, the male usually initiating the calling and the duet lasting up to as long as an hour. Such duets occur through the nest-building period and most of the incubation period. Eggs are typically laid early in the morning, normally on alternate days, with longer intervals sometimes occurring late in the clutch (Meng, 1951). Clutch records from 71 New York nests ranged from 3 to 6 eggs, with an average of 4.1 (Bull, 1974), and a general sample from eastern North America averaged 4.18 (Henny and Wight, 1972). Ontario records of 52 nests ranged from 1 to 7 eggs, averaging 3.63 (Peck and James, 1983). A smaller sample of 13 nests from Oregon averaged 3.8 eggs (Reynolds and Wight, 1978), or somewhat smaller than that of sharpshins but larger than goshawk nests from the same area. None of these data suggest any major geographic variations in clutch sizes.

Incubation typically starts with the laying of the third egg and is done almost entirely by the female, with the male sometimes taking over for short periods while the female is consuming food he has brought in (Meng, 1951). All three *Accipiter* species were found by Reynolds and Wight (1978) to have the same length of incubation, namely 30–32 days, although shorter estimates (to 21 days) as well as longer ones (to 36 days) have been reported for the Cooper's hawk in earlier literature. For example, Meng (1951) judged 36 days to be the normal duration, with pipping starting two days earlier.

Hatching in the nests studied by Reynolds and Wight (1978) was not greatly staggered, but instead occurred within a period of one or two days. The nestling period was determined by them to last 27–30 days, with females requiring longer periods for fledging than the smaller males. Males bring food to the nest three or four times a day during the incubation period, and do so increasingly frequently following hatching. During the first three weeks after hatching the male is seldom allowed to bring food directly to the nest; instead the female meets him on his approach and delivers the food herself. After three weeks the female is often absent from the nest while engaged in hunting, and at such times the male may take food to the nest directly. By 30 days the male chicks usually are able to leave the nest, while females may require about 34. Food may be brought to the nest for another ten days, and the young may also return to the nest to rest or sleep. They slowly start to hunt for themselves when about eight weeks old (Meng, 1951).

Evolutionary Relationships and Status

The Cooper's hawk and Gundlach's hawk (*A. gundlachi*) of Cuba belong to the same super-species (Brown and Amadon, 1968), and sometimes the bicolored sparrowhawk (*A. bicolor*) of Central and South America is also included in this group (Stresemann and Amadon, 1979). Reynard et al. (1987) concluded that the Cuban endemic *gundlachi* is indeed a valid species belonging to the *cooperii* and *bicolor* assemblage, and judged that it has been present in Cuba for a very long period. Wattel (1973) stated that it is futile to try to judge the relative relationships in this group, and simply suggested that the three forms evolved from a common stock of tropical American accipiters. The entire New World *bicolor* group may be related to the ancestral Eurasian sparrowhawk *A. nisus*, a population of which was able to invade North America and subsequently gave rise to *A. striatus*.

The Cooper's hawk was included in the Audubon Society's annual Blue List of apparently seriously declining species in 1971, remained on it through 1981, and was included again in 1986. In 1982 it was listed as a species of "special concern." Henny and Wight (1972) calculated an annual rate of population decline of 13.5 percent between 1941 and 1945, and rates as high as 25 percent annually after 1948, when DDT came into widespread use. However, since 1968 there have been improvements in the situation, and in at least some areas of the northeast reproduction rates have improved (Evans, 1982). In the western states, where pesticide levels were apparently also severe, the estimated number of fledged young per successful nest in a recent Oregon study of 24 nests (Reynolds and Wight, 1978) was 2.1. This average is less than estimates from the northeastern states of 3.53 fledged young per successful nest for the prepesticide years to 1945, and 2.67 for the postpesticide period of 1949–67 (Henny and Wight, 1972). During the period 1968–74 a continued improvement to 3.36 young per successful nest has been estimated (Braun et al., 1977).

Although no previous continental population estimates are available, I have very conservatively estimated on the basis of the Audubon Society's 1986 Christmas Counts that 19,400 birds were then present in the United States and Canadian provinces. Maximum calculated populations occurred in Arizona and California, which supported an estimated 3,250 and 3,200 birds respectively. An unknown additional number winter south to Central America.

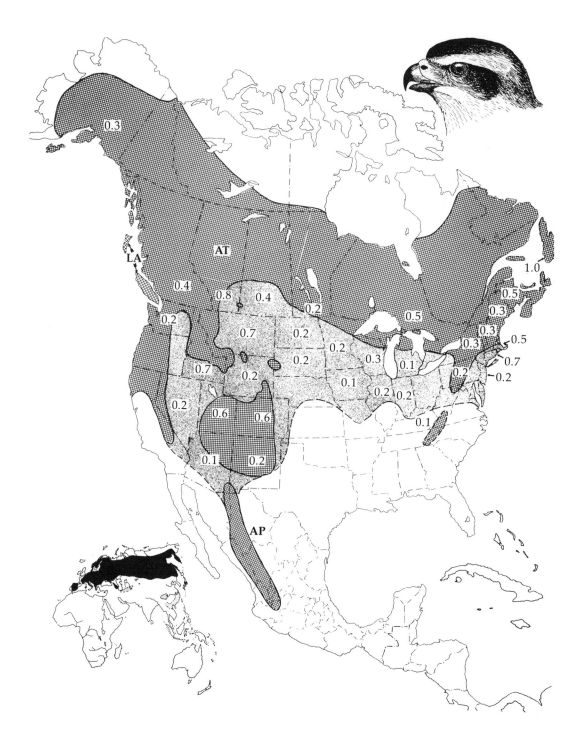

Figure 35. North American breeding (cross-hatching) and wintering (shading) ranges of the northern goshawk, including races apache (**AP**), atricapillus (**AT**), and langi (**LA**). Relative winter state or provincial density indices (average number seen per Audubon Christmas Count, 1986) are shown for major wintering areas. The breeding (inked) and wintering (shading) Old World ranges are shown on the inset map. Indicated range limits of contiguous races should not be considered authoritative.

Northern Goshawk *Accipiter gentilis* Linnaeus 1758

Other Vernacular Names

Goshawk; autour des palombes (French); gavilan pollero (Spanish)

Distribution

Breeds in North America from western and central Alaska, northern Yukon, eastern and southern Mackenzie, southern Keewatin, northeastern Manitoba, northern Ontario, central and northeastern Quebec, Labrador, and Newfoundland south to southern Alaska, central California, southern Nevada, southeastern Arizona, southern New Mexico, the eastern foothills of the Rockies and the Black Hills, central Alberta, central Saskatchewan, southern Manitoba, northern Minnesota, central Michigan, Pennsylvania, central New York, and northwestern Connecticut, and locally south in montane habitats at least to West Virginia and probably to eastern Tennessee and western North Carolina; local resident in the mountains of northwestern and western Mexico; also widely distributed in Eurasia.

Winters throughout the breeding range and in North America south regularly as far as southern California, northern Mexico, and Texas, occasionally to the northern portions of the Gulf states, and rarely to Florida; also in Eurasia. (See Figure 35.)

North American Subspecies

A. g. atricapillus (Wilson): Breeds in North America as described, except for areas indicated below. Includes *striatulus*.

A. g. langi (Taverner): Resident on Queen Charlotte and Vancouver islands.

A. g. apache (van Rossem): Resident from southern Arizona south locally in the mountains of Mexico to Jalisco.

Description (of *atricapillus*)

Adult male. Upperparts varying from deep gray to slate-gray and deep neutral gray, becoming darker on the interscapulars, nape, occiput, and crown, the top of the head often being blackish slate to slate-black and occasionally even a little blacker, the nape and sides of the crown so heavily mottled with white as to be more white than slate; upper wing coverts like the back; remiges fuscous, indistinctly broadly barred above with darker and fuscous on the inner webs; broadly barred below and mottled with whitish on the inner webs; five outermost primaries emarginated on their inner webs; rump and upper tail coverts like the back; tail similar but the rectrices tipped with whitish and crossed by four broad but indistinct fuscous bands; lores, cheeks, and auriculars varying from dirty white streaked with blackish to nearly pure black, the superciliary areas white more or less mottled with blackish slate; entire underparts white, thickly barred or vermiculated with gray to dark gray, the feathers of the chin, breast, and abdomen with blackish shaft streaks, which, like the vermiculations, are very variable in width; under wing coverts white barred with grayish fuscous. Iris orange-red, becoming deeper red to mahogany in older birds, bill bluish black; cere and inside of mouth yellow; tarsi and toes yellow, claws bluish black.

Adult female. Very similar to the male but usually very slightly browner above and more coarsely marked with dark gray below. Iris orange-yellow in older adults.

Subadult (sexes alike). Similar to adults by the second winter, but browner, the ventral markings coarser and heavier, the pectoral feathers with broad fuscous-gray shaft streaks and some fairly wide brownish-gray bars. The iris color remains bright yellow until about the fourth year, and some juvenal brown upper wing coverts may be retained until the second year. Initial breeding may occur at two or three years, even occasionally at one year.

Juvenile (sexes alike). Upperparts including wings fuscous to fuscous-black, with buffy white and cinnamon streaks, these broadest and most conspicuous on nape; scapulars with broad, mostly concealed bars of buffy white or cinnamon-buff; tail broadly barred with fuscous and light brown; remiges broadly barred inwardly with buffy white, drab or dull brown on outer webs; underparts, including wing lining, buffy white, washed with cinnamon-buff and heavily streaked with fuscous. The dark crown area is indicated only by broad black shaft streaks, but its boundaries are suggested by a whitish superciliary stripe, and the dark tail bars are narrowly bordered with whitish. Iris initially pale greenish gray, gradually turning pale yellow; bill dark brown toward tip, becoming yellow basally; cere and interior of mouth green to yellow; tarsi and toes dull yellow; claws brownish.

Measurements (of *atricapillus*, in millimeters)

Wing, males 303–354 (ave. of 27, 325.2), females 321–368 (ave. of 22, 333.6); tail, males 226.5–280 (ave. of 27, 245.7), females 250–301 (ave. of 22, 278.6) (Friedmann, 1950). Average egg size 59.2 x 45.1 (Bent, 1937).

Weights (in grams)

Males 735–1099 (ave. of 77, 912); females 845–1364 (ave. of 103, 1137) (Dunning, 1984). Males 677–1014, ave. 816, females 758–1214, ave. 1059 (sample sizes unspecified) (Clark and Wheeler, 1987). In *atricapillus* there is a clinal increase in average weights to the north (Cramp and Simmons, 1980). Estimated egg weight 59, or 5.2% of female.

Identification

In the hand. The large (at least 700 g), grayish (adults) to brown (immatures) body, with rounded wings (over 300 mm) and a long tail (over 225 mm) that is obscurely barred, slightly rounded, and about 75 percent as long as the wing, should serve to identify this species. The somewhat similar gray hawk (of considerably more arid habitats and a mostly more southerly distribution) is substantially smaller, and its tail is strongly banded with black and white.

In the field. When perched, the goshawk's large body size (body length 53–65 cm) and relatively long tail (the wingtips not reaching the midpoint of the tail) help to identify it, and in adults the dark gray upperparts, black crown, and pale grayish white underparts are distinctive. Immatures are less distinctive, but show heavy streaking below and a pale whitish stripe above the eye. (Juvenile red-shouldered and broad-winged hawks are quite similar, but have stronger tail banding and dark malar stripes.) In young birds the paired bars on each side of the tail feathers are normally slightly offset from one another (producing a zigzag pattern), whereas in

the Cooper's hawk they are more evenly aligned, and the white under tail coverts of young goshawks are more distinctly spotted.

In flight, birds of all ages show strong barring on the undersides of the flight feathers and tail, and they move swiftly through heavy woodlands with powerful wingbeats. They are more ponderous and buteo-like in flight than are the other two North American accipiters. They are roughly the size of a northern raven (*Corvus corax*), and adults appear rather silvery gray above, with a blackish crown, and grayish white below. The birds are usually quite silent except during the breeding season, when various screaming, shrieking, and wailing calls are uttered, often in association with alarm or territorial behavior.

Habitats and Ecology

Throughout its broad Holarctic range the goshawk varies considerably in its preferred nesting habitats, but nearly all can be described as consisting of a combination of tall trees having intermediate canopy coverage suitable for nesting, and small, open areas within the forest for foraging. At the northern limits of its range, it sometimes nests quite low, or even rarely on the ground. Otherwise, it is associated with such diverse habitats as dense coniferous taiga with scattered glades, tall coniferous, mixed, and deciduous woods, luxuriant riverine forests, cultivated coniferous plantations, and stands of birch and aspens or pines in steppe or steppe woodland. Coniferous forests are generally favored over deciduous ones, and this is perhaps especially true in the southern parts of its range. During the winter months the birds show a lower habitat specificity, often ranging into scrubby or parklike foothill areas, oak or other savannas, or other rather open environments including deserts (Wattel, 1973; Beebe, 1974).

In North America goshawks nest in nearly every kind of coniferous forest, from the northern subarctic spruce forests to the high-elevation coniferous (mostly pine) forests of the Mexican cordillera. Deciduous and mixed woodlands are also heavily used. Among 64 nests from various parts of North America, 44 percent were located in mixed woodlands, 34 percent in deciduous forests, and 22 percent in coniferous forests (Apfelbaum and Seelbach, 1983).

In Oregon the birds tend to select mature or

old-growth stands of conifers for nesting, typically those having a multilayered canopy with vegetation extending from a few meters above ground to more than 40 meters high. Generally nesting sites are chosen that are fairly near a source of water and are of moderate slope, usually having northerly aspects. This habitat type is quite similar to that used by the Cooper's hawk, but the trees tend to be older and taller and have a better-developed understory of coniferous vegetation (Reynolds, Meslow, and Wight, 1982). In a sample of ten nesting habitats from northwestern California the nests typically were located in small stands of dense, mature trees within larger dense communities of young conifers, having scattered large hardwoods and mature conifers and a parklike understory. The nesting stands were consistently sloped toward the northeast, although the degree of slope and elevation varied considerably (Hall, 1984).

Extensive home ranges are needed by this largest accipiter, and as a result large forest stands are favored for nesting habitats. A Wyoming territory/home range was judged to encompass 212 hectares (Craighead and Craighead, 1956), but a Minnesota study (Eng and Gullion, 1962) indicated a much larger breeding home range of about 13 square kilometers, encompassing an area extending about two kilometers out from the nest. In Oregon, the average distance between nests was determined to be 4.3 kilometers, with four nests located within an area of 117 square kilometers (Reynolds, 1978). Home ranges for a pair in Europe have been estimated at 30–40 square kilometers (Brown, 1976a), or even up to 50 square kilometers (Cramp and Simmons, 1980), so there is obviously a great deal of variation in population density for this species. In Europe, goshawk densities may locally be as high as a pair in less than 100 hectares; such densities were found in an area with only 12–15 percent woodland cover, suggesting that this is a species that is edge- or patchwork-adapted, rather than one associated with dense forests (Kenward, 1982).

Foods and Foraging

Virtually all North American studies indicate a high dependency on birds and mammals of moderate to large size for this species (Sherrod, 1978). The same pattern is true in Europe (Brown, 1976a; Cramp and Simmons, 1980), with a high degree of opportunistic feeding evident, depending upon the prey availability. To a greater extent than in the other North American accipiters there is a measurable difference in the average prey weight taken by the two sexes; Storer (1966) initially determined this for North America, calculating the average prey weight at 397 and 522 grams for males and females respectively, or nearly half the average weight of the adult birds. Comparable size differences have also been reported in Europe (Opdam et al., 1977).

In a summary of goshawk diets for North America, Sherrod (1978) found that mammals represented from 21 to 59 percent (numerically) of food intake in various studies and birds from 18 to 69 percent, with the remainder made up of reptiles (snakes) and invertebrates (mainly insects). Invertebrates may sometimes seem significant on a numerical basis, but do not contribute much to the diet from a standpoint of total mass. Gallinaceous birds such as grouse and ptarmigan are typically the most important avian prey for goshawks, as they often are found in comparable habitats and frequently can be captured in flight or on the ground. In boreal forests east of the Rocky Mountains, spruce grouse (*Dendragapus canadensis*), ruffed grouse (*Bonasa umbellus*), snowshoe hare (*Lepus americanus*), and red squirrel (*Tamiasciurus hudsonicus*) comprise the four basic foods of the goshawk, and two of these species (ruffed grouse and hare) are distinctly cyclic in their population traits. This means that the goshawk populations there must move south out of their breeding ranges periodically when their prey population crashes, especially during winter periods. Farther west and north, the birds have reliable food sources such as willow ptarmigan (*Lagopus lagopus*), blue grouse (*Dendragapus obscurus*), and ground squirrels such as the arctic ground squirrel (*Spermophilus undulatus*) to supplement their basic diets, and thus major winter irruptions into southern areas are not typical. Interestingly, on the Queen Charlotte Islands the birds specialize on the northwestern crow (*Corvus caurinus*), while on Vancouver Island they concentrate on two deep-forest passerines, the Steller's jay (*Cyanocitta stelleri*) and varied thrush (*Ixoreus naevius*) (Beebe, 1974).

In a Finnish study (Linden and Wikman, 1983), it was found that goshawks there preyed mainly on the hazel grouse (*Bonasa bonasia*), and a 500 square kilometer study area supported from 10 to 28 pairs of goshawks annually, depending on the density of the grouse population. However, brood size of goshawks was apparently not influenced by the relative annual density of hazel grouse, perhaps since other prey become important during the brood-rearing period.

Regardless of their specific prey, goshawks capture it by either or both of two means. The first is the perch-and-watch technique, followed by short attacks on prey when it is seen. Alternatively, they may conduct fast searching flights through forest edges, openings, or other vegetation. Concealed approaches, fast and low, are used when the birds hunt in woods where vegetation offers adequate cover, whereas when hunting from forest edges a small, silent accelerating glide down to the unsuspecting prey is typically employed. Should the prey see the approaching hawk in time an all-out chase invariably follows, with the goshawk doggedly chasing its quarry at full speed, often plunging into and through heavy cover in reckless pursuit. These flights, similar to those used by gyrfalcons, rarely last long or extend more than a kilometer before the prey is overtaken by the hawk's speed and determination (Beebe, 1974).

In an English telemetry study (Kenward, 1982), it was found that goshawks initiated hunting flights at about 3–4 minute intervals, and that these usually covered about 200 meters over open country. The birds remained in woodland areas for 50 percent of the time, and 70 percent of their prey was taken in or from woodlands, even though the area they occupied was only 12 percent woodland habitat. Most of their attacks were made from perches, with only 3 percent of their kills being made on prey in flight. Short-stay perched hunting was the most commonly used hunting mode, and there was an estimated 6 percent success rate for individual hunting attempts, or about one kill per 262 minutes of hunting time.

Social Behavior

Many goshawk populations are fairly sedentary, and such birds have been reputed to maintain life-long pair bonds (Brown and Amadon, 1968). However, those breeding in northern and northwestern areas of North America are more mobile, with birds nesting in British Columbia typically returning to their nesting territories in March (Beebe, 1974). Probably the pairs of the previous season associate little if at all on their wintering ranges, and if a prior mate is lost a new one may sometimes be obtained during the same season (Cramp and Simmons, 1980). The nesting territory is advertised by one or both of the resident birds by aerial display, which occurs before and during nest repair or construction. One such display, "high circling," consists of soaring high above the territory, often while calling and alternately extending and retracting the white under tail coverts.

Lone displaying birds may also perform "slow flapping" and "undulating flight" displays. These in turn may lead to a full "sky dance," which starts with high circling, followed by a slow-flapping and undulating-flight phase with a gradual loss of altitude; it may end with a plunge down to the trees or a new regaining of altitude and renewed high circling. These displays are performed mainly by the female, but the male sometimes joins her (Cramp and Simmons, 1980). A sequence may also begin by the male diving toward the female from a brief soaring flight above the territory, or may involve a direct chase below the level of the canopy. During such chases the female is usually in the lead but is closely followed by the male, both birds flying slowly with deep and measured wingbeats alternated with short glides while the wings are held in a steep dihedral angle. The under tail coverts are often spread so far that they extend out on either side of the tail; such "tail flagging" may also be used by perched birds (Beebe, 1974; Cramp and Simmons, 1980). The goshawk's aerial display is much like that of the Cooper's hawk, but the flights are generally longer and perhaps better developed, and the displaying birds typically call repeatedly.

The female may also attract the male by perching in the nesting area and calling, using the "perch-and-call" display (Brown, 1976a; Cramp and Simmons, 1980). This loud chattering call of the female also serves as a pair-contact call in a mutual duetting. A fast version of this may serve as a threat and mobbing call, and a soft and quick version is used by the male as a food call when bringing prey to the female and young. Wailing calls are used by the female when food-begging, sometimes also by males when bringing food, while a clicking note is uttered by males as a greeting call to the female at close quarters (Schnell, 1958).

Copulations occur mainly during the nest-building and egg-laying period, when the birds may mate several times a day, often on a branch close to the nest itself. Apparently the frequency of copulations (estimated as averaging more than 500 per breeding cycle) is one of the highest reported for all birds. This high frequency, beyond possible pair-bonding functions, may help assure parentage of the rightful male mate through successful sperm competition with other males that might copulate with the female (Moller, 1987). A crouching posture, with drooped wings, raised tail, and erected under tail coverts, is assumed by the female (see Figure 18A). The male may respond with a similar posture before flying toward the female and landing directly on her back. Much wing waving and calling occurs during treading. A small percentage of year-old birds may attempt to breed while in immature plumage; this is somewhat more common in females than in males, and indeed some birds may not attempt to breed until their third year (Reynolds and Wight, 1978; Cramp and Simmons, 1980).

Breeding Biology

Nest building may begin as early as two months before egg laying, but this would certainly be exceptional, and often an old nest is simply repaired. The nest site is typically in a large tree among fairly old-growth forest, and is usually placed on large, horizontal limbs either against the trunk or quite near the trunk. Nesting trees are often fairly close to water and typically have a plucking post within 50 meters of the nest (Shuster, 1980; Reynolds, Meslow, and Wight, 1982). The birds appear to select nesting trees that are fairly far from human habitation, but are often close to roads, swamps, or other forest openings. These openings in the forest may provide landmarks for locating the nest site, as well as convenient flight corridors. If large enough, they also often support a ground squirrel population, and may thus offer excellent hunting opportunities (Shuster, 1980). Southern slopes appear to be avoided for nesting sites in the northeast and in the western mountains south of Canada (Allen, 1978; Shuster, 1980; Moore and Henny, 1983; Hall, 1984; Speiser and Bosakowski, 1987), but in Alaska the birds exhibit a positive preference for them (McGowan, 1975), indicating the probable importance of microclimate in influencing specific nest site tendencies.

Although coniferous stands seem to be favored over hardwood stands, hardwoods offer large primary crotches or forks for suitable nest sites, and thus seem to be favored over conifers as specific nesting trees (Bull, 1974; Peck and James, 1983; Apfelbaum and Seelbach, 1983; Speiser and Bosakowski, 1987). Peck and James reported on 26 Ontario nests that had an average clutch size of 2.7 eggs. Bull's sample of 42 New York nests averaged 2.36 eggs. McGowan (1975) found average annual clutch variations of from 3.1 to 3.8 eggs in Alaska, and Reynolds and Wight (1978) reported an average clutch of 3.2 eggs for 48 Oregon nests. All told, it would seem that few if any significant regional variations in clutch sizes occur in North America, but perhaps clutches tend to increase slightly toward the north. Clutch sizes of about 3.5 eggs seem typical for temperate and northern Europe (Cramp and Simmons, 1980).

The eggs are laid at 2–3 day intervals, with a replacement clutch normally laid 15–30 days after loss of the initial clutch. As the female increasingly tends her clutch, she becomes wholly dependent upon the male for food. However, males may temporarily cover the eggs during egg-laying and early incubation stages, and while the female is eating prey that he has brought her. Incubation begins with the first egg or so, but this early incubation may not be intense, as the clutch typically hatches within about a 2–3 day period (Cramp and Simmons, 1980).

In a California study (Schnell, 1958), the brooding female was found to remain close to the nest both day and night during the early nestling period. She would leave the nest only to receive food from the male, to capture prey near the nest, to cache prey items, and to collect sprigs to bring to the nest. Sprig collecting reached a peak as the brooding ceased, and after the young were about 25 days old she would leave the nest area to hunt. About 85 percent of the prey eaten by the brood was brought by the male, but only the female actually fed the young. She also was responsible for defending the nest from humans. Nestling birds accounted for a considerable part of the summer diet, with the two nestling goshawks consuming an estimated total of 13 kilograms of prey during 49 days of posthatching development.

Typically the nestlings begin to move to nearby branches when about 35 days old, and males may fledge at 35–36 days, while the larger females usually take 40–42 days to fledge. They do not become independent of their parents until they are about 70–80 days old. Fledging success is of course highly variable, but European studies indicate that 2.7 to 3.1 young per successful nest represents a typical range (Cramp and Simmons, 1980). Reynolds and Wight (1978) reported an average of 1.7 young raised in 48 total nests (or 2.3 young per nest in 25 successful nests), and McGowan (1975) found that 1.8–2.5 young were raised per nest during various years of his Alaska study. Seemingly there are considerably greater annual and geographic differences in nesting and fledging success than in clutch sizes in this species.

Evolutionary Relationships and Status

This is the largest of the Holarctic accipiters, and it was suggested by Wattel (1973) that the species is of Old World origin, perhaps colonizing the New World comparatively recently. Part of his reasoning for this is that close relatives of *gentilis* exist in Africa in the form of *A. melanoleucus* (the black sparrowhawk) and in Madagascar, where *A. henstii* (Henst's goshawk) occurs.

In contrast to the other two North American accipiters, there is no hard evidence to indicate that the goshawk has suffered significant population declines in recent decades, and this is perhaps in part related to a seemingly rather low pesticide burden in the species (Snyder et al., 1972; Reynolds and Wight, 1978). In some areas of eastern North America the species has been able to extend its breeding range southward as forests have matured recently (Bull, 1974; Speiser and Bosakowski, 1987), while in other parts of the country such as the Pacific Northwest the destruction of old-growth forests has produced opposite population trends. Probably limited forestry, opening up some mature forests to a limited extent, has beneficial effects in promoting better nesting habitat, while more extensive forestry may well break the woodlands into areas too small to support nesting territories. Reynolds, Meslow, and Wight (1982) suggested that at least 8 hectares of habitat be left around goshawk nests as minimal protection.

A bibliography of the northern goshawk containing 98 references has been published (Shuster, 1977); Palmer (1988) cited 109 references.

TYPICAL BUTEONINE HAWKS

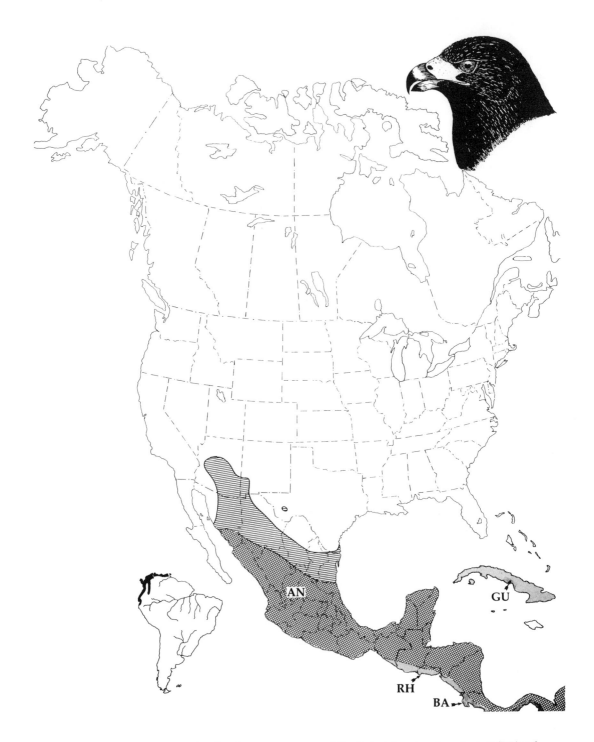

Figure 36. North and Central American breeding (hatching) and residential (cross-hatching) ranges of the anthracinus (AN) race of the common black-hawk. Residential ranges of the insular form gundlachii (GU) and the Central American forms bangsi (**BA**) and rhizophorae (**RH**) (the latter two currently considered specifically distinct and part of subtilis) are shown by shading.

Common Black-hawk *Buteogallus anthracinus* (Deppe) 1830

Other Vernacular Names

Black hawk, crab hawk, Cuban black-hawk (*gundlachii*), lesser black-hawk, Mexican crab hawk, Mexican black-hawk; aguililla cangrejera (Spanish)

Distribution

Resident (*anthracinus* group) in wooded riparian areas from central Arizona (formerly or casually to southwestern Utah, accidentally to Nevada), southwestern New Mexico, and western (and until at least 1937 also southern) Texas south through Mexico, Central America, and northern South America to Guyana. Also resident in Cuba and the Isle of Pines (*gundlachii* group, sometimes considered a separate species); perhaps also in the Pacific coastal lowlands of Central America and northern South America (*subtilis* group, here considered a separate species). Birds breeding in the southwestern United States usually migrate southward in the nonbreeding season, but some winter records for southern Texas and Arizona exist. Also seen irregularly in southern Florida from 1972 to 1976, the birds perhaps being vagrants from Cuba or possibly escapes from captivity. (See Figure 36.)

North American and West Indian Subspecies

B. a. anthracinus (Deppe): Resident or migrant from Arizona, New Mexico, and western (formerly also southern) Texas south to Panama. Casual nester in Utah; accidental in southern Nevada. (The 1983 AOU *Check-list* has been followed in regarding the Pacific coastal mangrove-adapted form *subtilis* as a separate species, including its Central American races *rhizophorae* and *bangsi*.)

B. (a.) gundlachii (Cabanis): Resident of Cuba and Isle of Pines, including coastal cays. Considered an allospecies by Stresemann and Amadon (1979) but regarded as part of *anthracinus* by Brown and Amadon (1968) and the AOU (1983).

Description (of *anthracinus*)

Adult (sexes alike). General color uniform slate-black on the back, neck, and breast; the head generally darker black; the feathers of the crown, occiput, and nape with white bases more or less tinged with pale tawny; upper wing coverts and remiges like the back, the inner webs and the concealed portion of the secondaries usually mottled with grayish brown to pale tawny-ochraceous; tail deep black, narrowly tipped with white and white at the extreme base, and crossed at about the middle by a broad, continuous white zone of variable width; upper tail coverts and under wing coverts more widely tipped with white. Iris brown; terminal half of the bill lead-black; the basal half, cere, lores, tarsi, and toes waxy bright yellow to orange-yellow; lower eyelids dull whitish; claws black.

Subadult (sexes alike). Like the adult by the spring following hatching, except that more concealed or partially concealed buff is present in the body feathers, especially on the hindneck, and the black feathers more sooty, less plumbeous than in older birds. Softparts similar to adults, but the cere, gape, lores, tarsi, and toes greenish yellow (Dickey and van Rossem, 1938).

Juvenile (sexes alike). Usually dark fuscous-brown above, variably spotted and streaked with white or buff and cinnamon; underparts pale buff to ochraceous, boldly spotted or barred with fuscous-brown; tibia closely barred; tail ochraceous, crossed with five to eight dark bars, the most posterior one the widest. (A lighter juvenal plumage variant also has been reported from Panama, in which the normally blackish coloration is diluted to brown or yellowish brown, and the pale areas of the upperparts are accentuated. The tail is dull white, crossed by five or six brownish bands.) Iris brown; bill black at tip, becoming olive-yellow to greenish yellow basally; cere and corner of mouth olive-green to yellowish green; tarsi and toes dull greenish yellow.

Measurements (of *anthracinus*, in millimeters)

Wing, males 337–400 (ave. of 29, 367.1), females 355–400 (ave. of 30, 381.5); tail, males 192–228 (ave. of 29, 206.1), females 203–244 (ave. of 30, 222.3) (Friedmann, 1950). Average egg size 57.3 x 44.9 (Bent, 1937).

Weights (in grams)

Males, ave. of 6, 793; females, ave. of 4, 1199 (Hartman, 1961). One female, 945 (Brown and Amadon, 1968). Range of weights 630–1300, ave. 950 (sex and sample size unspecified) (Clark and Wheeler, 1987). Estimated egg weight 63.8, or 5.3% of female.

Identification

In the hand. This species may be recognized by the combination of nearly naked bright yellow-orange lores, primaries that are only 40–70 mm longer than the secondaries, and (except in immatures) a nearly wholly black plumage except for a central white tail band.

In the field. When perched, adults of this species appear coal-black, save for the bright yellowish orange facial skin and cere (which are brighter and more extensively yellow than in the equally black zone-tailed hawk) and similar bright yellow legs. The tips of the wings nearly reach the end of the tail. Immatures appear mostly black, but have a buffy eyebrow stripe and underparts that are heavily streaked and spotted with buffy. The birds are usually found along rivers, swamps, and marshes where crabs are abundant.

In flight, the mostly black birds exhibit extraordinarily broad (front to back) wings, which in adults are almost entirely blackish below save for paler areas at the bases of especially the more anterior primaries, sometimes forming a light "window." A single broad white band crosses the mostly black tail, which is also tipped with white. Immatures are also blackish brown, heavily streaked with creamy to buffy on the undersides, and the tail is crossed with numerous (5–8) narrow dark and whitish bands, including a broader subterminal dark band. Immatures also show a more cream-colored and distinct "window" over most of the primaries, which is also visible from above. The usual flight behavior consists of extended eaglelike gliding on horizontally held wings, helping to distinguish the species from the similarly dark zone-tailed hawk, which has relatively longer and much narrower wings that are held in a distinct dihedral tilt when soaring. The typical in-flight alarm call is a series of high-pitched and rapid screaming whistles, lasting about four seconds, and consisting of 7 or 8 piercing notes that initially increase in intensity and then diminish.

Habitats and Ecology

In Arizona, this species is closely associated with and probably limited to areas having permanent streams, such as those along the base of the Mogollon Rim (Phillips, Marshall, and Monson, 1978). In that state it ranges up to about 1,500–1,800 meters, but is mainly found at lower elevations. In New Mexico it is limited to the southwestern corner of the state, but single nestings in Bernalillo County in 1971 and 1972 occurred along the Rio Grande river well to the north of its previously known range (Hundertmark, 1974). In Texas it was mainly found along cottonwood-lined rivers and streams, and formerly occurred in willow groves on the lower Rio Grande flood plain, probably until these were eliminated by timber cutting. It may also have occurred in marshy areas near the coast. Farther south, it once frequented sluggish rivers in adjacent Tamaulipas. The birds were discovered to be nesting in Jeff Davis County of western Texas in 1970, but the last known nesting in the Rio Grande area occurred in 1937, following a population collapse during the early years of this century (Oberholser, 1974). However, in recent years (since 1950) it has reappeared around the coastal prairie near Laguna Madre, and in the vicinity of Falcon Dam, Starr County (B. J. Rose, personal communication). It is also now regularly seen during summer in Big Bend National Park (*American Birds* 36:869).

Little can be generalized about its habitat in the U.S., which is clearly peripheral range, although Schnell (1979) identified some habitat traits of the species. Generally the birds prefer lowland areas, with a source of water where crabs, crayfish, or other aquatic foods are to be found, as well as trees for roosting or nesting. In Panama the birds are mainly found in coastal areas, such as swampy woodlands and mangroves. They range into the interior only along larger rivers, and there the presence or absence of crabs may determine the range of the species (Wetmore, 1965).

Foods and Foraging

According to early sources (Fisher, 1893; Bent, 1937), foods of this species include snakes, frogs, fishes, young birds, and land crabs. Brown and Amadon (1968) considered crabs (land or beach), frogs, snakes, fish, insects, rodents, and rarely birds to be their foods. Insects such as grasshoppers and caterpillars are sometimes also eaten, and the birds may follow grass fires to eat such prey. In Cuba the birds hunt at dawn and again in early evening hours, pouncing on crabs along the mangroves. Thomas (1908) indicated that large land crabs are the primary if not almost exclusive food of the birds in Belize.

Snyder and Wylie (1976) estimated that invertebrates comprise nearly 80 percent of this species's prey numerically, with lower vertebrates adding another 15 percent and birds 6 percent, based on literature records available to them. Glinski (unpublished observational data cited by Sherrod, 1978) found that spring and summer foods in Arizona consisted of 40 percent fish, 37 percent reptiles and amphibians, 12 percent invertebrates (crayfish and insects), 3 percent each of birds and mammals, and 5 percent miscellaneous foods. The rather marked differences in these percentages suggest that considerable opportunistic feeding may occur, and probably there is no real dependence upon crabs except perhaps in more tropical areas.

The foods of two Arizona pairs were reported upon by Millsap (1981), who found a nearly complete dependence upon the riparian habitats for foraging, with nearly 80 percent of the diet consisting of fishes, amphibians, and reptiles. Of the fishes, which represented 59 percent of the total estimated biomass and 35 percent of the items, the Gila sucker (*Catostomus insignis*) was most important, comprising 32.5 percent of the biomass and 20 percent of the items. Among reptiles, which made up 14 percent of the biomass and 21.5 percent of the items, garter snakes (*Thamnophus cyrtopsis*) were most important, comprising 9 percent of the biomass and 14 percent of the food items. The only amphibian represented was the leopard frog (*Rana pipiens*), which represented 18 percent of the biomass and 32 percent of the items. Nesting was not observed to occur where catostomid fishes such as suckers were lacking, nor in areas with low densities of amphibian and reptilian prey.

The birds typically wait for prey while perched on a low, well-vegetated tree limb near water, or stand quietly on a sand bar or mud bar (Oberholser, 1974). From these situations they can easily capture crabs, and they are also notably agile on the ground (Wetmore, 1965).

Social Behavior

No detailed information is available on pair bonding, territoriality, or most other aspects of social behavior. Soaring and calling above the nesting area by the pair is probably a part of courtship (Wauer and Russell, 1967), and the display flight of the male reportedly consists of a series of undulating climbs and dives, often while the bird is dangling its feet and calling. The pair also often will fly together, with their wings fluttering and held at a strong angle of dihedral (Clark and Wheeler, 1987). The male may also plunge from a high altitude almost to the ground, and then check its momentum near the surface and glide away. Loud and hoarse whistles, sounding something like those of a night-heron (*Nycticorax*) are common during the nesting season (Wetmore, 1965; Oberholser, 1974).

Copulation occurs about 15–90 meters from the nest, on a branch or prominent rock. The male may swoop down and land directly on the female, or may perch beside her for a time before mounting. Up to four copulations per day occur as the egg-laying period approaches (Jay Schnell, in Palmer, 1988).

Breeding Biology

Nests of this species are often built (in Panama) at the edges of mangrove swamps, usually about 12 meters high (Wetmore, 1965); more generally they are placed in diverse trees including palmettos and cypresses, at heights up to about 30 meters (Brown and Amadon, 1968). In Belize they are typically in pines, usually about 7–10 meters high, and they have also been reported built in rocky recesses (Thomas, 1908; Bent, 1937). In Arizona the birds have been found to nest in rather large trees in mature, vigorous stands, and all of 24 nests were within 120 meters of permanent flowing water. Nineteen of these nests were in cottonwoods, with average diameters of 82 centimeters, while the rest were in alder, sycamore, and willow of similar size; the nests averaged about 15 meters above ground.

The clutch size is relatively small in this species; Bent (1937) reported that 13 sets he found ranged from 1 to 3 eggs, averaging 1.9. Thomas (1908) reported that 26 of 27 nests in Belize had a single egg, and the remaining one had two. Of 86 sets in the collection of the

Western Foundation of Vertebrate Zoology the range is 1–3 eggs, averaging 1.6. Bent (1937) noted that 22 clutches from Texas and Central America were found between February 8 and May 30, with half of them found between March 31 and May 5. Fifteen sets in the Western Foundation of Vertebrate Zoology from Texas and Arizona were obtained from February 8 to June 3, with 9 from May 8 to 18. Seventy-two sets from Mexico are from March 12 to May 13, with 47 from April 2 to 15. Seventy of these sets were from nests situated 2.5–42 meters above ground, with 43 located between 12 and 18 meters. Among 63 sets in which the identity of the tree supporting the nest was mentioned, 11 were in mangroves, 12 were in "higo" (apparently a kind of *Opuntia* cactus), 6 in "higuera" (a kind of *Ficus* fig tree), 7 in cottonwoods, 5 in cypress, and 22 in 14 other kinds of trees. Eggs are probably laid at 2- or 3-day intervals (Jay Schnell, in Palmer, 1988).

Millsap (1981) noted that in west-central Arizona the birds arrived on breeding territories in late March and laid between April 17 and May 20, with half of the clutches completed by about May 5. Incubation lasted approximately 37 days at one nest, and one renesting attempt was observed after an initial attempt failed. About half of the eggs had hatched by June 11, with nestlings remaining in two nests for an average of 40 days, and with half of the nests fledged by July 21. The average clutch size in 22 nests was 1.92 eggs, and the average initial brood size was 1.5 young, with an 80 percent hatching success. The average number of young fledged per nest was 1.3 young, with a fledging success rate of 84 percent. Somewhat lower reproductive success rates were reported by Schnell (1979, and in Mader, 1982) for southeastern Arizona. He found an average clutch of 1.67 for 12 nests, and an average 1.0 young fledged per nesting attempt (58 nests), with an overall nesting success of 81 percent. He also estimated an incubation period of 38–39 days, a fledging period of 43–50 days, and a period of postfledging dependence lasting 6–8 weeks (Newton, 1979). A much longer period of postfledging dependence may be typical of the closely related great black-hawk (*B. urubitinga*) of tropical America, where broods are small and infrequent (Mader, 1982).

There does not appear to be any information on other details of the nesting cycle, although

Thomas (1908) noted that up to three sets may be laid in a single season if the clutches are removed. According to him, the adults are very bold when the nest is disturbed, and the male may often make plunging dives toward the intruder, swerving away just in time to avoid striking.

Evolutionary Relationships and Status

This is clearly a very close relative of *B. subtilis*, and the two forms have often been considered conspecific. Amadon (1982) has most recently treated the two as separate species, since they seem to be in contact without intergrading. The total number of species recognized in the genus has ranged recently from three (Brown and Amadon, 1968) to five (Stresemann and Amadon, 1979; Amadon, 1982), indicating something of the taxonomic uncertainty of the group. Perhaps the nearest relatives of *Buteogallus* are found in the genera *Parabuteo* and *Leucopternis*, the latter being a tropical New World group of hawks that includes some species once placed in the genus *Buteogallus*, and that occupy similar ecological niches.

The U.S. status and prospects of the black-hawk are rather bleak; as already mentioned, the population nesting in southern Texas was apparently eliminated early in the century, and for more than 30 years the birds were not known to nest anywhere in the state (Oberholser, 1974). However, at least three pairs were nesting in the Davis Mountains during the early 1970s (Porter and White, 1977). Nesting also occurred during 1988 in Val Verde County (*American Birds* 42:458). The best U.S. population is perhaps in Arizona, where the birds are nevertheless only locally present in riparian areas, and they have an even more geographically restricted distribution in New Mexico. In both cases they are mainly limited to cottonwood groves along the Rio Grande and Gila rivers, and these groves are rapidly disappearing as a result of clearing for agricultural purposes (Hubbard, 1965). Perhaps 6–12 breeding pairs were still present in New Mexico in the mid-1970s, and at that time there were at least 11 known nest sites in Arizona (Porter and White, 1977). They have nested at least twice in southern Utah as far north as the Virgin River valley (Wauer and Russell, 1967;

Gifford, 1985). Black-hawks were seen during the early 1970s in southern Florida, perhaps as a result of birds that have come from Cuba (Abramson, 1976), but are not known to have nested there. As of the early 1980s an estimated 200 pairs were present in the United States (J. Schnell, in Palmer, 1988). A long-term study of breeding success and prey of black-hawks is under way in Aravaipa Canyon, Arizona, under the direction of Jay Schnell.

Figure 37. North and Central American residential range of the Harris' hawk, with recent historical range in *Arizona indicated by dashed line. The South American residential range is shown on the inset map.*

Other Vernacular Names

Bay-winged hawk, dusky hawk, eastern Harris hawk (*harrisi*); aguilla cinchada (Spanish)

Distribution

Resident of semiopen habitats from northern Baja California, southern Arizona, southern New Mexico, and central Texas (one 1963 breeding in southern Kansas) south through Central America and South America to Chile and northern Patagonia. Previous resident of southeastern California (to the early 1970s) and adjacent southwestern Arizona, where extirpated but more recently (1979–85) being reintroduced along Colorado River. (See Figure 37.)

North American Subspecies

P. u. harrisi (Audubon): Resident from Arizona and Texas south to western Ecuador. Includes *superior* (Stresemann and Amadon, 1979).

Description (of *harrisi*)

Adult (sexes alike). Entire head and most of body dark sooty brown to fuscous (with fuscous predominating in the northernmost parts of the range), usually darker on the scapulars and upper back than on the head or underparts; the lower back and rump similar but tinged or edged with dark chestnut; lesser upper wing coverts entirely rich chestnut; the median upper wing coverts fuscous, broadly edged with chestnut; greater upper wing coverts dark fuscous; remiges dark fuscous becoming fuscous-black on the primaries; upper tail coverts white, sometimes with concealed fuscous bars or spots; rectrices basally white and widely tipped with white; otherwise black, the lateral rectrices paling very slightly to fuscous on the inner webs; under wing coverts and thighs rich chestnut; under tail coverts pure white. Iris hazel-brown; cere, lores, edge of gape, and orbital ring bright yellow; bill light bluish ash, dusky toward the tip; tarsi and toes orange-yellow to dull yellow; claws black.

Subadult. The adult plumage is apparently attained within a year, during a prolonged, gradual molt and a wearing away of buffy feather edges (Bent, 1938). Some subadult birds, usually females, may breed successfully (Whaley, 1986).

Juvenile (sexes alike). Similar to the adult, but with the feathers of the upperparts generally narrowly edged with dark chestnut; the chin, throat, and abdomen streaked with white or buffy white, the lower abdomen nearly plain tawny whitish; thighs pale tawny narrowly barred with chestnut; the rectrices indistinctly multiple-banded (the subterminal band the broadest and most conspicuous) with dark fuscous-black and grayish fuscous bands; the outer under wing coverts light gray, with narrow fuscous barring and varying amounts of rufous near the bend of the wing. Iris chestnut-brown; cere yellow or yellowish green; tarsi and toes pale yellow.

Measurements (of *harrisi*, in millimeters)

Wing, males 318.3–331.5 (ave. of 3, 323.2), females 325–370 (ave. of 15, 358.4); tail, males 215–262 (ave. of 3, 234), females 213–243 (ave. of 15, 232.5) (Friedmann, 1950). Average egg size 53.7 x 42.1 (Bent, 1937).

Weights (in grams)

Males 634–877 (ave. of 37, 735); females 913–1203 (ave. of 14, 1047) (Dunning, 1984). Adult males in winter 550–829 (ave. of 220, 690); adult females in winter 825–1173 (ave. of 177, 998) (Hamerstrom and Hamerstrom, 1978). Estimated egg weight 56.1, or 5.4% of female.

Identification

In the hand. This species is identified by the combination of nearly naked pale yellow lores and rusty chestnut on the wings (the basis for the vernacular name "bay-winged hawk"), including both the upper and under wing coverts (the latter much less evident in immatures than adults).

In the field. When perched, the chestnut-rufous upper wing coverts are variably evident in both young and adult birds, and contrast with the otherwise generally dark brownish black upperparts. In adults the white tail is very broadly banded with contrasting black; white is otherwise evident only on the tail coverts.

In flight, the bright chestnut under wing coverts are evident both in immatures (where they are less apparent) and adults; in young birds the primaries are nearly white basally, becoming increasingly barred with darker toward the tips, whereas in adults the undersides of the flight feathers are uniformly medium gray. In adults the tail has a distinctive broad white tip, an even wider central black band, and is also white basally and on the upper and lower tail coverts. Immatures have a much narrower white tail tip and inconspicuous gray and black barring on the rest of the tail. Gliding is done on slightly cupped wings. The alarm call is a prolonged *iirrr* note, lasting about three seconds.

Habitats and Ecology

This species has a broad range in tropical parts of the New World, but a very limited one in the United States, where it is now restricted to Texas, New Mexico, and Arizona. In Arizona its traditional breeding habitats varied from upland deserts dominated by saguaros (*Carnegeia*) to inundated trees in the Colorado River valley (now extirpated) and arid mesquite (*Prosopis*) and paloverde (*Cercidium*) woodlands (Phillips, Marshall, and Monson, 1978). In Texas it favors mesquite woodlands with an *Opuntia* cactus understory as prime habitat, but also ranges into the yucca-cactus-creosote bush deserts, and locally extends into juniper-oak habitats on the Edwards Plateau (Oberholser, 1974).

In Mader's (1975a) study area of Pima and Pinal counties, Arizona, the birds occurred exclusively in hot Sonoran Desert flatlands, dominated by trees such as paloverde, mesquite, ironwood (*Olneya*), and various large cacti (*Carnegeia, Opuntia*), with an understory of bur sage (*Franseria*) bushes. Mader believed the presence of these saguaro-paloverde flatlands to be indispensable for the survival of the Harris' hawk in southern Arizona. Whaley (1979, 1986) did a more general Arizona study, finding that the now-extirpated (but being reintroduced) Colorado River population nested near water in flooded mesquite and in willows and cottonwoods. The major and still extant population occurs in upland habitats between about 400 and 1,000 meters elevation, mainly nesting in paloverde-saguaro cactus habitats or in the more localized and narrow strips of paloverde-ironwood associated with large arroyos and rarely in large riparian cottonwoods. These generally structurally complex communities support the greatest prey diversity and density of any Sonoran habitats. In

Texas, live oak–persimmon stands were found to be preferentially used by breeding adults, these trees providing the highest hunting perches and preferred nesting sites (oaks) in that area (Brannion, 1980).

Home ranges during the breeding season ("nesting ranges") typically are fairly large; in Arizona these average about 5 square kilometers per active nest in dense populations (Mader, 1975a). These areas are well separated from those of conspecifics, and may contain two or three old nests, suggesting considerable territorial site tenacity. Whaley (1986) confirmed this, estimating an 84–91 percent reoccupancy rate of territories that had been used the previous year. He also reported a nesting density of one nest per 2.5 square kilometers in two years, with an average distance between nests of 1.8 kilometers, suggesting some degree of territorial overlap. Brannion (1980) estimated home ranges of from 2.13 to 3.2 square kilometers for three breeding pairs.

Compared with Swainson's hawks in the same area, Harris' hawks tend to locate their nests in areas with greater tree densities and more exposed ground, but habitat overlap is considerable, as was dietary overlap in a recent study (Bednarz, 1988b). On the other hand, red-tailed hawks occupy a broader ecological range of habitats than do Harris' hawks in Arizona, including more arid habitats, but apparently avoid some of the very spiny or densely canopied trees that the Harris' hawk uses for nesting (Mader, 1978).

Foods and Foraging

Considerable difference of opinion as to the foods of this species has occurred in the past, with authors variously describing the species as a "sluggish carrion-eater" and a "dashing predator" (Brown and Amadon, 1968). Reviewing the literature, Snyder and Wylie (1976) estimated mammals to comprise about 53 percent of the diet numerically, birds about 40 percent, lower vertebrates about 7 percent, and invertebrates only about 1 percent. Pache (1974) found a much higher incidence of invertebrates (mostly beetles, ants, ticks, and grasshoppers) than did earlier studies, perhaps because he used pellet analysis rather than other more "coarse-grained" techniques such as counting prey remains at nests. Based on 641 prey items found at nests, Whaley (1986) found that mammal remains comprised 61 percent of the numerical total, birds 28 percent, and reptiles 11 percent. The most frequently encountered mammals, desert cottontails (*Sylvilagus audubonii*), comprised 22 percent of the total, while the most commonly encountered bird was the Gambel's quail (*Callipepla gambelii*) at 9 percent, and the most common reptile was the desert spiny lizard at 10 percent. By comparison, Brannion (1980) found that 66 percent of the prey he analyzed from nest sites in Texas were mammals, and most of these were cottontails. Fairly large mammals such as cottontails, ground squirrels (*Spermophilus*), antelope squirrels (*Ammospermophilus*), and woodrats (*Neotoma*) seem to be taken frequently, and even black-tailed jackrabbits (*Lepus californicus*) are sometimes used as prey. Cottontails made up the largest frequency as well as biomass components of prey remains found in and near nests in New Mexico (Bednarz, 1988a), and cottontail populations typically peaked when most second nestings were initiated.

The Harris' hawk is apparently a highly versatile predator, shifting its prey as their availability changes. Harris' hawks are perhaps more flexible in their prey exploitation abilities than are red-tailed hawks in this regard (Mader, 1978). This flexibility may help to explain the considerable dietary differences that have been reported for the Harris' hawk by various authors. This species is similarly flexible in adjusting its breeding season locally to an extended and perhaps variable food supply (Bednarz, 1988a). As compared with the sympatric Swainson's hawk, the Harris' hawk is somewhat more of a specialist on larger mammals and some birds, whereas the Swainson's hawk is more of an opportunistic generalist (Bednarz, 1988b).

In a radio-telemetry study, Bednarz and Hayden (1987) found that during winter Harris' hawks often hunt cooperatively (average group size 4.7 hawks) and frequently thus kill both cottontails and jackrabbits. They found that surprise pounces, flush and ambush, and relay attack methods were all used, and that cooperative hunting is not only advantageous to individual hawks but may also help explain the tendency for group living in this species. Bednarz (1988a) observed that such cooperative hunting allowed the birds to kill larger prey than might be handled by individual hawks, and improved the success rate per amount of time. He also noted that such groups regularly consist of a pair of adult breeders, up to two "auxilliary" younger

birds in adult plumage, and up to three immature-plumaged birds of the year. The whole group thus presumably consists of close relatives among whom kin selection is most likely to be operative.

Social Behavior

Mader (1975a) observed two apparent courtship flights during his studies. Both times an adult male went into a nearly vertical stoop, with outstretched feet, from an altitude of 150–200 meters, ending it with an upward swooping glide to the back of an adult female perched on the top of a saguaro, where copulation followed immediately. However, not all copulations he observed were preceded by such courtship flights, nor were such flights always followed by copulation attempts. Thus, two males were observed making identical flights in the presence of a female during the nestling phase, and on several occasions all three hawks stooped together, either in single file or as a loosely formed group. Similarly, Brannion (1980) observed slanting or vertical stoops toward the ground by males, which were often followed by copulations.

Pache (1974) observed copulation occurring repeatedly during a short period, including five times in a two-and-a-half-hour period and three times in a three-hour period. All of these were in above-ground situations. However, Parmelee and Stephens (1964) reported that copulation on the ground was seen twice. Mader (1979) observed copulation only on elevated sites, and noted that

it was sometimes preceded by the male standing on the female's back for a fairly extended period (up to 170 seconds for the entire sequence). Brannion (1980) observed 44 copulations, all of which were on elevated sites (including rock outcroppings).

Although some earlier writers have questioned the existence of exclusive breeding territories in the Harris' hawk, Dawson (1988) found that individual breeding groups resided on and defended their core nesting areas year round, and these groups also seasonally maintained and defended exclusive-use breeding territories that included both nesting and foraging areas. Aggregations that included members of nearby groups and transient hawks occurred only in fall and winter during nonbreeding periods.

In 50 active nests studied by Mader (1975a, 1975b), 46 percent had three adults present, and in the cases where sex of all could be determined the ratio was always two males and a female. Most trios remained intact through the nesting cycle. Immature birds were sometimes also seen at nests occupied by adults, but in only one case was an immature observed to help in brooding or shading the young. Where two adult males were present, one obtained prey and fed the chicks directly, as well as staying near the nest when the female was absent, while the other only caught prey and transferred it to the female or to the other male. In later observations (1979), Mader established that both males copulated with the female at approximately the same frequency, although one male was larger and dominated the smaller male. Whaley (1986) also confirmed that color-marked juveniles of earlier nesting attempts were allowed in the immediate vicinity of, and often on, the nest containing eggs or young of subsequent nesting efforts. On one occasion a bird from a brood that fledged four months earlier was observed bringing prey to the nest, and two juveniles that had fledged only a few weeks earlier were observed helping incubate their parents' second clutch. Similarly, Brannion (1980) observed recently fledged juveniles bringing prey to the nests of their parents' second breeding attempts.

Harris' hawks are fairly tolerant of most raptors' intruding into their nesting range, such as immatures of their own species (including their own offspring) and even nonresident adult Harris' hawks (Mader, 1975a).

Much the most interesting aspect of this species's breeding biology is the tendency for simultaneous polyandry that it often exhibits. Mader (1975a, 1975b, 1979) was the first person to report on this in detail, noting that it had otherwise been observed among hawks only in the Galapagos hawk (*Buteo galapagoensis*). He observed 27 nests tended by pairs as compared with 23 tended by trios, and found that the percentage of nests that were successful, the number of young raised per nesting attempt, and the number of young per successful nest were all higher in nests having trios rather than pairs present. He concluded (1979) that the higher breeding success of trios may have been the result of the extra male's assistance in nest attendance, food procurement, and nest defense. Additionally, the extra male might insure against total nest failure should one adult die during the breeding season. Presumably, an initially monogamous male accepts a second male and its associated copulation efforts because the former may thereby gain in inclusive fitness through increased overall breeding success, while it may be advantageous to the second male to remain mated to such a pair if it is likely to increase his own fecundity later in life, as for example after the first male has died and needs to be replaced. The situation is apparently most common in areas of rich habitat quality, whereas in very poor (arid) habitat there is less chance of successful reproduction even with trios available for hunting. Polyandry may also be more common in Arizona, where the sex ratio is seemingly strongly skewed toward males, than in Texas (Griffin, 1976), where the adult sex ratio is closer to equality (Hamerstom and Hamerstom, 1978).

Recent observations in Arizona (Dawson, 1987) shed more light on this difficult sociobiological problem. He found that of 53 nesting groups, the group sizes ranged from 2 to 7 individuals, averaging 3.8 and most often consisting of 3 birds. These groups included a female breeder, an alpha adult male, a beta male helper, and a variable number of adult or immature "gamma" male or female helpers. The alpha males excluded helpers from the nest, but did tolerate beta males for short times. Helpers typically provided food to the alpha pair, who would then feed it to the nestlings. Most groups apparently consisted of a monogamous pair with one or more nonbreeding helpers.

In a recent effort to understand the values of cooperative breeding, Bednarz and Ligon (1988) have compared ecological variables of breeding pairs versus larger breeding groups in New Mexico. Their data did not provide clear support for any of the currently available hypotheses concerning cooperative breeding. Instead, the species's group living tendencies seem to be related to its dependence on fairly large and elusive prey (largely lagomorphs), in a manner analogous to the social systems of wolves (*Canus lupus*) and wild hunting dogs (*Lycaon pictus*). Its social system thus seems to be related to the species's extended parental care, in which young birds remain with older group members for a year or more, and to its cooperative behavior, in which group hunting and prey sharing provide critical survival advantages to individual birds.

Breeding Biology

Nest sites in Arizona are mostly in saguaro cactus and paloverde trees, at heights typically of about 4 to 6 meters, but recorded sites have also included pines, palms, cottonwood, mesquite, ironwood, and electrical towers (Mader, in Palmer, 1988). In Texas, the average height of 19 nests was 3.4 meters, with all but one of these situated in hackberry, mesquite, or Spanish dagger (Griffin, 1976).

The nesting season in this species is quite prolonged. Whaley (1986) reported that eggs were laid in his Arizona study area from mid-January to mid-August, with half of all first clutches (284) laid between February 20 and March 22; some of those laid during and after April were second clutches. Double or triple clutches within a single season were reported in 22 percent of the nests he studied, with 64 percent of these after a successful first nesting attempt, and 36 percent after an unsuccessful first nesting. In some cases second clutches were begun before the young of the first clutch had fledged, the eggs sometimes even being laid in a nest still containing unfledged young. Brannion (1980) reported an average interval of 28 days between the fledging of the first brood and the completion of the second clutch. Bednarz (1987) also reported successive nesting in New Mexico, noting that a second breeding attempt was frequently made in late summer or autumn, often after successful spring breeding.

The egg-laying interval is uncertain, but Mader (1975a) established an average clutch size of 2.96 eggs for 50 Arizona nests. By comparison, Whaley (1986) reported an average clutch size of 3.07 eggs for 71 Arizona nests. In Texas the clutch size may be somewhat smaller, as Griffin (1976) determined an average clutch of 2.85 eggs for 20 nests, and Brannion (1980) found an average of only 2.33 eggs in 24 nests. The incubation period lasts about 35 days, with the male sometimes sharing incubation duties. Fledging requires another 40–45 days, with the young birds usually remaining in the area of the nest for at least two or three months after hatching (Mader, 1975b).

Although Mader (1975a) provided useful comparative data on the success of nesting by pairs versus trios, the best data on nesting success are those of Whaley (1986). Of 319 nesting attempts, 72 percent were judged by Whaley to be successful (the young surviving at least 28 days). The average number of young fledged in 319 nesting attempts was 1.62, while the average number fledged in all successful attempts was 2.26. A fairly high rate of egg loss was attributed to egg infertility, nest abandonment, and predators, while losses of nestlings were the result of diverse causes, including human disturbance, disease, and accidents. The productivity estimates of Mader (1975a, 1975b) are almost identical to Whaley's (2.35 young fledged per nest for 34 successful nests), while those of Griffin (1976), Brannion (1980), and Bednarz (1988b) indicate rather lower productivity estimates. Brannion reported an overall hatching success rate of 66 percent for 24 nests, and a fledging success of 89 percent, but an average of only 2.06 young fledged per successful nest. Predation appeared to be the major cause of mortality in both eggs and young in that study. Repeat nesting attempts were more successful than initial attempts, with larger clutch sizes and higher fledging success rates typical. Bednarz (1988b) found that, although productivity for 114 nests averaged only 1.25 young fledged per breeding cycle and had a nesting success rate of only 68 percent, double brooding resulted in a somewhat higher annual overall productivity of 1.73 fledged young per nest.

Evolutionary Relationships and Status

According to Brown and Amadon (1968), *Parabuteo* is very closely related to *Buteo*, although its adult and immature plumages suggest a possible relationship with such genera as *Buteogallus* and also *Heterospizias meridionalis* (the South American savanna hawk). Amadon (1982) more recently has suggested that *Parabuteo* may be closely related to the savanna hawk (which he included in the genus *Buteogallus*), but the two should probably be maintained as separate genera.

The distribution and status of the Harris' hawk has been very well summarized by Whaley (1986). He said that the extirpated Colorado River population was probably dependent upon the riparian woodland community that began to be affected by such factors as dam construction in the 1930s, a spreading of saltcedar (*Tamarix*) and its associated competition with cottonwoods, disturbance from dredging operations as well as increased use of the river for recreational purposes, and nest destruction. Of these factors, drastic habitat alteration and increased recreational impact may have been the most significant. Between 1979 and 1985 a program of attempted reintroduction of the birds to the Colorado River area was undertaken, with 105 birds released at hack sites and cross-fostered to red-tailed hawks. The long-term results of this program are uncer-

tain, but at least a few of these birds have nested and produced young (Stewart and Walton, 1985).

The remainder of the Arizona population is also shrinking, and many of the known nest sites are in areas that are vulnerable to human activity. Urban sprawl has also had its effects near Tucson and Cave Creek, and other areas are similarly threatened (Whaley, 1986).

The Texas component of Harris' hawks is the largest of any state's, and at least in the mid-1970s the winter population there may have consisted of as many as 10,000 birds. However, rapid habitat alterations occurring at that time were believed to pose a serious threat to the birds (Porter and White, 1977). A recent (early 1980s) estimate of the total U.S. winter population was of 5,500 birds, or 38 percent below the estimated population of the early 1970s (Anonymous, 1986). The Harris' hawk was on the Audubon Society's Blue List of apparently declining species from 1972 to 1981 and was listed as a species of "special concern" in 1982 and of "local concern" in 1986.

Gray Hawk *Buteo nitidus* (Latham) 1790

Other Vernacular Names

Mexican goshawk, shining buzzard-hawk, Sonora gray hawk (*maximus*); gavilan gris (Spanish)

Distribution

Breeds in riparian woodlands and forest edges from extreme southern Texas (where very rare and limited to lower Rio Grande valley) and southeastern Arizona (mostly Santa Cruz valley and Nogales area) southward through northern Mexico (Sonora, Jalisco, Hidalgo, and Tamaulipas), Central and South America to northern Argentina. Formerly (and perhaps still very rarely) also a breeder in southwestern New Mexico (Gila and Mimbres rivers). Breeding populations in Arizona are migratory, and most recent records for New Mexico and Texas are for winter; more southerly populations are resident. (See Figure 38.)

North American Subspecies

B. n. plagiatus (Schlegel): Breeds from southern Arizona south across central Mexico (Nayarit, Jalisco, and Tamaulipas) to Honduras and Costa Rica. Local in extreme southern Texas. Northernmost populations are migratory. Includes *micrus* and *maximus* (Stresemann and Amadon, 1979).

Description (of *plagiatus*)

Adult male. Deep, rather dark ash-gray or slate-gray dorsally, becoming paler and narrow on the head, where the feathers have the shaft streaks black; wings with faint lighter bars; rump almost black; upper tail coverts immaculate pure white, tail deep black tipped with pale grayish brown; the subterminal black band bounded by a narrow white band, anterior to which is a second black band interrupted by a second narrower and less perfect white one; under tail coverts and greater upper tail coverts also white; primaries approaching black toward tips; the tips edged with dull white, as are also the secondaries; head uniformly delicate ash-gray, becoming white on chin and throat, and approaching the same on the forehead; black shafts on head and neck feathers; neck with faint pale transverse bars, these most distinct on chest; underparts regularly barred transversely with ash gray and pure white; under wing coverts white, with sparse, faint, zigzag bars on the axillaries and on larger coverts; under surface of primaries white anterior to their emargination, beyond which they are more silvery. Iris dark brown; cere, edge of gape, tarsi, and toes bright waxy yellow; bill and claws bluish black.

Adult female. Similar to the male, but the slate-gray upperparts darker, the barring of the wings hardly observable; front and throat scarcely whitish; rump almost pure black; second tail band much broken and restricted; ashy prevailing on the chest; ashy bars beneath broader.

Subadult. The juvenal plumage is evidently normally held for a full year, but some individuals may breed when a year old, and these birds may carry some adult gray feathers by that time (Dickey and van Rossem, 1938).

Juvenile (sexes alike). Mainly fuscous-brown above, with buff or russet intermixed, sides of head mostly whitish with contrasting fuscous streaking through the eye and malar areas, upper tail coverts buffy with brown spotting, underparts buffy to cinnamon-tinted, striped or spotted with fuscous; tail crossed with six to seven blackish bands, the subterminal one the broadest. Iris light brown to grayish brown; bill bluish black at base; cere, edge of gape, tarsi, and toes dull greenish yellow; eyelids and lores cobalt blue.

Measurements (of *plagiatus*, in millimeters)

Wing, males 232.5–252 (ave. of 12, 244.6), females 254–259 (ave. of 6, 256.7); tail, males 146–163 (ave. of 12, 155.9); females 161–167 (ave. of 6, 164.3) (Friedmann, 1950). Average egg size 50.8 x 41 (Bent, 1937).

Weights (in grams)

Males (of *plagiatus*), 364–434 (ave. of 3, 401), females 572–655 (ave. of 3, 621) (Brown and Amadon, 1968). Ave. of 5 males, 416; ave. of 4 females, 637 (Snyder and Wiley, 1976). Range of weights 378–660, ave. 524 (sex and sample size unspecified) (Clark and Wheeler, 1987). One male ca. 454, one female ca. 567 (Oberholser, 1974). Estimated egg weight 47.1, or 7.5% of female.

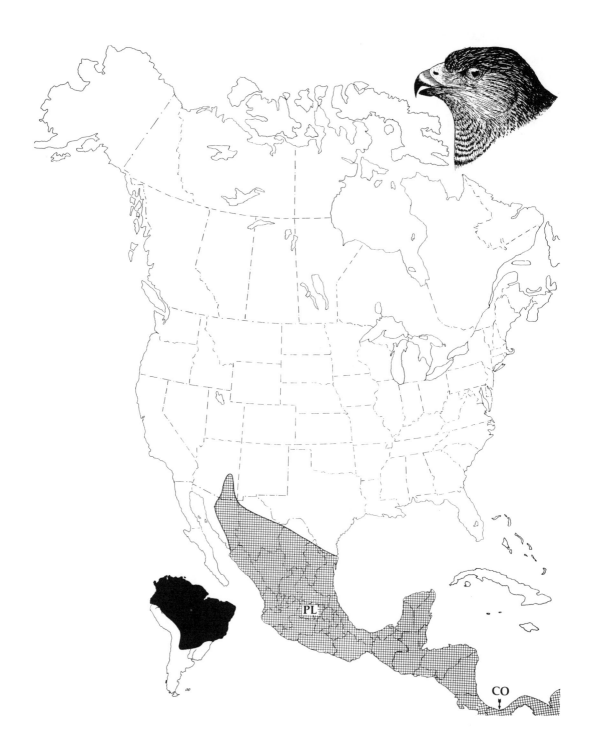

Figure 38. North and Central American residential range of the gray hawk, including races costaricensis (**CO**) and plagiatus (**PL**). The South American residential range is shown on the inset map. Indicated range limits of contiguous races should not be considered authoritative.

Identification

In the hand. This buteo may be identified by the combination of its rather small size (wing length under 260 mm) and a relatively long black tail (about two-thirds the length of the wing) that is strongly banded with white (one broad central band and one narrow anterior band in adults) or gray (numerous brownish gray bands in immatures), bounded anteriorly by a narrow white crescent formed by the upper tail coverts.

In the field. This small buteo is usually found in riparian woodlands, in isolated clumps of trees or along forest edges near water. Adults appear almost uniformly gray on the head and upperparts, with a conspicuous yellow cere and the underparts narrowly barred with gray and white. When perched, the wingtips reach about the midpoint of the tail, where a broad white band crosses. Immatures are dark brown above and streaked with brown and white below, but also have a bright yellow cere and show a dark eye-streak, bordered above by a contrasting buffy eyebrow line and below by a whitish cheek area that is bounded anteriorly by a dark malar "mustache."

In flight, both adults and young show nearly white to pale grayish underwing surfaces, including both the wing coverts and undersides of the flight feathers, and a tail that is either mostly black, crossed by two white bands of varying width (in adults), or shows multiple banding of black and gray (in immatures). In both age classes there is a narrow white crescent formed by the upper tail coverts at the base of the tail. The species's normal flight is rather accipiter-like, consisting of a series of rapid flaps followed by gliding, but it also does some typical buteo-like soaring on horizontally held wings. The usual call is a loud and clear double whistle, with the second syllable higher, which is loudly and frequently uttered upon disturbance.

Habitats and Ecology

Throughout its range, the gray hawk is associated with riverine forests, scattered groves, and extensive tracts of thin xerotic or deciduous tropical forests, but rarely extends into humid heavy forests (Brown and Amadon, 1968). In Panama this species occurs mostly in the tropical zone, typically remaining in the higher level of forest trees, but favoring rather open cover rather than dense forest vegetation (Wetmore, 1965). In the American southwest it occupies mature riparian woodlands and nearby semiarid scrub grasslands. In Texas it has been historically confined as a nesting species to cottonwood- and willow-lined areas of the lower Rio Grande (Oberholser, 1974), and a similar dependence upon riparian woodlands is probably typical of its Arizona range (Zimmerman, 1965; Phillips, Marshall, and Monson, 1978). In Arizona, where perhaps only 45 breeding pairs existed in the mid-1980s, the birds usually nest only where permanent running water is to be found (Clark and Wheeler, 1987). Most Arizona birds apparently favor areas where the southwestern desert shrub and plains grasslands form a transitional environment. There is no certain evidence that the species has ever bred in New Mexico (Hubbard, 1974a), although it probably rarely does so, and an occupied territory on the Mimbres River was found in the 1970s (Porter and White, 1977; Zimmerman, 1976a).

Foods and Foraging

Early summaries of this species's foods (Fisher, 1893; Bent, 1937) indicated that a rather general array of mammals, small to medium-sized birds, and reptiles, especially lizards, represent its basic foods. Brown and Amadon (1968) concluded that the gray hawk apparently prefers lizards and small snakes, and in a summary of data available to them Snyder and Wiley (1976) judged that lower vertebrates numerically comprise about 75 percent of the total diet, with birds adding about 20 percent and mammals the remaining 5 percent. This is the same general breakdown as provided by Stensrude's (1965) observations on an Arizona nesting pair, for which the Clark's spiny lizard *Sceloporus clarkii*, an arboreal lizard common in mesquite forests, was the favorite prey. Birds were secondary in importance, such as adult brown towhees (*Pipilo fuscus*) and unidentified fledglings, and rodents were sometimes captured. Unpublished data of Glinski (cited by Sherrod, 1978) indicated that reptiles comprised about 80 percent of the foods, including a variety of lizards (*Sceloporus, Cnemidophorus, Urosaurus,* and *Holbrookia*) as well as some snakes (*Masticophis* and *Thamnophis*). Bird prey totaled 11 percent, including adult and nestlings of Gambel's quail (*Callipepla gambelii*), doves (*Zenaida* spp.), and kingbirds (*Tyrannus* spp.). Mammals made up the remaining 10 percent, and included mice (*Peromyscus*), woodrats (*Neotoma*), and cottontail rabbits (*Sylvilagus*).

Hunting is done by two basic methods. Most commonly the birds perch within the foliage of a thickly foliaged tree, watching and listening for prey on the ground or in the trees. The capture involves a long, flat descent with accelerating speed toward the prey, often lizards that are plucked off the tree while the bird scarcely slows in flight. Only a single effort is made in capturing prey by this manner. A slow soaring flight at rather low altitude is the second mode of hunting, which apparently is successful in hunting for rodents. The preferred hunting habitat is apparently mesquite forest, in which scaly lizards are common, but cultivated fields are also searched (Stensrude, 1965).

Social Behavior

Stensrude (1965) observed a possible sky-dancing type of courtship flight in April. A bird was observed gliding at high altitude; it was then joined by a second individual, and both climbed to about 300–450 meters. They then dived rapidly downward, their flight paths crossing. One then climbed up again and, with tremendous speed and closed wings, plummeted back toward earth. A similar behavior was observed again in early May. He also observed copulation on several occasions, always during early morning hours. Typically the female would land on a horizontal and dead limb of a live cottonwood, followed a few moments later by the male. Almost immediately the male would mount her, but only for a few seconds. This apparently was not directly associated with courtship feeding. Little else is known of display behavior, although early observers reported aerial circling by pairs, with the male closely following the female and the birds uttering flutelike calls.

The most frequent call Stensrude (1965) heard was the "peacock call," which was a loud peacocklike scream that seemed possibly to serve to strengthen the pair bonds and perhaps also as a location call. Other loud, penetrating *creee* calls were used mainly by the female but to some extent also by the male in defense of nest and territory, and lasted for several seconds. Finally, low-pitched calls were sometimes heard of unknown function, in which the bird would utter short notes at about three-second intervals, lifting the head and bill straight up with each vocalization.

Breeding Biology

Very little has been written on the breeding biology of this species, but the birds build rather small and inconspicuous nests, usually in fairly large trees and often situated fairly close to the top of the canopy. Bent (1937) listed 48 nesting records from Arizona and Mexico extending from March 16 to July 2, with half of them between April 19 and May 31. A total of 21 Arizona clutches in the Western Foundation of Vertebrate Zoology are from February 2 to June 19, with 12 of these between May 10 and 24. A larger sample of 71 Mexican clutches extends from March 11 to July 3, with 37 of these occurring between April 11 and May 14. The heights of 67 of these nests were from 6 to 27 meters, with half of these between 10 and 12 meters. Of the associated trees that were identified, mesquites and oaks were represented with 11 nests each, cypress with 8 nests, 6 were in cottonwoods, and 6 were in other kinds of trees.

Clutch sizes in this species are quite small; Bent (1937) and Brown and Amadon (1968) indicated that most clutches contain only two eggs, but range from one to three. Of 93 clutches in the collection of the Western Foundation of Vertebrate Zoology 71 are of two eggs, and the average clutch size is 2.16 eggs. The incubation period has been reported as 32 days, and the fledging period as 42 days (R. Glinski's observations, cited by Newton, 1979).

Little information exists on productivity, but of 13 nests observed in Arizona, the average numbers of chicks fledged in 1973 and 1974 were 2.4 and 1.1 respectively. In 1975, 23 nests were monitored, and 16 of these fledged a total of 34 young (data of R. Glinski, cited by Porter and White, 1977).

Evolutionary Relationships and Status

Johnson and Peeters (1963) reviewed the biology and morphology of this species, concluding that it belongs within the genus *Buteo* rather than being separated (mainly on the basis of molt pattern) in a distinct genus *Asturina*. Brown and Amadon (1968) also grouped it in *Buteo*, but mentioned that it and *magnirostris* (the roadside hawk) are both aberrant in various respects and provide apparent evolutionary links with *Buteogallus* and *Leucopternis*. However, Stresemann and Amadon (1979) recognized *Asturina* as a monotypic genus situated between the two just-mentioned genera. Amadon (1982) likewise considered *Asturina* as monotypic (including *plagiata* in *nitida*), but Palmer (1988) has recently expanded the genus to include the roadside hawk and red-shouldered hawk. He also regarded the northern (north of Costa Rica) form *plagiata* as specifically distinct from *nitida*, and thus identified the U.S. population as belonging to *A. plagiata*.

This species is barely a part of the U.S. breeding hawk fauna at present. Zimmerman (1976a) offered evidence of the species's continuing rare occurrence and probable nesting in New Mexico (which was questioned earlier by Hubbard, 1974), and mentioned that a principal prey group, *Sceloporus* lizards, are fairly common in southwestern New Mexico. He believes that peripheral populations are bound to fluctuate in time, and that periodic breeding in the state is not out of the question. Zimmerman (1965) also

suggested that the future of the gray hawk (as well as the black-hawk) in the American southwest largely depended upon the preservation of the mature cottonwood woodlands along rivers, which are rapidly being eliminated. In the mid-1970s there were probably only 2–4 nesting pairs present in New Mexico, and 39 active nesting territories have been found in southeastern Arizona since 1973. Seven additional territories were abandoned between 1963 and that period. Most of these Arizona nests were in dense cottonwood groves along permanent streams, a habitat type that has mostly been destroyed as a result of human settlement (Porter and White, 1977). As of the mid-1980s, the U.S. nesting population was about 50 pairs (Palmer, 1988), making it (along with the aplomado falcon and hook-billed kite) one of the three rarest North American hawks, and yet one of the most neglected from a conservation standpoint. In 1988 there were 11 nests in the San Pedro River valley of Arizona, which produced 18 fledged young (*American Birds* 42:1325).

Figure 39. Breeding (hatching) and residential (cross-hatching) ranges of the red-shouldered hawk, including races alleni (**AL**), elegans (**EL**), extimus (**EX**), lineatus (**LI**), *and* texanus (**TE**).

Red-shouldered Hawk *Buteo lineatus* (Gmelin) 1758

Other Vernacular Names

Florida red-shouldered hawk (*alleni*), northern red-shouldered hawk (*lineatus*), red-bellied hawk (*elegans*), Texas red-shouldered hawk (*texanus*); buse à epaulettes (French); aguililla listada (Spanish)

Distribution

Breeds in woodlands near water from western Oregon south to extreme northern Baja California; and from eastern Kansas, western Iowa, central Minnesota, northern Wisconsin, northern Michigan, southern Ontario, southwestern Quebec, and southern New Brunswick south to central and southern Texas, eastern Mexico south to the Valley of Mexico, and east along the Gulf coast to Florida including the Keys.

Winters, at least sporadically, through the breeding range, but in eastern North America primarily from eastern Kansas, central Missouri, the Ohio valley, northwestern Pennsylvania, southern New York, and southern New England southward. (See Figure 39.)

North American Subspecies

B. l. lineatus (Gmelin): Breeds in the eastern U.S. and Canada south to Texas and Nuevo Leon.

B. l. alleni (Ridgway): Resident from Texas and Oklahoma east to South Carolina and Florida.

B. l. extimus (Bangs): Resident in extreme southern Florida and the Keys.

B. l. texanus (Bishop): Resident from southern Texas south to Veracruz and the Distrito Federal.

B. l. elegans Cassin: Resident in California, recently locally to western Oregon and northern Baja California, south at least in winter to Sinaloa.

Description (of *lineatus*)

Adult (sexes alike). Crown, nape, scapulars, interscapulars, back, and rump fuscous to fuscous-black, the feathers edged with pale cinnamon to tawny and to a lesser extent with whitish; upper wing coverts dull fuscous to brown, barred with grayish white; the lesser coverts at the bend of the wing broadly edged with cinnamon or orange-cinnamon; remiges dull fuscous becoming whitish on the inner webs, especially on the outer primaries, narrowly tipped with white and barred with grayish white, these bars becoming more grayish on the inner webs of the secondaries; upper tail coverts barred broadly dark brown and whitish, the lateral ones usually with more white, the median ones often washed with pale tawny; rectrices fuscous-black, crossed by four or five narrow grayish white bars and narrowly tipped with white; lores and cheeks grayish tinged with cinnamomeous; chin and middle of throat whitish to pale drab-gray, the feathers broadly tipped with brown; sides of throat like the nape and interscapulars; breast and upper abdomen usually ochraceous-buff, barred with buffy white (this area is paler in the southeastern races *extimus* and *texanus*, and much brighter cinnamon in the western forms *elegans* and *alleni*); lower abdomen, thighs, and under tail coverts pale buffy white, barred with light pinkish cinnamon to pinkish cinnamon on the thighs and upper abdomen, the bars disappearing posteriorly; axillaries and under wing coverts barred and marked with light buff and cinnamon. Iris brown; orbital ring, cere, and inside of mouth light yellow to dark yellowish green; bill bluish horn, becoming blacker at base; tarsi and toes dull cadmium-yellow to yellowish green, generally darker on the toes than the tarsi.

Subadult (sexes alike). The juvenal plumage is apparently worn for about 15 months without much change except for wear and fading, with the adult plumage apparently attained gradually during the second fall of life (Bent, 1938). Birds in juvenal plumage have been known to breed, sometimes successfully (Henny et al., 1973; Apanus, 1977).

Juvenile (sexes alike). Upperparts as in adult, but with the reddish and whitish feather edges much reduced, giving them a darker, more uniformly fuscous-black appearance; less chestnut on bend of wing; underparts buffy, heavily spotted with hazel to fuscous and the under wing coverts more spotted with fuscous; tail olive-brown to fuscous, with narrow and more numerous bars of dull gray, brown, tawny, and cinnamon. Iris variable, from whitish or pale yellow to sepia; cere pale lemon-yellow; bill bluish at base and black at tip; tarsi and toes variable, from dirty light greenish yellow to nearly white.

Measurements (of *lineatus*, in millimeters)

Wing, males 309–346 (ave. of 26, 320.8), females 315–353 (ave. of 22, 339.1); tail, males 192–228.3 (ave. of 26, 207); females 206.9–236.6 (ave. of 22, 219.9) (Friedmann, 1950). Average egg size 54.7 x 43.9 (Bent, 1937).

Weights (in grams)

Males, ave. of 10, 475; females, ave. of 14, 643 (Dunning, 1984). Males, ave. of 25, 555; females, ave. of 24, 701 (H. Snyder and J. Wiley, in Palmer, 1988). Range of weights 460–930, ave. 629 (sex and sample size unspecified) (Clark and Wheeler, 1987). Estimated egg weight 58.2, or 9.0% of female.

Identification

In the hand. This medium-sized buteo may be separated from other North American species by its rounded wings (wing 309–353 mm) that have the inner webs of the outer four primaries strongly notched, and have all the primaries and secondaries distinctly barred with white (the bars buffy and less distinct in immatures) on their outer webs.

In the field. When perched, the rusty brown anterior wing coverts of this woodland species are typically evident, as are the "spangled" larger and more posterior coverts and secondaries, which are heavily spotted and barred with white. The strongly banded black-and-white tail is usually also visible; the wingtips of perched birds usually reach about the midpoint of the tail. Immatures lack such conspicuous wing markings, but often at least show tinges of rusty in the coverts and some white spotting on the upper wing surface.

In flight, the upper wing surface shows a strong black-and-white pattern on the flight feathers, which combination is repeated in the strongly banded black-and-white tail patterning. From below, the wing lining appears rather reddish brown or cinnamon (especially in adults), and a crescent-shaped white (buffy in immatures) "window" is formed near the middle of the primary feathers where the black banding is reduced. The intensity of the reddish underparts varies considerably throughout the species's range, as noted above, and may be very pale in the extreme southeast. Flight is often accipiter-like, consisting of a series of rapid wingbeats followed by gliding. The typical and fairly distinctive call is a loud scream that drops strongly in pitch and volume. It is more down-slurred than the red-tail's, and unlike that species's call is usually rapidly repeated several times. However, the blue jay (*Cyanocitta cristata*) has a nearly identical and apparently mimicking call.

Habitats and Ecology

This species is associated with mixed coniferous-deciduous woodlands, rather moist hardwood forests, swamps, river bottomlands, and wooded marshy margins, and rather rarely with pure coniferous forests. A sample of 155 North American nesting habitats included 69 percent in decid-uous woodlands (including many riparian woodlands), 24 percent in mixed woodlands, and 5 percent in coniferous woodlands (Apfelbaum and Seelbach, 1983). Morris (1980; Morris et al. 1982) found that in Quebec the birds select the most mature stands of sugar maple (*Acer sac-charum*) or beech (*Fagus grandifolia*) that are avail-able for their nesting territories. The borders of lakes and streams or other wetlands seem to be especially favored habitat. Bednarz and Dinsmore (1981, 1982) found that wetland and forested habitats were used exclusively by breeding birds in Iowa, with 84 percent of observed usage occurring in forests, including 64 percent in flood-plain forests and 20 percent in upland forests. Marshes and wetland edges comprised the remaining 16 percent of usage. Large, con-tiguous tracts of forests are apparently necessary to support breeding birds, perhaps those ap-proaching 250 hectares in size, according to these authors. Such areas, interspersed with small marshes, may be required habitat in Iowa, which is at the western edge of the species's range in the Great Plains area. In California the race *elegans* also favors river-bottom forests at fairly low altitudes, while in Florida the resident race *alleni* occupies both flat pine woodlands and dense live-oak hammocks (Bent, 1937).

Craighead and Craighead (1956) estimated breeding home ranges in Michigan as averaging about 63 hectares, but with considerable variation (7.7 to 155 hectares for 42 ranges), and winter home ranges averaging about 339 hectares, also with substantial variation (127 to 503 hectares). McCrary (1981) estimated maximum breeding home ranges of males as 61.8 hectares, and of females as 36.8 hectares, with adjacent pairs overlapping in their ranges from 6 to 11 percent. In Maryland, 51 breeding territories averaged 194 hectares, with an average population density of a pair per 2.1 square kilometers (Stewart, 1949). Similar breeding densities for Florida were reported by Bent (1937). A considerably denser population of breeding birds (of about a pair per 50 hectares or 120 acres) was reported in three study areas of the same flood-plain forest area of Maryland by Henny et al. (1973).

Foods and Foraging

The foods of this species are highly varied, and include an array of small mammals, snakes, lizards, small birds including nestlings, frogs, toads, and also insects (Bent, 1937; Brown and Amadon, 1968). Snyder and Wiley (1976) estimated that invertebrates comprise about 55 percent of the food intake numerically, as compared with 21 percent from lower vertebrates, 20 percent from mammals, and only about 3 percent from birds. In an analysis based on the percentage of stomachs in which particular items were found, red-shouldered hawks were found to have mammals in 65 percent, reptiles and amphibians in 29 percent, arthropods in 21 percent, and birds in 7 percent (Johnson and Peeters, 1963). Although this is probably a somewhat more realistic appraisal than a simple numerical tally of food items, it still does not deal with differing weights of the individual prey types, which would further diminish the relative importance of invertebrates such as insects. Foods brought to nestlings in Michigan were nearly all vertebrates, including a wide diversity of small rodents, shrews, snakes, small birds, rabbits, frogs, and toads, plus other prey represented in diminishing quantities (Stewart, 1949).

Like other woodland hawks, this species employs various hunting methods, but probably mainly uses direct searching while in flight as a primary method, as well as some still-hunting from perches (Johnson and Peeters, 1963). Parker and Tannenbaum (1984), using radio-telemetry methods, found that the birds foraged in open habitats as well as in wooded areas; an estimated 22 percent of their time was spent in forest habitats that were within 25 meters of water. Their apparent foraging association with water, and the associated effects of fluctuating water on prey movements and densities, appeared to these authors to be primary factors affecting both habitat use and foraging behavior in that study area of Missouri.

Social Behavior

Although a common and widespread species, the social behavior and especially pair-forming behavior of the red-shouldered hawk is still only very poorly documented. Circling over the territory by one or both birds, followed by a series of dives accompanied by calls, seems to be a method of courtship display and/or territorial advertisement that is commonly used (Bent, 1937; Brown and Amadon, 1968). Bent has quoted an account by L. O. Shelley of a soaring male that had been circling up to about 300 meters, screaming periodically. Shortly after a female arrived the male began lunging downward with drops of several hundred feet at a time, then checking its fall and making a wide spiral before descending again. Finally, from about 60 meters it made a last dive that ended with him landing on the female's back, after which copulation followed. The female had spread her wings and crouched down in the last few seconds prior to contact. Afterward the male again took off, and was soon climbing beyond the range of the unaided eye.

In contrast to this behavior is an account of a male engaged in talon locking and downward whirling with another conspecific bird that intruded into the territory. This whirling descent apparently occurred following an aerial attack but near miss on the intruding bird. When they had nearly reached the ground the two birds separated, with one chasing the other. The female of the resident pair landed in a tree, where she was joined by her returning mate about five minutes later (Kilham, 1981).

Since this species is residential over most of its range, it seems likely that pair bonds might be maintained more or less continuously. Bent (1937) described an apparently strong territorial constancy in the species, based on his own observations covering 50 years in Massachusetts. In one case a territory was occupied more or less continuously over a period of 41 years, during which the nest was located 20 times. Another territory was occupied continuously for at least 42 years, and the nest located during 31 of those years, while a third territory was occupied for 45 years, during which time the nest was found 29 times, and a fourth was occupied 47 years, and the nest found 15 times. Most nests were used only a single year, but some were used two or even three years before being abandoned and a new one built or some older one refurbished. A high level of nest side fidelity (37 percent of 54 nests used the following year) was observed by Jacobs, Jacobs, and Erdman (1988). Bent believed that the birds mate for life, but that if one of the pair members is killed a new mate is promptly taken.

Breeding Biology

Nests are built in the spring over a fairly prolonged period of about four or five weeks; if an old nest is reused it is marked with a fresh sprig of coniferous vegetation, apparently to indicate "claimed possession" (Bent, 1937). About 90 percent of the 283 nesting sites summarized by Apfelbaum and Seelbach (1973) from throughout North America were in deciduous trees, with oaks (*Quercus*) and beeches (*Fagus*) the most commonly used genera. Bull (1974) and Peck and James (1983) also indicated a preponderant use of deciduous trees for nest supports. In southern Florida the cabbage palmetto is apparently a favored nest location (Bent, 1937). Woodrey (1986) determined that the major factors accounting for nest placement in a study of 15 Ohio nests were canopy height, distance to nearest permanent water, distance to nearest forest opening, and the percentage of total canopy cover. In short, the birds typically place their nests in mature riparian forests where they use large-diameter trees for nesting and have large open-habitat hunting grounds nearby. A study in Iowa (Bednarz and Dinsmore, 1982) similarly indicated that the birds use dense woodlands, typically flood-plain forests, with a large amount of edge and numerous small nearby hunting areas, and place their nests in trees with good canopy cover. Their nest sites differed from those of red-tailed hawks of the same region in nearly all of the measured environmental variables, which probably tends to reduce direct competition between them.

Eggs are laid at normal intervals of 2 or 3 days, and clutch sizes in northern areas consistently average about three eggs. Peck and James (1983) reported on 152 Ontario nests that had from 1 to 6 eggs, averaging 3.07, and Bull (1974) listed 156 New York nests as having 2–5 eggs, averaging 3.1. A sample of 29 recent nests from California averaged 2.69 eggs (Wiley, 1975)

although a larger sample (322) of pre-1961 nests averaged 3.08 eggs. Finally, the Florida races *alleni* and *extimus* most frequently have only two eggs (Bent, 1937). Birds in Maryland and Texas have clutch sizes similar to those in California, being intermediate between the northern and southern extremes (Stewart, 1949). The incubation period has frequently been estimated about 28 days (Bent, 1937), although currently it is believed to average 33 days per egg (Palmer, 1988). Incubation is mainly by the female, with the male participating regularly in at least some pairs. Fledging probably normally requires about 45 days, although with disturbance the birds may leave the nest at about 39 days after hatching.

Breeding success rates have been extensively analyzed by Henny et al. (1973), who considered data collected in Maryland over a period of nearly 30 years. Of 74 nesting attempts during that time, 68 percent were successful in fledging at least one young. The average number of birds fledged per breeding pair was 1.58, and the average number of young fledged per successful breeding pair was 2.34. This compares with 1.87 young fledged per nesting pair during two years in Michigan (Craighead and Craighead, 1956), 1.1 fledged per nest in 287 Wisconsin nests (Jacobs, Jacobs, and Erdman, 1988), and an average of only 1.34 young fledged per nesting attempt in a California study (but 1.82 in an area unaffected by human disturbance), where the nesting success was 65.6 percent (Wiley, 1975). Wiley believed that the rather low productivity he observed in California might be partly pesticide-related, although this factor was not believed by Henny et al. (1973) to have influenced reproductive performance in the Maryland population they studied.

Evolutionary Relationships and Status

According to Johnson and Peeters (1963), this species is part of an array of several New World woodland hawks, whose relationships are close but uncertain in detail. They believed that the nearest relative of *lineatus* is *ridgwayi* (Ridgway's hawk of Hispaniola), which is similar in juvenal and adult plumages as well as in some measurement ratios.

Although the data of Henny et al. (1973) do not document a reduction in the breeding population of red-shouldered hawks in central Maryland over a 29-year period from 1943 to 1971, other data do suggest that a general decline occurred after World War II. Brown (1971) analyzed nearly 10,000 Audubon Christmas Count reports for the years 1950–1969, and concluded that winter populations had declined in all states except California, where no significant change was evident, and in West Virginia, where increases may have occurred. Cohen (1970) judged that the species has declined nationally, probably as a result of habitat changes. Wiley (1975) believed that the California population might be experiencing reproductive problems associated with pesticide contamination and eggshell thinning, and suggested close monitoring of its status. Wilbur (1973) judged that the California population is maintaining itself. Adding credence to this position is the fact that the species has

recently expanded its range into southwestern Oregon (Rogers, 1979), and has since been reported across most of its coastal length (Henny and Cornely, 1985). Analysis of Audubon Christmas Bird Counts made in Florida between 1948 and 1982 indicate that that state's population of red-shouldered hawks declined rather sharply in the early 1950s and has shown no clear trend since, whereas the red-tailed hawk has gradually increased during that same time span (Kiltie, 1987). The red-shouldered hawk has been on its Blue List of apparently declining species continuously from 1972 through 1986. It was listed as endangered, threatened, or a species of special concern in 17 states as of 1988.

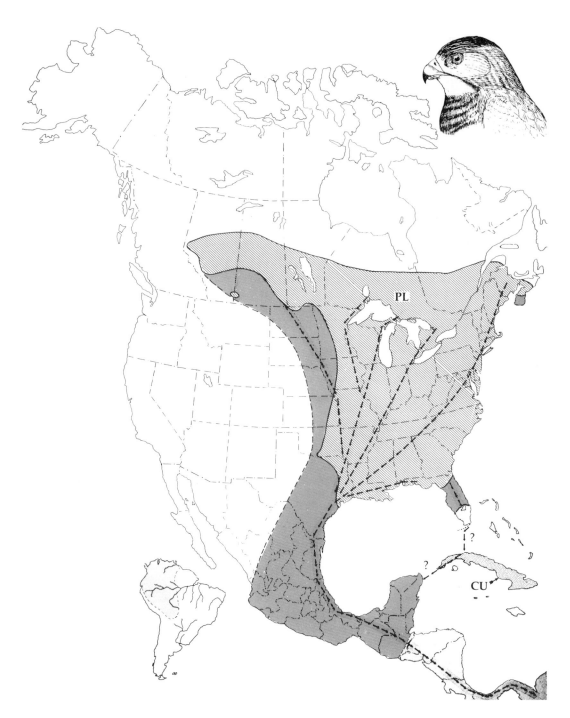

Figure 40. North and Central American breeding (wide hatching) and wintering (shading) ranges of the platypterus (**PL**) race of the broad-winged hawk. Also shown are approximate major fall migration routes (arrows and narrow hatching) of this race (based in part on Smith, 1980, and Palmer, 1988), and the residential range (stippled) of the race cubanensis (**CU**). The South American wintering range is shown on the inset map. Indicated range limits of contiguous races should not be considered authoritative, and Smith's suggested migratory route across the Gulf of Mexico is still unproven and considered doubtful by some (e.g., Palmer, 1988).

Broad-winged Hawk *Buteo platypterus* (Vieillot) 1812

Other Vernacular Names

Broadwing; petite buse (French); aguililla alas anchas (Spanish)

Distribution

Breeds from west-central Alberta east through central Saskatchewan, southern Manitoba, southern Ontario, southern Quebec, New Brunswick, and Nova Scotia south through northern North Dakota, Minnesota, Iowa, eastern Kansas, and eastern Oklahoma to eastern Texas, the Gulf coast, and northern Florida. Also resident in the Greater and Lesser Antilles.

Winters from Guatemala south through Central America to Peru and Brazil. Occasionally also winters in small numbers locally in southern Florida, sometimes north to northern Florida, on the lower Mississippi Delta, and in southern coastal Texas, these birds being nearly all juveniles. (See Figure 40.)

North American and Greater Antillean Subspecies

B. p. platypterus (Vieillot): Breeds in continental North America as indicated above, wintering south to South America.

B. p. brunnescens Danforth and Smith: Resident on Puerto Rico, where very rare.

B. p. cubanensis Burns: Resident on Cuba.

Description (of *platypterus*)

Adult (sexes alike). *Light morph:* A narrow forehead line (sometimes lacking) buffy white to buffy; crown, nape, scapulars, interscapulars, back, rump, and upper wing coverts varying from brown to fuscous-black, the feathers of the occiput and nape with fairly conspicuous white bases and edged with whitish to brown; the feathers of the back more uniformly dusky; upper tail coverts brown, tipped with white; secondaries smoky brown becoming white on the inner webs, the inner webs barred with five or six narrow blackish bands, these bars incomplete on the four or five outermost feathers; rectrices dull black, with an indistinct terminal band of dull brown, this fading terminally into whitish; across the middle of the tail a broad band of dull light umber to nearly dull white, about as far anterior

to the main band as this is from the tip is another much narrower and less distinct light band; lores whitish to dirty buff; cheeks brown to fuscous; a broad but inconspicuous fuscous to fuscous-black "mustache" running downward across the cheek; chin and throat white, sometimes tinged with buff; rest of underparts variably whitish, the breast feathers heavily streaked and broadly barred with dull to rufescent-brown; upper abdomen, sides, and flanks like the breast, but with the shaft stripes reduced; lower abdomen immaculate or only very slightly and lightly marked with brownish; under tail coverts usually pure white, thighs like the sides, but with the brown marks paler and narrower; under wing coverts whitish marked very sparsely with brown. Iris brownish yellow to light hazel or dark brown; bill slate to black, becoming lead-colored basally; cere greenish yellow to chrome-yellow; tarsi and toes light yellow; bill and claws bluish black.

Dark morph (very rare and known mainly from western Canada, especially Alberta): Like the light morph, but the feathers of the sides of neck and upper back edged with cinnamon or brown, and the bars below with a rufescent cast and sometimes confluent on the breast to the near exclusion of the white. Tail banding as in the light morph.

Subadult (sexes alike). The juvenal plumage is carried for nearly a year, with no change except for wear and fading. A nearly adult plumage is attained by the second fall of life (Bent, 1938).

Juvenile (sexes alike). Similar to adults but with pale feather edges and the dark marks on the breast variable longitudinal fuscous streaks, not rufous bars. Tail with five to six obscure dark bars, the subterminal one the widest. Iris pearl-gray to pale gray, bill black, becoming lead-colored basally, cere yellow to greenish yellow; eyelids olive-green, inside of mouth yellowish green; tarsi and toes ochre-yellow to greenish yellow.

Measurements (of *platypterus*, in millimeters)

Wing, males 244–277 (ave. of 17, 262.8), females 265–296 (ave. of 17, 282.8); tail, males 148–173.5 (ave. of 17, 159), females 155–185.4 (ave. of 17, 171.2) (Friedmann, 1950). Average egg size 48.9 x 39.3 (Bent, 1937).

Weights (in grams)

Males, 310–400 (ave. of 9, 357), females 389–460 (ave. of 10, 440) (Mosher and Matray, 1974). Males 325–384 (ave. of 40, 361), females 394–559 (ave. of 37, 443) (Palmer, 1988). Estimated egg weight 41.7, or 9.5% of female.

Identification

In the hand. This is one of the smallest North American buteos, and has relatively pointed but short wings (wing length 244–283 mm) relative to the tail (about 60 percent of the wing length). The combination of its small size, wings with only three outer primaries having notched inner webs, and the broadly banded tail (two white bands alternating with black ones in adults, about five light brown bands alternated with black in immatures) serve to identify the species.

In the field. Perched birds (which are usually found in woodland habitats) appear quite dark brown above (lacking the white upperpart spangling found in red-shouldered hawks) and lighter below, with dull brown underpart spotting (immatures) or narrow rufous barring (adults). The strongly banded black and white (adults) or light brown (immatures) tail, which is shorter than that of a red-shouldered hawk but has broader white banding, provides the best distinguishing marks of this very small (330–430 mm long) buteo. The wingtips reach about the midpoint of the tail in perched birds. Juvenile broad-winged and red-shouldered hawks are very difficult to distinguish when perched.

In flight, the very light underwing surface is usually conspicuous, including both the under wing coverts (but which are dark brown in the rare dark morph) and the flight feathers, which become nearly pure white at the base of the dark-tipped primaries, forming an indistinct pale "window" in most birds when seen overhead. Flying birds typically exhibit one complete black tail band (the second, more anterior one being broken by the rump and tail coverts) if adults, or several less distinct pale bands if immatures. The birds glide and soar on flattened wings, and during migration often gather in rather large flocks ("kettles"). The usual alarm call is a "melancholy" two-syllable whistle, somewhat like the "pee-wee" call of an eastern wood-pewee (*Contopus virens*), with the first note slightly higher in pitch, and both notes unusually high-pitched for a hawk. As typical of many hawks, the male's call is noticeably higher than that of the female.

Habitats and Ecology

This is one of the most common and typical woodland hawks of eastern North America, one that is associated with dry to wet deciduous, mixed, or occasionally coniferous forests. Peck and James (1983) noted that of 65 nests in Ontario, 31 were in mixed woods, 27 in deciduous woods, and 1 was in a pure coniferous stand. The birds often nest in fairly continuous rather than extensively broken woodlands, although edges (such as road openings or wetland edges) are important for hunting. The usual presence of both woodland openings and wet areas of habitat was indicated by both Keran's (1978) and Rosenfield's (1984) studies. Keran found that woodland trails were found near most of 29 nests in Minnesota and Wisconsin (average distance 22 meters), while paved or gravel roads or woodland fields were variably near the rest. The average distance to an opening of one of these types from the nest was 42 meters. He also found that swamps, marshes, pools, or lakes were on average 67 meters from the nest, and never more than 143 meters away. All of these had standing water and frogs present. In that area an oak-aspen sere was the preferred forest habitat for nesting, while in a New York study (Matray, 1974) the birds seemed to select areas of yellow birch (*Betula alleghaniensis*), a tree that is associated with moist sites such as near streams, lakes, and swamps. In Maryland, broad-winged hawks similarly selected nesting areas that were closer to water and to forest openings than predicted by random forest sampling, but otherwise showed little predictability in their habitat selection.

Breeding densities were estimated by Keran (1978) to average a pair per 5.2 square kilometers in Minnesota. This area centers on a breeding, nesting, and feeding territory that is defended until the young are fledged and the nest abandoned. Rosenfield (1984) estimated a density of a pair per 2.4 square kilometers in Wisconsin. Burns's (1911) data indicated a Massachusetts density of a pair per 5.2 square kilometers, while at the extreme northwestern margin of the species's range in Alberta the maximum density was only about a pair per 23.3 square kilometers. Actual home ranges and territory measures are not yet available for this species.

Foods and Foraging

The early summaries of broadwing foods by Fisher (1893) and Burns (1911) tended to be somewhat misleading, as the numerical incidence of small arthropod food biased the overall results in that direction. Thus, Fisher indicated that invertebrates comprised 78 percent of the total foods, while Burns (1911) calculated this component as 55 percent. Estimates of the mammalian component for both surveys were about 12 percent, and the reptilian and amphibian components were estimated to total from 8 (Fisher) to 24.5 (Burns) percent. The numerical importance of birds was minor in both studies.

These problems were resolved in more recent studies, in which relative biomass contributions were estimated. Thus, Rusch and Doerr (1972) estimated that nestlings were fed on mammals for about 56 percent of their food by weight, with hares and voles the most numerous, and about 38 percent on birds, mainly ruffed grouse (*Bonasa umbellus*). Amphibians and insects made up a minor part of the biomass total.

Similarly, Fitch (1974) estimated prey biomass from three nests, of which various birds contributed 27 percent, lizards about 16 percent, cottontails 14 percent, voles 12 percent, and snakes about 12 percent. The average prey weight was only 19 grams, compared with 106 grams for prey taken in the same area by redtailed hawks.

In a Wisconsin study, Rosenfield and Gratson (1984) estimated that mammals comprised 62.5 percent of the prey biomass brought to nestlings, with eastern chipmunks (*Tamias striatus*) the most important single prey, followed by shrews and voles among the mammals. Birds comprised an estimated 21.5 percent, with 25 of 30 such prey being nestlings or fledglings. Amphibians made up about 13 percent of the biomass; the broadwing's use of this food source as nestling food may help explain the tendency of the birds to nest near water, according to these authors. Thus, Mosher and Matray (1974) reported that amphibians comprised 29 percent of the estimated biomass of foods brought to four nests in New York, while mammals made up about 42 percent (mostly consisting of chipmunks, red squirrels, and mice, shrews, or voles), reptiles about 23 percent, and birds an estimated 6 percent. In all of these studies insects and other invertebrates were absent or insignificant in relative importance.

On their South American wintering areas the birds evidently do feed largely on insects and other arthropods, as well as perhaps some lizards and frogs (Burns, 1911). Apparently individual foraging territories are maintained on wintering areas, although this is still only poorly documented.

Johnson and Peeters (1963) judged that nearly all vertebrate groups as well as arthropods are used as prey, based on the then-available literature. Mosher and Palmer (in Palmer, 1988) estimated that at least 30 species of mammals, 25 species of birds, 11 species of snakes, 4 lizards, 5 amphibians, 1 turtle, and a fish have all been

reported as prey. Still-hunting from perches is evidently the most commonly employed hunting technique, with direct searching in flight being of secondary significance and soaring or hovering followed by stooping on prey of tertiary importance (Johnson and Peeters, 1963).

Social Behavior

Apparently courtship in this species is much like that of the other *Buteo* hawks, involving territorial advertisement by soaring and swooping behavior, often by the paired birds simultaneously, together with the usual loud whistling typical of the species. Mueller (1970) described the copulation attempts of a tame, hand-reared male broadwing toward him. Prior to mounting his arm, the bird would assume a nearly horizontal posture, raising the back feathers, flattening and spreading the breast feathers, and lowering and spreading the tail. In this posture the bird would pace about calling with a short and emphatic *whee-oh* note that was quite different from the species's usual calls. When trying to copulate the toes were completely balled, so that the talons were not exposed and thus could not be used for gripping. Bent (1937) quoted a description of a normal copulation that occurred in a tree near the pair's nest, during which a similar two-toned courtship call was uttered by the male. Bent believed that pair bonding in this species probably lasts for life, although there does not seem to be any real information to support this attractive idea. Matray (1974) did report one case of a banded pair maintaining a pair bond over two successive breeding seasons. Certainly maintenance of pair bonds during the long period of migration is unlikely, and it seems probable that any real permanence of pair bonding is likely to be the result of both sexes tending to return to their prior year's territory. However, the birds rarely use the same nest for two consecutive years, and never for more than three or four, according to Bent.

Breeding Biology

Nests are typically located in a crotch next to the trunk of the tree, and a very wide variety of deciduous trees seem to provide suitable nest sites (Burns, 1911; Bent, 1937). Of 113 Ontario nests, 83 were in deciduous trees, and 11 in coniferous ones, with birches, poplars, and maples most commonly used (Peck and James, 1983). These nests were typically placed in crotches and primary forks, usually near the trunk and rarely near the crown of the tree. Forty of 86 nests were located from 7.5 to 12 meters above ground. Two of the nests were reused in successive breeding seasons.

Nest building is normally done over a fairly extended period of three to five weeks after the birds return to their breeding areas, and only infrequently is the previous nest used again. However, old raptor, crow, or even squirrel nests are sometimes renovated and used (Bent, 1937). Matray (1974) noted that at least two of 10 nests he studied had been renovated rather than newly built. The delivery of sprigs to both old and new nests is common behavior, and Lyons, Titus, and Mosher (1986) suggested that in addition to possibly serving as an ectoparasite repellent, the addition of greenery to the nest may lengthen the period that a nest remains usable. Other possible functions attributed to the addition of greenery include its use in shading or hiding nestlings, territorial advertisement, and nest sanitation functions.

Clutch sizes are fairly consistent in this species, with 2–4 eggs typical. Peck and James (1983) reported that 64 Ontario nests had 1–4 eggs, with 56 of them having 2 or 3 eggs, and the average clutch size was 2.71 eggs. A smaller sample of 24 nests from New York had an average of 2.67 eggs (Bull, 1974).

There is a clear division of labor during incubation, according to Matray (1974), with the female incubating and the male hunting, the male only covering the eggs during times that the female is eating food brought by the male. Matray also observed that males never brooded the young and that their visits were always of short duration. He also determined an incubation period of not less than 28 days, in contrast to frequently quoted earlier literature suggesting a 21–25 day incubation period. Probably 31 days is closer to the normal period (Mosher, in Palmer, 1988).

The young are brooded fairly intensively for the first few weeks after hatching, and through the night until they are about 21–24 days old. They are fed by the female until they are about 29–30 days old, and at that time begin venturing out of the nest. They begin to produce an adultlike whistle at 30–36 days of age, and become able fliers during their sixth week after hatching (Matray, 1974). Fitch (1974) reported a similar fledging period. Fledglings continue to receive food from their parents until about 50-56 days of age, and may remain a few hundred meters from the nest for a short time longer. They evidently do not return to their immediate natal area as yearlings (Matray, 1974). Fitch (1974) and Lyons and Mosher (1987) have also provided useful information on the behavioral and physical development of young broad-winged hawks.

Productivity estimates for this species are not numerous, but 36 nests in the Appalachians studied by Janik and Mosher (1982) had an 86 percent nesting success rate and an average of 1.7 young fledged per nest. In a larger sample of 70 nests from Wisconsin (Rosenfield, 1984) the average clutch size was 2.4, the average initial brood size was 1.8, and the average number of young fledged per nest was 1.5, with 79 percent of the nests studied fledging at least one young. Mammalian and avian predation appeared to be the major cause of egg and young mortality, with raccoons (*Procyon lotor*) and American crows (*Corvus brachyrhynchus*) suspected predators.

Evolutionary Relationships and Status

This is seemingly a species with no obviously close relatives in the genus *Buteo*, but Johnson and Peeters (1963) suggested that it be placed in linear sequence between *lineatus* and *nitidus,* to indicate their probable relative relationships. Palmer and Mosher (in Palmer, 1988) said that "no useful information" is available on the species's evolutionary affinities.

This is easily one of the most abundant of the North American hawks and almost certainly the most abundant *Buteo,* judging from the numbers seen during migration counts. Thus, Thiollay (1980) reported that this species alone comprised 77 percent of more than 262,000 hawks observed during a spring migration period of 23 days in eastern Mexico, and indeed peak numbers had probably occurred prior to the start of the study. Earlier counts noted by him suggest that up to 100,000 birds may pass through in a single day and that the total North American population may amount to more than a million birds. There is no good evidence of significant population changes in this species during recent years; it seems to have avoided major population losses during the era of hard pesticides.

Figure 41. North and Central American breeding (hatching) and residential (cross-hatching) ranges of the short-tailed hawk. The South American breeding or residential range is shown on the inset map.

Short-tailed Hawk *Buteo brachyurus* Vieillot 1816

Other Vernacular Names

Little black hawk, short-tailed buzzard; aguililla cola corta (Spanish)

Distribution

Resident locally in peninsular Florida, and from Sinaloa and Tamaulipas south through Central America and South America to Chile (the Andean population *albigula* sometimes being considered a separate species from the lowland nominate form *brachyurus*). (See Figure 41.)

North American Subspecies

B. b. fuliginosus Sclater: Breeds in peninsular Florida (most birds retreating south of Lake Okeechobee in winter, and to the Keys), and from northern Mexico south to Panama.

Description

Adult (sexes alike). *Light morph:* Upper surface nearly uniformly fuscous to blackish brown, darkest and most uniform on the head, which, with the exception of the anterior half of the lores, the anterior malar region, chin, and throat, is solid fuscous-black; the scapulars and wings dull grayish brown with the feathers darker centrally; sides of the rump strongly tinged with rufous; tail grayish brown, very narrowly tipped with dull white, crossed near the end by an indistinct band of dusky, and the basal portion crossed by up to five narrower and often broken dark bands; lateral upper tail coverts lighter brownish gray, with broad but rather indistinct dusky bars or spots; a white area on each side of the base of the bill, covering the anterior half of the loral and malar regions; chin, throat, lower throat, breast, and remaining lower parts all pure white, the thighs washed with pale ochraceous or light buff; lining of the wing and axillaries pure white, the under primary coverts, however, with a large comma-shaped patch of dusky near their tips. Iris brown; bill black, becoming bluish basally; cere, tarsi, and toes bright yellow.

Dark morph: Uniform black or dusky, varying from dark sooty brown to almost a coal-black, freshly molted specimens usually having a chalky or glaucous cast on the back, and a more or less distinct purplish reflection to the general plumage; forehead usually more or less distinctly white, but this sometimes wholly absent; occipital feathers pure white beneath the surface; primaries plain black above; tail brownish gray or grayish brown, crossed by about six or seven narrow black bands, of which the subterminal is much the broadest, the grayish bands becoming gradually narrower toward the base of the tail; under surface of the tail appearing silvery white, with slate cross-bands. Under surface of the primaries chiefly white anterior to their emargination, but this broken by irregular bars, or confused mottlings of grayish; rest of under surface of wing uniformly dark brown or black, the under primary coverts sometimes spotted or barred with white.

Subadult: Not yet well described, but apparently the juvenal plumage is carried much of the first year, with the buffy underparts of light-morph birds gradually disappearing. The pale edges of the dark upperpart feathers are also lost, but the juvenal tail is retained (Bent, 1937). The age of sexual maturity is still unknown, but probably occurs within two years.

Juvenile (sexes alike). *Light morph:* Upperparts as in adult, but the feathers edged with white, buff, or rufous; dark tail bands more numerous (seven or eight, of which three are usually exposed), more nearly equal in width, and more distinct than in adults; whitish to tawny below, sometimes streaked or spotted with fuscous-brown. A few fine dark streaks present on the breast, and also a buffy to rufous wash in this area among younger birds. Iris beige; bill dusky; cere, tarsi, and toes pale yellow.

Dark morph: Uniformly dark brown on the breast, becoming more mottled with white posteriorly, the white mottling also extending to the dark under wing coverts and present on the under wing coverts. The tail has more dark bands than in the light morph (nine or ten, of which five are usually exposed).

Measurements (in millimeters)

Wing, males 265–310 (ave. of 8, 284.7), females 300–335 (ave. of 7, 322.2); tail, males 133–166 (ave. of 8, 145.1), females 144–199 (ave. of 7, 166.8) (Blake, 1977). Average egg size 53.4 x 42.8 (Bent, 1937).

Weights (in grams)

Males (of *brachyurus*), 1 immature 450, 1 adult 470, 1 female 425 (Brown and Amadon, 1968); 1 female 530 (Haverschmidt, 1968). Range of weights 342–560, ave. 426 (sex and sample size unspecified) (Clark and Wheeler, 1987). Estimated egg weight 60.6, or 11.4% of female.

Identification

In the hand. This very small buteo is of comparable size to the broad-winged hawk, but differs from it in that there are at least five indistinct light bands (three exposed) crossing the tail in front of the broad subterminal dark band (in adults), or up to about ten (five exposed) more definite bands crossing the tail (in immatures). A white area on the forehead and anterior lores is typical of all age classes, although some dark-morph birds may lack this feature. The two plumage morphs do not intergrade.

In the field. When perched, this small (380–430 mm) buteo in the more common (at least in Florida) dark morph appears sooty brown both above and below, with white evident only on the forehead and the tail. The contrasting bright yellow legs and cere are also conspicuous. Light-morph adults are very different, having pure white underparts and uniformly dark brown upperparts, and also having a distinctively set off white forehead area on an otherwise blackish or "hooded" head. Light-morph immatures are very similar to those of young broad-winged hawks, and the two may not be separable in the field where they occur together. Dark-morph immatures are solid blackish brown on the breast and are mottled with white more posteriorly. In perched birds (which are rarely seen because they perch in inconspicuous sites) the wingtips reach or nearly reach the end of the relatively short tail.

In flight, dark-morph adults show a nearly black underwing lining and body underparts, contrasting with their bright yellow feet and cere coloration, and with paler whitish gray "windows" near the bases of the dark-tipped primaries. Immature dark-morph birds have more heavily spotted underwing linings. Light-morph adults have nearly immaculate white underwing linings with darker wrist spotting and primary barring, and the bases of the primaries are also somewhat paler than those of the secondaries. Light-morph immatures sometimes have some brown spotting or indefinite streaking on their breast and abdomen. The birds often soar with horizontally held wings and upswept wingtips, sometimes flying at considerable height and occasionally diving falconlike toward prey. Their usual alarm call is a high-pitched nasal squeal that drops in pitch and volume. It is similar to that of the red-tailed hawk, but generally higher in pitch than other buteos except the broad-winged hawk. The male's alarm call also somewhat resembles the scream of the blue jay (*Cyanocitta cristata*).

Habitats and Ecology

In Florida this species nests and roosts in stands of mature cypress (*Taxodium*), riparian hardwoods, mangroves, or pines, especially in mixed hardwood-savanna edge habitats, as where such woodlands are adjacent to broad areas of open prairie or marshes. The most important nesting areas are where mature woodlands border lakes, creeks, or rivers situated in prairies, pastures, or marshlands. They have also been reported to favor swamp woodlands for nesting, especially cypress or mangrove swamplands. In one summary of 39 habitat use records, the birds occurred in mangrove areas 11 times and in cypress swamps 13 times (Moore, Stimson, and Robertson, 1953). The Fisheating Creek region of Glades County apparently is ideal nesting habitat, with the river bordered by stands and patches of cypress, pines, swamp hardwood, and various evergreen oaks. The adjoining areas are pastures and prairies dominated by saw palmettos (*Serenoa repens*) and native grasses and herbs. The birds hunt over open country areas or the interface zones between woodlands and open areas, but not over clean pastures or clean-cut woodland edges. Woodland cover at sites used by the birds during spring and summer varies from about 20 to 60 percent, with the rest generally open plant communities (Ogden, 1974; Kale, 1978).

In Mexico the birds often occupy pine-oak ridges at moderate elevation, and elsewhere in Central America and Trinidad they seem to prefer woodland edge habitats or open-country areas, but also hunt over low thorn scrub, agricultural fields, and even over areas of extensive forests. There they may be able to exploit the relatively greater numbers of birds associated with canopy and forest edges in these tropical habitats (Ogden, 1974).

Foods and Foraging

Although the information on the foods of this species is rather limited, they are known to include a diversity of rodents, birds, lizards, snakes, frogs, and insects (Bent, 1937; Brown and Amadon, 1968). The observations of Ogden (1974) provided the first information suggesting that the species is a specialized bird hunter. He identified 66 birds as prey items in seven nests, and saw 29 birds and 3 rodents taken during periods of hunting. Two species, the eastern meadowlark (*Sturnella magna*) and the red-winged blackbird (*Agelaius phoeniceus*), made up over half of the identified bird prey. Other birds that occurred more than once among either captured or intended prey included northern bobwhite (*Colinus virginianus*), yellow-billed cuckoo (*Coccyzus americanus*), tree swallow (*Iridoprocne bicolor*), northern cardinal (*Cardinalis cardinalis*), rufous-sided towhee (*Pipilo erythrophthalmus*), seaside sparrow (*Ammospiza maritima*), and American goldfinch (*Carduelis tristis*).

The hawks spend most of their time attempting to capture adult birds, and no nestlings or fledglings were found among prey remains. They search almost exclusively from the air, hunting over low terrain at heights of from 50 to 350 meters, but usually between 75 and 250 meters. Three basic hunting techniques are used. The first is aerial "still-hunting" by hanging on updrafts, as are common along woodland edges, gradually working along the length of the edge. The same technique is used when thermals are available or on windy days. Secondly, the birds use very slow soaring while working into the wind, then sometimes go back with the wind to re-search ground just covered, or off to one side, then turn again and sail still farther back. Soaring in tight circles is also used, but mainly to maintain or gain altitude while hunting. Occasionally other methods are used, such as "direct searching" while flying in straight routes above the treetops of wooded areas, or perching on the very topmost branches of trees, the birds possibly scanning for prey while also preening or sunning themselves.

Aerial hunting normally begins a few hours after sunrise, when thermals are developing, and the birds stop hunting late in the afternoon as cooling begins. They typically catch their prey by making steep stooping dives, descending at angles of between 45 and 90 degrees, with the primaries folded back falconlike. Or the birds may initially make a parachutelike descent with the wings spread and the body held horizontally, dropping thus up to 50 meters before folding back the wings for a final rapid attack. When about 15 meters or less from its prey, the hawk extends its talons and spreads its wings to slow its descent. It makes no effort at a second strike if the prey is initially missed, but instead regains height by spiraling upward. Of 160 stoops, 87 percent were toward prey in trees or shrubs, 9 percent toward prey in open fields, and 4 percent toward low-flying birds. Of 107 stoops, only 60 percent were carried through to striking distance, and only 11 percent resulted in actual prey captures.

Maximum diameters of three hunting ranges were estimated by Ogden at 1.9–2.6 kilometers (averaging 2.3 kilometers), or more than twice as large as the hunting range diameters he estimated in the same general region for red-shouldered hawks. The commonest open-country bird species in these areas (eastern meadowlark and red-winged blackbird) made up 33 percent of the available avian prey, but 68 percent of the prey collected at nest sites one year, whereas three woodland edge birds that made up 32 percent of the available prey comprised less than 13 percent of the prey identified at the nest. Ogden believed that this concentration on open-country birds reflected the relative ecological efficiency of the hawk's hunting methods and the relatively greater conspicuousness of such prey in open country.

Social Behavior

Almost nothing is known of the social life of this species. As it is essentially sedentary it is likely to maintain permanent pair bonds, and pairs have been known to maintain the same breeding territory for up to three years. Breeding pairs have been known to consist of light-morph females and dark-morph males in several cases, but there are also numerous examples of birds mated to like-morph birds. Ogden (1974) mentioned a dark-morph male mated to a light-morph female in 1966, but the same male mated to a dark-morph female in 1970 and 1971. At least in Florida, pairings between morphs are considerably more common than those involving birds of the same morph, implying that assortative mating tendencies may be lacking. Of 12 specimens identified as males, all were dark-morphs, while of 11 females 8 were light-morphs (J. Ogden, in Palmer, 1988), suggesting that a possible sexual bias may exist in morph color inheritance. It would be useful to have some more detailed data on possible assortative mating tendencies as well as on the inheritance of color morph by sex in this species. In Florida the dark morph is probably several times more common than the light morph, but in the Andean portion of the species's range the dark morph is unknown, and this taxon (*albigula*) is sometimes considered a distinct species.

Social displays of the species include a shallow roller coaster–like form of sky dancing by the male, which at times is interspersed with swift, downward dives of undulating spirals as the wings are alternately spread and closed. Such displays may be terminated with a steep head-first dive ("tumbling"). These flights often cover large areas, with undulations continuous for up to 350 meters, and the male may also carry nesting materials or prey, ending his flight with a dive to a perch near the female. Copulation has been observed to occur almost immediately after a male landed beside or even on the back of a perched female, with the male calling during copulation (J. Ogden, in Palmer, 1988).

Breeding Biology

Only a relatively few nests of this species have been described, nearly all of which are from Florida (Ogden, 1974). Most nests are placed in fairly tall trees, sometimes as high as about 30 meters from the ground, and the nests tend to be relatively large. They are often located in forks of large branches, or on larger branches high in the canopies of such trees. They have been reported in cypresses, magnolias, gums (*Nyssa*), mangroves, and cabbage palms (Brown and Amadon, 1968). Of 12 nest records in the egg collection of the Western Foundation of Vertebrate Zoology, 7 were in cypresses, 4 in oaks, and 1 in a pine. Fifteen nests were at heights of 7.5 to 19.6 meters, with 9 located between 9 and 12 meters. Seven Florida nesting records were from March 2 to May 1, with a median date of March 30. Probably most eggs in Florida are laid between mid-March and mid-April. Nine Mexican egg

records are from February 12 to May 10, with a median date of April 7. Bent (1937) listed 14 records from both areas between February 12 and June 10, with half between March 15 and May 1.

Many pairs build new nests each year, although sometimes a nest may be used for two or more consecutive years. Nest building is done mostly or entirely by the female, as the male brings in sticks carried in his talons or beak. After the nest is completed and until after egg laying begins, both birds, but mainly the female, line the nest with green materials (J. Ogden, in Palmer, 1988).

Clutch sizes in this species are rather small; Bent (1937) indicated that the most typical number of eggs is two, with a range of one to three. Twelve clutches in the collection of the Western Foundation of Vertebrate Zoology range from 1 to 3, with 9 having 2 eggs and the average being 1.9. The incubation period at two Florida nests was about 34 days, with apparently all of the incubation being done by the female. While the young are downy, apparently all the hunting is done by the male, but later both sexes participate in hunting. The age of fledging is still unknown, and there is also almost no information on productivity or other details of the nesting biology. Of six Florida nests observed during the nestling period, five had one young and only one contained two, suggesting a rather low productivity rate (J. Ogden, in Palmer, 1988).

Evolutionary Relationships and Status

The evolutionary affinities of this species are not readily apparent, although the rufous-thighed or white-rumped hawk (*B. leucorrhous*) of the woodlands of South America is distinctly similar in having light and dark morphs that closely resemble the short-tailed hawk in adult body plumage. The rufous-thighed hawk mainly differs in having broader tail banding and rufous thighs, the latter a condition also being found in the taxonomically uncertain form *albigula* that is usually included in the present species.

The population status of this species is also somewhat uncertain. Kale (1978) estimated that in the 1970s the Florida population might number only a few hundred birds that were widely and sparsely distributed across peninsular Florida, but were more concentrated in the southern part of the state during winter. According to Kale, one important nesting site (Fisheating Creek, Glades

County) needed protection from development, as did habitats associated with various other rivers such as Crystal, Hillsborough, Kissimmee, St. Johns, and Suwannee, as well as lake habitats in Brevard, Glades, Lake, Okeechobee, Orange, Osceola, Seminole, Volusia, and other central Florida counties. Recently the Florida population has been estimated as possibly less than 500 birds (J. Ogden, in Palmer, 1988). If this is the case, it certainly warrants listing as an endangered species. Too few appeared on the 1986 Audubon Christmas Bird Count to estimate the total population.

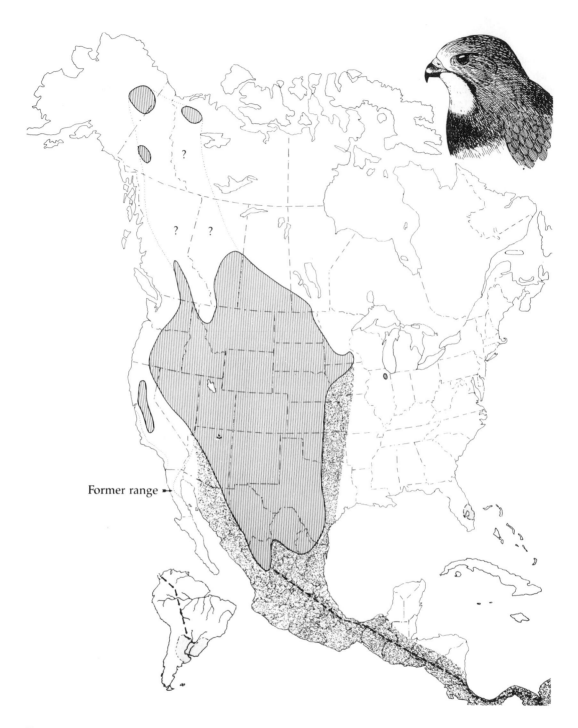

Former range ▸

Figure 42. North American breeding (hatching) and minor wintering (shading) ranges of the Swainson's hawk. Dotted lines indicate probable former breeding range limits in California and northern Baja, and circumscribe possible areas of local breeding in northwestern Canada. The general migratory corridor between North and South America is shown by stippling, and arrows indicate probable primary Central American migration route (Smith, 1980). The source of the mostly immature birds wintering in Florida is unknown. The wintering area in South America is shown on the inset map.

Swainson's Hawk *Buteo swainsoni* Bonaparte 1838

Other Vernacular Names

Black hawk, brown hawk, grasshopper hawk; buse de Swainson (French); aguilila de Swainson (Spanish)

Distribution

Breeds locally in east-central Alaska, Yukon, and Mackenzie, and from central British Columbia, central Alberta, central Saskatchewan, southern Manitoba, and western and southern Minnesota south through the eastern parts of Washington and Oregon locally to California's central valley (formerly to Baja California), Arizona, Sonora, Chihuahua, Durango, and central and southern Texas. The usual eastern breeding limits in the Great Plains are in Kansas and Nebraska, but breeding occurs periodically in Iowa and has rarely occurred in northwestern Missouri. Also breeds or has bred locally in northern Illinois and southwestern Wisconsin.

Winters primarily on the pampas of southern South America, casually north to the southwestern United States, with a few (mainly immatures) regularly wintering in southeastern Florida, some also present on the Texas coast, and a scattering elsewhere (Browning, 1974). (See Figure 42.)

North American Subspecies

None recognized.

Description

Adult (sexes nearly alike). *Light morph:* Forehead whitish or pale buff, crown, occiput, nape, auriculars, sides of throat, and most upperparts more or less uniform dark grayish brown to grayish black (often darker in females than males), the feathers edged with grayish tawny-olive to cinnamon; a concealed white patch on the nape; secondaries fuscous to grayish black, becoming somewhat paler basally on their inner webs where obscurely banded with grayish black; primaries fuscous-black to grayish black, also becoming slightly lighter basally on their inner webs where banded with grayish black; rectrices whitish basally, otherwise light gray to gray more or less tinged with brownish, especially on the median pair, tipped with grayish or buffy white,

subterminally obscurely and broadly banded with grayish black, and crossed with nine or ten blackish bars; lores, chin, and middle of throat white; a very broad area covering the entire breast pinkish cinnamon to light russet (often dull buffy brown in females), the feathers with narrow dusky shaft streaks; rest of underparts white or buffy white, the feathers of the sides of the upper abdomen incompletely barred with brown or tawny-olive, sometimes almost immaculate; the under wing coverts with a few small fuscous bars, otherwise distinctly lighter in tone than the undersides of the remiges. Iris dark brown; bill blackish; cere and corner of mouth pale greenish yellow; naked eyebrow olive-yellow; tarsi and toes wax-yellow.

Dark morph (sexes alike): Remiges and rectrices as in light morph; all the rest of the body and head, except the under tail coverts, fuscous-black to grayish black, the feathers of the upper back and the upper wing coverts with slightly paler edges; under tail coverts buffy whitish heavily marked and barred with brown to fuscous; the dusky bars on the rectrices broader than in the light morph. This morph is generally very rare, but in California it constitutes about a third of the population (Clark and Wheeler, 1987). A "rufous" or erythristic variant is somewhat intermediate in darkness, but the underparts are heavily washed with pale brown to hazel and obscurely but narrowly and abundantly barred with russet, and the under wing coverts are pale tawny marked with brown and fuscous.

Subadult. Although some early observers believed that a period of up to as long as four years might be required before the adult plumage is reached, it is now clear that, like other buteos, this species attains its definitive plumage in two years. It begins to molt into its adult plumage about one year after hatching, when the juvenal wing and tail feathers are gradually molted and the streaked underparts are lost. Yearling birds generally resemble juveniles, but are somewhat paler.

Juvenile (sexes alike). *Light morph:* similar to the light-morph adult dorsally, but more spotted or streaked with white above and buffy white and dark brown below, the breast and sides rather heavily spotted with fuscous. *Dark morph:* similar to the light morph, but more heavily streaked with brown on the underparts and under wing coverts, and with reduced buffy feather-edging dorsally. *Rufous variant:* similar to the dark morph, but with rufous spotting on the underparts. Iris brown to grayish; bill dull black, becoming dull blue or pale olive-buff at corner of mouth; cere greenish yellow to yellow; tarsi and toes creamy to light grayish green.

Measurements (in millimeters)

Wing, males 362–406 (ave. of 20, 383.6), females 375–427 (ave. of 38, 404.6); tail, males 185–214 (ave. of 20, 204.6); females 193.6–235 (ave. of 38, 214.6) (Friedmann, 1950). Average egg size 56.5 x 44 (Bent, 1937).

Weights (in grams)

Males, ave. of 5, 908; females, ave. of 7, 1069 (Brown and Amadon, 1968). Males 693–936 (ave. of 69, 808); females 937–1367 (ave. of 50, 1109) (J. Schmutz, in Palmer, 1988). Range of weights 595–1240, ave. 849 (sex and sample size unspecified) (Clark and Wheeler, 1987). Estimated egg weight 60.4, or 5.6% of female.

Identification

In the hand. This medium-sized buteo is separable from other North American species of that genus by the combination of having only the outer three primaries notched on their inner webs, a wing length of over 360 mm, and the tail approximately 60 percent as long as the wing.

In the field. Perched birds are usually seen in open surroundings (often standing on fence posts or telephone poles in grasslands), where at least adults typically appear rather uniformly dark above. They are variably white below, but normally have a dark breast band and white or at least lighter lower underparts, and often have a white forehead patch. The wingtips extend slightly beyond the end of the tail in perched birds. Immatures sometimes closely resemble immature red-tailed hawks, including a similar dark malar stripe and white dorsal spotting. The rather rare (except in California) dark morph appears uniformly sooty brown both above and below, and unlike light-morph birds the under wing coverts may be darker rather than lighter than the undersides of the flight feathers. Confusing individuals are sometimes identifiable

when they take flight, by determining whether the flight feathers are as light as or lighter than the under wing coverts (if red-tailed) or distinctly darker (if Swainson's), and, if the upper tail surface is mostly grayish, whether they have a whitish basal area and tail coverts (if Swainson's), or at least somewhat tinged with rusty (if red-tailed).

In flight, the distinctive two-toned underwing color, with the undersides of the flight feathers appearing darker than the more anterior wing linings, provides the most reliable means of identifying Swainson's hawks in any plumage or age class, including most dark-morph birds. When seen from above, the tail appears generally grayish (rather obscurely banded with blackish), grading to a whitish area at the tail base and the adjacent tail coverts. The wings are relatively longer and more pointed than those of red-tailed hawks, and appear to be slightly more uptilted and swept-back when gliding. The typical alarm call is a long and plaintive whistle that is somewhat higher-pitched and weaker than the red-tail's. As in many buteos, the call of the female is somewhat lower in pitch than that of the male.

Habitats and Ecology

This is a hawk that, although similar in general appearance to the red-tailed hawk and often confused with it, differs substantially from it in many ways. Beebe (1974) has described it as the western plains and desert ecological equivalent of the rough-legged hawk of the arctic, inasmuch as it is largely a mouse eater, at least during the breeding season. However, nonbreeders are insectivorous to a degree not found in either rough-legged or red-tailed hawks, and because of this tendency it (along with the similarly distinctly insectivorous broad-winged hawk) is among the most migratory of the North American buteos.

During the summer, the birds seem to prefer mixed to short grassland habitats with scattered trees, where occur both foraging opportunities for small rodents and insects and nesting opportunities in the form of isolated trees. Dunkle (1977) determined that 49 nesting pairs in Wyoming nested in available and apparently suitable habitats (dry grasslands, irrigated meadows, and edges between these two situations, all having tree or bush nesting sites and not subject to much human interference) at rates that did not indicate any preferences for any of these habitats over the others. Woodbridge (1987) reported that in California the birds strongly favor agricultural areas such as irrigated alfalfa fields during the nesting season, with productivity in these habitats over four times greater than that of pairs in rangelands, apparently because of the much greater prey base of voles and ground squirrels. Olendorff (1972) found that 61 percent of 71 Swainson's hawk nests in his Colorado study area were in creek bottoms, while 25 percent were in pure grasslands and 14 percent were on cultivated lands. Trees served as nest sites for all of these nests. In a North Dakota study (Gilmer and Stewart, 1984), pasture and haylands comprised the predominant (75 percent) land use within a kilometer of each of 10 occupied nests.

The small population of mostly immature birds that regularly winters in North America in southern Florida is almost entirely associated with ploughed fields. There the hawks apparently feed on insects and/or mice, moving from area to area as the fields are ploughed (Brown and Amadon, 1968). They have also been reported to forage on locusts in Argentine wintering areas.

Maximum breeding densities were estimated by Dunkle (1977) as a pair per 6.7 square kilometers in suitable nesting habitats of eastern Wyoming, based on estimated home ranges and average distances (1.77 kilometers) observed between active nests. He believed that the home ranges of adjoining nesting pairs rarely overlapped, thus essentially representing territories. This population density is similar to that estimated by Craighead and Craighead (1956), who estimated the density of five pairs in Wyoming to average a pair per 6.2 square kilometers (but only 0.032 nests per square kilometer over their entire study area), with an estimated home range averaging 2.6 square kilometers. Fitzner (1978) estimated the average home range of males in Washington to be 9.08 square kilometers, and females' to be 3.55 square kilometers. A rather similar average nesting home range of 4.1 square kilometers per pair was estimated in Utah by Smith and Murphy (1973). In their area the nesting density was far lower than the figures just given, perhaps because of the large study area utilized, which contained considerable habitat apparently relatively unsuited to Swainson's hawks. Gilmer and Stewart (1984) also estimated breeding densities in North Dakota grassland, with yearly averages of from 0.036 to 0.079 nests per square kilometer over their entire study area, and nests averaging 2.3 kilometers apart in all years.

Foods and Foraging

Early analyses of food intake in this species (Fisher, 1893; Bent, 1937) indicated that a combination of small mammals and insects represent the primary food sources, with the mammals primarily mice, voles, ground squirrels, and rats, and the insects mainly coleopterans and orthopterans. On a strictly numerical basis the insect component may indeed comprise over 90 percent of the total intake (Snyder and Wiley, 1976), but of course such numbers are not nearly as useful as biomass calculations. Dunkle (1977) estimated that even numerically insects comprised only trace amounts of the foods on his Wyoming study area, with 68 percent of the prey items being mammals (mostly young rodents and lagomorphs), and 25 percent being birds of considerable variety, ranging in size from sparrows to sage grouse (*Centrocercus urophasianus*).

Smith and Murphy (1973) estimated that insects comprised about 31 percent of the prey numerically, with birds totaling about 17 percent and mammals 52 percent, but on a biomass basis the insect component represented only trace quantities, and the birds only about one percent. Over 98 percent of the total biomass was made up of jackrabbits and cottontails during all three years of the study. Similarly, Fitzner (1978) found insects to be an insignificant component of spring and summer foods in Washington, making up less than one percent on a numerical basis; there was a similar substantial component provided by jackrabbits and cottontails, with a variety of rodents, birds, reptiles, and toads represented as well. Fitzner believed that serious competition with red-tailed hawks and ferruginous hawks was reduced by the Swainson's hawk taking generally smaller and more diverse prey than either of these two species.

In a North Dakota study, Gilmer and Stewart (1984) found that Richardson's ground squirrel (*Spermophilus richardsonii*) was the most important single prey brought to nests, making up nearly half of the estimated prey biomass, with northern pocket gophers (*Thomomys talpoides*) of secondary significance, and no insect remains present at all. Likewise, Schmutz (1977) found that nesting Swainson's hawks in Alberta consumed insects at a numerical incidence of less than one percent and birds at about 15 percent, the rest of the foods being of mammalian origin. Ground squirrels (*S. richardsonii* and *S. tridecemlineatus*) made up the majority of these items and represented about three-fourths of the total estimated food biomass, while white-tailed jackrabbits (*Lepus townsendi*) were of secondary importance from a biomass standpoint.

The apparent discrepancy between this species's reputed insect dependence and its actual high level of mammal consumption while on the breeding grounds was discussed by Johnson, Nickerson, and Bechard (1987), who observed flocks of nonbreeding Swainson's hawks feeding almost exclusively on grasshoppers during summer in Saskatchewan and Idaho. Individual pellets contained up to 120 grasshopper remains, and thus the birds individually probably consumed about 100 grasshoppers per day, assuming that a single pellet is regurgitated per day. They suggested that breeding requires a shift to larger mammalian food items, but that when the constraints of breeding are removed, as when on migration and in wintering areas, the birds become essentially insectivorous. When feeding on grasshoppers the hawks foraged in recently plowed or mowed alfalfa fields, using a "pounce and peck" hunting strategy that is quite different from the species's perch-and-soar style of hunting for small mammals.

Foraging and habitat partitioning among the Swainson's, red-tailed, and ferruginous hawks were investigated in Oregon by Cottrell (1981), who found dietary overlaps among these three species to range from 80 to 95 percent during two different years. However, habitat overlap was considerably less, with the overlap between Swainson's and red-tails estimated as 52 and 54 percent during two years, and between Swainson's and ferruginous as 41 and 47 percent. Redtails and ferruginous were estimated to overlap 33 and 37 percent. The habitat overlap of the Swainson's with the other two species was apparently minimized by its selection of nesting trees having a different configuration from those used by the others.

Schmutz (1977) judged that the food intake of breeding Swainson's hawks in his Alberta study area had a 95–99 percent dietary overlap with that of ferruginous hawks in terms of prey species, and there was no evidence that the two species partitioned their common food resource on the basis of prey size or hunting periods. There was also a substantial (73–82 percent) overlap in the nestling and postfledging periods of the two species, and similarly red-tails nesting in the same area had a very comparable prey base and nesting seasonality. Schmutz judged that different interspecific nesting habitat preferences (red-tails in tall stands of trees, ferruginous on open plains, and Swainson's on areas supporting scattered trees or shrub- and tree-bordered rivers and lakes) helped to reduce food competition, and that the rodent prey population was probably sufficiently abundant to make it an unlikely factor in limiting the breeding populations of all three buteos.

Nonetheless, apparent reduction of reproductive success may result from interspecific interactions brought about by proximate nesting of Swainson's hawks near other buteos, judging from the studies of Schmutz (1977). Although Janes (1984a) found no such reduced productivity among red-tailed hawks nesting near Swainson's hawks, he did observe prolonged and intense fighting over territories, and noted that red-tailed hawks, which had established their territories early in the season, sometimes lost parts of their territories to the Swainson's hawks when they arrived later in the spring.

Social Behavior

Although it is certainly a monogamous species, the length of pair bonds in Swainson's hawks is questionable, especially considering the long migratory route of the species and consequent low likelihood of a pair remaining together throughout the year. Bloom et al. (1985) reported that recent California research indicates a high degree of mate and territorial fidelity, but details are still lacking.

Olendorff (1974) described a courtship flight during which a pair soared separately near a tree containing a nest, with the flight paths describing circles of about a half-kilometer diameter as the birds increased altitude to about 90 meters. Then one bird, perhaps the male, soared over the nest, set its wings, and glided downward away from the nest, losing about 70 meters altitude before starting to soar up again in circles. One also began a rapid flight when above the nest, followed by a closing of its wings and a brief dive, which was followed by a short circular climb and another dive. This occurred twice in rapid succession, and was then followed by another climb and stall, which in turn was terminated by a long parabola-like dive that ended with the bird lighting on the edge of the nest. The presumed female soon joined the other bird at the nest, but no further display occurred.

Fitzner (1978) observed three courtship flights quite similar to this, one of which consisted of five stooping dives and ended with a landing in a tree where Fitzner had often observed copulation to occur. Two others were seen with single and triple dives, all of which were similar to those described by Olendorff but differed in the mode of recovery after the dive, the birds usually using their own momentum to regain lost height, rather than using flapping flight. He observed copulations to occur mainly during morning and evening hours, on dead limbs of trees. The female typically assumed a receptive posture without apparent prior display by the male, who then quickly landed on her back. One bird, presumably the female, called during treading.

Breeding Biology

Although nest site fidelity is not well documented in this species, five birds marked by Fitzner (1978) returned the following year to use the same site. Of 59 nests he studied, about half were freshly built, with the remainder being old nests of the species, or of magpies, crows, ravens, etc. The average height of 57 of these nests was 5.6 meters. The majority of the nests were situated in trees in broken grasslands or cultivated lands, with a few situated in unbroken grasslands or "juniper grasslands." Nest construction began 7–15 days after the birds arrived, and required 1–2 weeks to complete. The males brought most of the nesting materials and did most of the arranging of the nest. Both sexes brought green sprigs, both before and after hatching.

Egg laying occurs at approximate two-day intervals, with the laying normally occurring during morning hours. Fitzner found the clutches of 39 nests to average 2.18 eggs, or very close to the average reported by Olendorff (1972) of 2.34 eggs in 95 Colorado nests and by Dunkle (1977) of 2.55 eggs for 31 Wyoming nests. The female did all the incubation while the male brought her food. Only while the female was off the nest to feed did the male cover the eggs. Incubation of 10

eggs in three nests required 33–36 days, averaging 34.5. The average fledging period was 42.9 days, but varied from 38 to 46 days. Typically the young began to leave their nest at 33–37 days, fledging soon afterward. During the entire postfledging period prior to departure on migration the young remained in the adults' territory and were largely dependent upon them for food. This postfledging period averaged 29.2 days, ranging from 22 to 38 days. By about 30 days after hatching the young were beginning to kill insects and snakes (Fitzner, 1978).

Productivity estimates have been provided by Dunkle (1977), Olendorff (1973), Fitzner (1978), and Schmutz (1985b). All of these reported a moderately high rate of nesting success, with the number of fledged young per successful nest varying from 1.19 to 2.0, and the number of fledged young per adult pair ranging from 1.18 to 1.54. Schmutz reported an adult survival rate to the following season of 75 percent.

Evolutionary Relationships and Status

There are no obvious close relatives of this species. Although it is traditionally placed with a group of species (such as the white-tailed hawk) that have only the three outermost primaries notched, it is unlikely that this trait has any real phyletic significance.

The Swainson's hawk has been a species of increasing concern to conservationists, and in many areas it has certainly declined drastically since the 1940s, as in California, where it may have declined as much as 90 percent (Bloom et al., 1985; Schlorff, 1985). The species has also declined significantly in southeastern Oregon (Thompson, Littlefield, and Johnstone, 1985). Clark (1985a) judged that perhaps some 40,000–53,000 breeding pairs existed in the mid-1980s, assuming a collective population of about 320,000 birds, an estimate derived from migration counts. Smith (1985) reported seeing a maximum of 344,400 migrating birds in Panama during 1972, which seems to provide the only quantitative basis for estimating the continental population. Clark (1985b) was unable to establish any evidence for population declines in the Great Plains states, at least excluding peripheral populations such as Iowa and Illinois. A small disjunct population breeds in northern Illinois (Keir and Wilde, 1976), and a similar remnant population exists near Nogales, Arizona (Glinski, 1985b).

Figure 43. North and Central American residential range of the white-tailed hawk. The South American residential range is shown on the inset map.

White-tailed Hawk *Buteo albicaudatus* Vieillot 1816

Other Vernacular Names

Sennett's white-tailed hawk (*hypospodius*), white-tail; aguililla cola blanca (Spanish)

Distribution

Resident from Sinaloa (historic records north to southern Arizona), Chihuahua, Coahuila, and southeastern Texas (mainly coastally, formerly rarely to central Texas) south through Mexico and Central America to southern South America. Some winter sightings outside the breeding area (Padre Island, southern Arizona, and along the upper Rio Grande in Texas) indicate probable limited winter wandering occurs. (See Figure 43.)

North American Subspecies

B. a. hypospodius Gurney: Breeds from Sonora and southeastern Texas south to Venezuela.

Description (of *hypospodius*)

Adult (sexes nearly alike). *Light morph:* Crown and sides of head and neck dark ashy gray to neutral gray, each feather more or less distinctly edged with deep neutral gray; most upperparts neutral gray edged with deep gray, basally white; lesser upper wing coverts cinnamon to russet-brown (females averaging more rufous than males), the scapulars also tinged with the same; these feathers with black shaft streaks and some irregular dusky bars; middle and greater wing coverts dark neutral gray to blackish slate, as are also the outer webs and tips of the primaries; outer secondaries tipped with whitish, secondaries and inner webs of primaries gray, paling internally to grayish white; lower back, rump, and upper tail coverts white, narrowly barred with dark gray; rectrices white, darkening to pale neutral gray, the tail feathers tipped with white, with a broad subterminal black band, and sparsely barred anteriorly with up to ten narrow, wavy dusky bars; underparts white, more or less barred with narrow dark gray or brownish gray bars on the sides, flanks, and thighs, but sometimes almost immaculate white; sides of throat and breast gray; axillaries and under wing coverts white, narrowly barred with dark grayish or brownish gray near the bases of the primaries. Iris dull yellow (in younger birds?) to sienna or

hazel; cere pale green to yellowish or yellowish olive; orbital skin light yellowish green; bill black at tip, bluish gray at base; tarsi and toes bright yellow; claws blackish.

Dark morph (fairly rare in the U.S., apparently continuous variation exists between light and dark extremes): Similar to the light morph but with the entire underparts neutral gray to slate, the feathers of the abdomen and thighs sparingly edged or barred with white, the thighs sometimes barred with rufous and white, and the abdomen and smaller upper and under tail coverts somewhat suffused with rufous, remaining tail coverts white barred with slate; rectrices mostly white, with eight narrow dark bars and a broad black subterminal band.

Subadult (sexes alike). Similar to adult of the same morph by the end of the first year, but with the head, chin, and throat deeper neutral gray; the upperparts generally darker, from dark neutral gray to dusky neutral gray, and with only limited russet edging evident in the scapulars and inner wing coverts; rump and upper tail coverts buffy white to cinnamon, barred with brown; underparts variably white to heavily streaked or barred with rufous and blackish, except on the anterior breast, which is white; under wing coverts buffy white to cinnamon-buff, heavily marked with brown or blackish brown, the undersides of the flight feathers barred with dark gray as in adults.

Juvenile (sexes alike). Upperparts, including wings, usually sooty brown to mouse-gray, becoming darker anteriorly and dull buff behind, forming a pale whitish U-shaped patch in front of the tail. Underparts similarly dark (even in light morphs), except for variably large creamy breast patch; remainder of underparts mottled dark and light, with pale under tail coverts, the under wing coverts similarly dark brownish, darker than the barred gray undersides of the flight feathers. Rectrices washed with light gray, evenly barred with up to 11 fuscous bars, and often with an ill-defined subterminal black band. Iris buffy gray; bill black above and lighter below, tarsi and toes flesh-colored, claws light brown.

Measurements (of *hypospodius*, in millimeters)

Wing, males 400–432 (ave. of 9, 416.9), females 390–450 (ave. of 7, 424.3); tail, males 178–212 (ave. of 9, 195.6); females 192–211 (ave. of 7, 201.1) (Blake, 1977). Average egg size 58.9 x 46.5 (Bent, 1937).

Weights (in grams)

Average weight (sex and sample size unspecified) 884 (Dunning, 1984); one female 984 (Howell, 1972). Six adults, sexes not stated, 920–1351, ave. 1111 (Palmer, 1988). Range of weights 880–1235, ave. 1022 (sex and sample size unspecified) (Clark and Wheeler, 1987). Estimated egg weight 70, or about 7% of female.

Identification

In the hand. This is the only North American buteo with the combined traits of having only the outer three primaries notched on their inner webs and a tail length that is less than half that of the wing. There is a continuous variation between the light-morph and dark-morph plumages, but dark-morph birds are rare at the northern edge of the range.

In the field. Associated with open, often semiarid or coastal grassland habitats. When perched, the wingtips of nonjuveniles reach beyond the tip of the tail, and nonjuveniles also have extensive white areas on the rump, basal half of the tail, and under tail coverts. In juveniles the tail is more grayish, and the entire bird appears quite dark, but white is variably present on the upper and lower tail coverts and as a breast patch that is mostly or entirely surrounded by blackish. Birds at least a year old show some rufous on the scapulars and anterior wing coverts, are white on the upper breast and throat but may be barred with darker markings on the lower breast and abdomen, and have a broad black band near the tip of the otherwise nearly white tail.

In flight, the typical light-morph birds appear almost completely white on the undersides of the body, the wing linings, and the tail anterior to the broad blackish subterminal band (thus closely resembling the light-morph short-tailed hawk, but with the primaries appearing slightly darker than the secondaries, rather than the reverse). First-year juveniles appear very dark from below, including the wing linings, but typically have a conspicuous whitish breast patch and pale under tail coverts. Unlike the short-tailed hawk, soaring is done with the wings held at an uptilted angle. In any plumage, the wings appear unusually long relative to the tail. The usual call is a series of high-pitched, musical notes, and a harsh *ack kehack'*, *kehack'* call has also been described.

Habitats and Ecology

In Texas this species occupies coastal grassland, including saltgrass flats near Gulf beaches and dry grassy mesquite–live oak savanna farther inland. Its current breeding range corresponds very closely to that of the native distribution of the Gulf prairies vegetation type (Kopeny, 1988). Farther south it occupies similar open-country habitats but extends up on open hillsides to at least about 700 meters, and may occur about clearings in dry woodlands or even in grasslands, as on the northern pampas of Argentina (Brown and Amadon, 1968).

Population densities are almost unstudied, but Stevenson and Meitzen (1946) remarked that a group of 16 adults and 4 immatures that were attracted to a prairie fire probably represented the entire population in an area of about 10 miles radius. Farquahar (1986) estimated that in a Colorado County (Texas) study area (Attwater Prairie Chicken National Wildlife Refuge) males had average home ranges of 32.7–34.3 hectares during two years, and females had home ranges averaging 12.6–13.9 hectares. He reported mean breeding density of 0.18–0.21 pairs per square kilometer during these two years. Kopeny (1988) found a similar density in his Kleberg County study area of 0.11–0.17 pairs per square kilometer in two adjacent pastures during one year. Nests in savanna-chaparral habitat were situated slightly closer together (averaging 3.5 kilometers between nests) than those in open savanna (4.0 kilometers between nests).

Foods and Foraging

This is one of the lesser-known species of North American hawks in terms of its food requirements and foraging behavior. Stevenson and Meitzen (1946) summarized the available literature and added their own observations. It would seem that this species is highly opportunistic, feeding on whatever is available conveniently. Based on the prey brought to the young at eight nests, a variety of reptiles, birds, and mammals are exploited, including lizards (*Sceloporus* and

Eumeces), snakes (*Coluber, Pituophis,* and *Thamnophis*), northern bobwhite (*Colinus virginianus*), clapper rail (*Rallus longirostris*), king rail (*R. elegans*), eastern meadowlark (*Sturnella magna*), northern mockingbird (*Mimus polyglottos*), seaside sparrow (*Ammodrammus maritima*), cottontails (*Sylvilagus*), pocket gophers (*Geomys*), and mice (*Peromyscus*). Other data indicate the feeding of nestlings on blue-winged teal (*Anas discors*), greater roadrunner (*Geococcyx californicus*), and a mole (*Scalopus*). Observations further indicate that the birds take other vertebrates such as snakes (*Opheodrys, Crotalus*), jackrabbits (*Lepus*), cotton rats (*Sigmodon*), pocket mice (*Perognathus*), and fox squirrels (*Sciurus niger*), as well as larger insects such as grasshoppers (Stevenson and Meitzen, 1946). Ditto (1983) noted that the young in one nest were fed snakes (*Thamnophus, Coluber*), horned lizards (*Phrynosoma*), ground squirrels (*Spermophilus*), cottontails, crabs (*Callinectes*), and an unidentified rodent.

In a more recent study, Farquahar (1986) judged the prey biomass found at 10 nests to be comprised of about 45 percent mammals (mostly cottontails), 34 percent birds (largely Attwater's prairie chickens), and 16 percent reptiles, with the prey choice being largely opportunistic. Considerable numbers of grasshoppers were also eaten; these represented 36.3 percent of the prey numerically, but did not contribute significantly to ingested food biomass. Kopeny (1988) found that in his study area small mammals, especially cotton rats, and reptiles, especially snakes, were the primary foods, emphasizing the opportunistic feeding tendencies of the species.

The birds typically hunt at altitudes of about 15–50 meters, using a combination of gliding or soaring, hovering, and flapping flight. When winds are favorable the birds often hover while hunting or will hang motionless on thermals, then glide for a distance before repeating the process. This sort of hunting is often used above grassy hillsides, where updrafts caused by declivity winds are common (Brown and Amadon, 1968). The birds also regularly congregate at prairie fires to take advantage of the availability of rodents and insects that are thus exposed (Stevenson and Meitzen, 1946; Tewes, 1984; Farquahar, 1986).

Heredia and Clark (1984) estimated that the birds select foraging habitats that have from less than 10 to about 40 percent tree cover, and grasses of from less than 20 to 80 centimeters.

Like the sympatric black-shouldered kite, they hunt mostly from the air using searching techniques, but also hunt from perches. Much of their hunting occurs late in the day. They have also been observed stealing food from black-shouldered kites.

Social Behavior

Almost nothing can be said of the territorial, pair-bonding or early courtship phases of behavior in this species, although high-circling behavior by pairs and single birds has been observed. Farquahar (1986) observed a behavior by pairs that may function as pair bonding or courtship, involving the placement of long woody stems or branches on nests. He observed copulation to occur as early as January 30 and as late as April 12. Flight displays resembling territorial behavior usually (65–71 percent of occurrences in two years) preceded copulation, and similar aerial displays followed copulation in a few cases (5–8 percent of occurrences). Often these involved chasing an intruding bird, and may have served both for pair bonding and territorial defense. Usually both members of the pair chased intruders; there was no significant sexual difference in territorial behavior during incubation and brooding periods. Pairs spend most of the year within their breeding territory, and at least some defend their territory throughout the year.

Along the central Gulf Coast the breeding birds are paired and have occupied territories by early January, at a time when many wintering raptors such as red-tailed hawks and northern harriers are still abundant in that area. Some interspecific aggression has been seen among these species. The birds probably initially breed at two years of age, but in one case a full-plumaged male was seen paired with a female in apparently immature plumage, and in another case both members of a breeding pair were in immature plumage (M. Kopeny, in Palmer, 1988).

Breeding Biology

Like ferruginous hawks, the birds seem to select areas with wide vistas for their nesting territories, nesting on perhaps almost any tall bush, yucca, or tree that is available. Seven of eight nests observed by Stevenson and Meitzen (1946) were between 2.3 and 3.8 meters above ground in the tops of blackjack or live oaks, the other being some 9 meters up in a live oak within an oak grove. All these nests were at the extreme top of a tree, usually in the center of the crown. They were sometimes refurbished and used in subsequent years, being repaired in February, or well before egg laying. Thirty nests found by Burrows (1917) were from 0.5 to 4.4 meters high, averaging about 2.3 meters, and were usually in thorny bushes or in small trees, such as mesquite, huisache (*Acacia*), and hackberry. Morrison (1980) reported that 124 nest sites from Texas ranged from 0.9 to 12.2 meters, averaging only 2.97 meters. Nest supports listed in the egg collection of the Western Foundation of Vertebrate Zoology include 37 in oaks, 18 in Spanish dagger (*Yucca treculeana*), 17 in mesquite (*Prosopis*), 16 in huisache (*Acacia*), 8 in other kinds of trees, and 13 in bushes.

Kopeny (1988) noted that on his study area the birds nested in three habitat types: open mesquite savanna, scrub live oak, and mixed savanna-chaparral. Of 44 nests there, 48 percent were located in granjeno (*Celtis pallida*) or ebony (*Pithecellobium flexicaule*) trees, with 39 nest trees averaging 3.2 meters in height. In Farquahar's (1986) study area all but two of 20 nests were placed at or near the tops of Macartney rose (*Rosa bracteata*) bushes, the most common tall shrub in that area.

There is no information on the rate of egg laying, but based on the two-day hatching pattern observed by Farquahar (1986) the eggs are probably laid at 48-hour intervals. Clutch replacement is frequent if the first clutch is lost, in two cases occurring 10 and 15 days after the loss. Egg laying in Texas occurs from late February through May, with a peak between March 11 and 20, and with nests started during and after April probably being renesting efforts. Such renesting has been reported by several observers (Farquahar, 1986). The average clutch size of 139 clutches was 2.26, with a range of 1–4 eggs (Morrison, 1980). There is a record of one pair raising two broods in a single season (*American Birds* 23:673–75).

The incubation period was 31.2 days, based on a sample of 13 nests (Farquahar, 1986). Stevenson and Meitzen (1946) noted that the young are hatched about a day apart. They also observed that two young left a nest prematurely at 35 days when they were banded, but a third remained until it was 47 days old. In one case crested caracaras were found to be the cause of nestling mortality, and great-tailed grackles (*Quiscalus mex-*

icanus) have been blamed for the loss of eggs. Kopeny (1988) reported a 47–53 day fledging period in one nest. However, the young are fed by their parents for their first seven months of life, which is a longer period of immature dependency than that reported for any other North American buteo (Farquahar, 1986).

Some information on productivity has recently become available. Thus in one study 71 percent of 76 nesting attempts were successful, with 1.11 young fledging per nesting attempt. Productivity averaged 1.74 young per successful nest for 54 nests (Kopeny, 1988). An even higher rate of success was found by Farquahar (1986), who estimated that 18 of 20 nest attempts (90 percent) were successful. Further, 92 percent of all 39 eggs hatched, and 92 percent of all 36 hatched young survived to fledging, or an average of 2.0 fledged young per successful nest.

Evolutionary Relationships and Status

According to Brown and Amadon (1968), this is one of a group of large open-country American buteos that have only the outer three primaries notched, and that is otherwise represented among North American forms only by the Swainson's hawk. The red-backed buzzard (*Buteo polyosoma*) of South America is perhaps the most similar of this group to the white-tailed hawk. Kopeny (in Palmer, 1988) has suggested that this species as well as another South American buteo, the puna hawk (*B. poecilochrous*), may even be conspecific with the white-tailed hawk. All of these forms have reddish upperparts and mostly white tails with broad subterminal banding as adults. The AOU (1983) noted that the relationship among these three forms "needs clarification."

The small U.S. range of this species is now limited to coastal Texas, and in that state it apparently suffered from the development of brushy vegetation that may have originated from grazing effects or from the elimination of range fires (Oberholser, 1974). However, some isolated trees or shrubs are needed for nest sites, and complete removal of woody plants for prairie restoration can have undesirable effects. Indications of eggshell thinning have been reported since the late 1940s in Texas, presumably as a result of pesticides, but recent productivity data do not indicate a current reproduction problem. The total Texas population during the 1970s may have been only 250–500 pairs (Porter and White, 1977). Kopeny (1988) has recently reported that at least 32 pairs were present in Kleberg County in 1986, in the heart of its Texas range. Other important areas for the species are in the lower coastal portions of Kenedy, Brooks, and Willacy counties. Audubon Christmas Count data summarized by Kopeny suggest that a significant population increase may have occurred since about 1968–70, and that more than the approximate 200 pairs estimated by Morrison (1978) are now probably present in Texas. Based on the Audubon Society's 1986 Christmas Counts, I estimated a total U.S. winter population of 1,040 birds, all in Texas. Assuming that the population consists of about 80 percent adults (Kopeny, 1988), this might suggest a potential population of around 400 breeding pairs.

Figure 44. North and Central American breeding (hatching) and residential (cross-hatching) ranges of the zone-tailed hawk. The South American residential range is shown on the inset map.

Zone-tailed Hawk *Buteo albonotatus* Kaup 1847

Other Vernacular Names

Band-tailed hawk, zonetail; aguililla cola cinchada (Spanish)

Distribution

Resident in northern Baja California, and from Arizona, New Mexico, and western Texas (where mostly migratory) south locally through Mexico (where mostly resident) and Central America to Paraguay and southern Brazil. Casual north to southern California, where breeding was attempted in Santa Rosa Mountains from 1979 to 1981 (Weathers, 1983). (See Figure 44.)

North American Subspecies

None currently recognized; South American birds average slightly smaller and have at times been racially separated.

Description

Adult (sexes nearly alike). Plumage uniformly black or black with a faint brownish to grayish cast, the feathers pure white basally; the upper back and breast with a slight slate tinge in some specimens; lores and a narrow forehead line whitish; inner webs of remiges paling to deep gray and to grayish white, barred with fuscous-black; the outer webs of all but the outermost primary barred with fuscous-black and deep gray; both webs of the middle pair of rectrices and the outer webs of the other tail feathers slate-color, narrowly tipped with white and very broadly subterminally banded with black and irregularly banded with black and white on the basal three-fifths (males having broader banding than females, and two rather than three white tail bands), the inner webs of all but the median pair whitish; entire body, under wing coverts, and tail coverts black. Iris dark brown to reddish brown; cere and edge of mouth clear yellow, bare lores grayish; inside of mouth pale yellow; bill black, paling to light grayish white at base; tarsus and toes bright lemon-yellow; claws black.

Subadult. The juvenal plumage is apparently worn for most or all of the first year. Later stages of molt are still undescribed.

Juvenile (sexes alike). Similar to adults, but more brownish throughout, with small irregular white spotting below and above, tail whitish below and brownish above, with up to seven black bands (of which four or five are exposed beyond the coverts), the subterminal band the broadest; flight feathers whiter below than in adults, and with fewer dark bars. Iris brown; bill black, tinged basally with dull green; cere, tarsi, and toes yellow.

Measurements (in millimeters)

Wing, males 380–393 (ave. of 7, 387.5), females 409–438 (ave. of 9, 419.6); tail, males 216–222.8 (ave. of 7, 219.9), females 224.6–234.5 (ave. of 9, 228.9) (Friedmann, 1950). Average egg size 55.6 x 43.5 (Bent, 1937).

Weights (in grams)

Males 607–667 (ave. of 3, 628), females 845–937 (ave. of 4, 886) (Dunning, 1984); 1 female ca. 1020 (Oberholser, 1974). Range of weights 610–1080, ave. 830 (sex and sample size unspecified) (Clark and Wheeler, 1987). Estimated egg weight 58.1, or 6.5% of female.

Identification

In the hand. This species may be separated from the other North American buteos by the combination of having the four outer primaries notched on their inner webs, a wing length of 380–438 mm, and a moderately long (at least 50 percent of wing length) tail that is crossed by three or four white bands alternating with black (these numbering about seven and less contrastingly patterned in immatures).

In the field. Perched birds appear almost coal-black, with wingtips reaching the end of the tail, and more grayish (rather than bright yellow, as in the black-hawk) lores. The tail band is more grayish and thus less conspicuous than in the black-hawk, and in that species the wings do not reach the tip of the tail.

In flight, this rather narrow-winged species flies with distinctively uptilted wings (compared with horizontally held wings in the more broadly winged black-hawk), and this plus its generally gray to blackish body and underwing coloration makes it closely resemble a turkey vulture. This similarity is reduced by the strongly banded tail, showing a single white continuous cross-band, plus one (males) or two (females) more anterior white bands broken by the tail coverts. Young birds have generally more whitish undersides of the flight feathers (and thus a more two-toned wing pattern), usually have scattered white spotting on the otherwise black body, and have a tail that is closely banded with silvery white (from below) or brownish (from above) and blackish bands. Vocalizations include a long, drawn-out and rather catlike whistle uttered as an alarm call.

Habitats and Ecology

The habitats of this species in Texas are similar to those elsewhere in the U.S. and northwestern Mexico, the birds occupying rough and deep, rocky canyons and streamsides within semiarid mesa, hilly, and mountainous terrain (Oberholser, 1974). In western Texas the birds breed in the Davis Mountains on steep north-facing slopes and among open stands of ponderosa pines, while in the Chisos Mountains they have been found nesting on north-facing slopes dominated by oaks and junipers. They have also been found nesting in Boquillas Canyon, Brewster County, on north-facing slopes dominated by such typical hot-desert plants as creosote bush (*Larrea*), mesquite (*Prosopsis*), and lechuguilla (*Agave*) (Matteson and Riley, 1981). In Arizona the birds nest in tall trees along the main rivers and canyons, but do most of their soaring and no doubt also their foraging over the higher mountains (Phillips, Marshall, and Monson, 1978). Millsap (1981) found the birds to be most abundant at about 1,100 meters in Arizona, with a wide range of plant communities utilized. Typically the birds nested in steep canyons with riparian hardwoods or open pine stands on steep hillsides, foraging mostly over adjacent uplands but also over fairly level grasslands and gently rolling paloverde-saguaro habitats. All of the nests were found on hillsides or immediately adjacent to cliffs or steep talus, and most were within a kilometer of surface water.

In Mexico, Guatemala, and Nicaragua the species is largely associated with the pine-oak belt of montane woodlands, while in Costa Rica it is mainly found in the dry forested lowlands. It is quite rare in Panama, and in South America is also relatively rare and local in distribution. It may there be attracted to dry deciduous woodlands, a habitat that it uses preferentially in Venezuela, especially where the terrain consists of eroded badlands (Brown and Amadon, 1968).

Foods and Foraging

There is still only rather limited information on the foods of this species, which have in the past been characterized as consisting of lizards, frogs, and small fishes for the most part, but with some small birds such as nestlings and mammals such as rodents also eaten (Bent, 1937). However, Willis (1963) suggested that the species may be an aggressive mimic of the turkey vulture (*Cathartes aura*), not only in general appearance but also in its mode of flight. He believed that lizards, squirrels, and chipmunks constitute its major prey, and that all of these prey types might readily habituate to the sight of turkey vultures, thereby making themselves more vulnerable to the zone-tailed hawk, which typically flies vulturelike at about 20–30 meters above ground in a tilting, irregular flight, with its wings held up in a distinct dihedral. Willis later (1966) described seeing the capture of prey (a small bird) following a vulturelike soaring terminated by a shallow dive to an isolated tree at the edge of a forest, without the hawk having stimulated alarm calls from the birds using the forest edge or pasture.

Mueller (1972) suggested that although the zone-tailed hawk may indeed resemble a vulture in its shape and manner of soaring, their similarities are more likely to be the result of aerodynamic adaptations required by birds that tend to glide close to the substrate, such as vultures and harriers. However, he did agree that the color pattern of the species may be the result of aggressive mimicry evolved subsequent to the evolution of this mode of flight. Zimmerman (1976b) added to the discussion by pointing out that small mammals do not altogether ignore turkey vultures, nor do migrant birds such as shorebirds. He noted that hunting zonetails often cruise parallel to elevated ridges, usually between 15 and 20 meters above ground, but sometimes only about 10 meters above, concentrating on the bases of the hills. In the process they have been observed to stimulate strong aggressive reactions from a resident American kestrel as well as a pair of western kingbirds (*Tyrannus verticalis*), whereas at least the kestrel typically recognized and regularly ignored passing turkey vultures. Zimmerman judged that the idea of vulture mimicry is unproven but "an intriguing concept" deserving of additional consideration. Helen Snyder (cited in Palmer, 1988) observed that 2 of 6 capture attempts made by zone-tailed hawks in the presence of vultures were successful, whereas only 1 of 14 attempts made in their absence was successful, suggesting a potential beneficial hunting effect of associating with vultures.

Snyder and Wiley (1976) judged from available data that birds numerically comprised the single largest food component (47 percent), with lower vertebrates making up 33 percent and mammals the remaining 20 percent. Sherrod (1978) summarized several unpublished analyses, including observations by R. Glinski from Arizona, in which 16 birds as large as Gambel's quail (Callipepla gambelii) and as small as the horned lark (Eremophila alpestris) were taken, as well as four lizards (Sceloporus, Crotaphytus, and Sauromalus). Other unpublished data from Arizona by H. A. and N. F. R. Snyder included a variety of birds to the size of quail and as small as warblers (Dendroica) and house sparrows (Passer domesticus). The largest sample size came from Texas observations by S. W. Matteson, J. O. Riley, and J. T. Harris (in an unpublished report of the Chihuahuan Desert Research Institute), in which ground squirrels (Spermophilus) and antelope squirrels Ammospermophilus) comprised the mammalian component of 1.5 percent, at least 19 species of birds comprised the avian component of 27.5 percent, and reptiles made up 71 percent numerically of the total, nearly all of these being 97 crevice spiny lizards (Sceloporus poinsettii).

Millsap (1981) provided a numerical and biomass analysis of Arizona food, representing a sample of 88 prey items. Of these, mammals comprised an estimated 41 percent of the biomass and 32 percent of the items, with Harris' antelope squirrels (Ammospermophilus harrisii) the most important prey species. Birds comprised 37 percent of biomass and 15 percent of the items, with the Gambel's quail (Callipepla gambelii) responsible for nearly the entire amount. Reptiles comprised 20 percent of the biomass and 43 percent of the items, with the collared lizard (Crotaphytus collaris) representing about three-fourths of the total. The average prey weight (98 grams) was relatively small as compared with that of other buteos of comparable size (260 grams for red-tailed hawk).

These data strongly suggest that the zone-tailed hawk is an effective predator on small,

agile lizards, on a variety of open-country birds (meadowlarks, horned larks, etc.) as well as woodland edge or woodland birds (jays, wrens, kingbirds, bluebirds, warblers, woodpeckers, etc.), and on terrestrial rodents such as ground squirrels. Many of these are apparently caught by short, swift plunges to the ground, or even by snatching nestlings out of nests in full flight (Brown and Amadon, 1968).

Social Behavior

Very little has been written on the social behavior of this species. Oberholser (1974) described an apparent aerial display, in which the half-spread wings were raised in the form of a triangle so that their tips met behind, the bird dropping from a great height almost to the earth with great velocity, then pulling out and regaining height, only to repeat the performance. Hubbard (1974) observed aerial display in February, when he saw three birds soaring nearly 300 meters above a valley floor. Two of them were initially higher than the third, when suddenly one of the higher two stooped on the lower one, which at the moment of contact turned over and locked talons with the stooping bird. The two birds then somersaulted downward for about 100 meters. At that point they released and both returned to nearly their original positions before all three birds gradually moved off and disappeared in the same direction. Although Hubbard speculated that this behavior represented courtship, it seems equally likely that this was actually an agonistic encounter. Clark (1984) made such an interpretation on a similar encounter he observed, in which the resident male of a territorial pair dived on an intruder judged by Clark also to be a male, and the two birds locked talons for a short time.

Breeding Biology

Nests of this species are rather bulky, and are placed at various heights in a variety of trees. Of 21 nest records in the files of the Western Foundation of Vertebrate Zoology, 13 were located 9–15 meters above ground, and all were between 4.6 and 23 meters. Eight were placed in cottonwoods, two each were in oaks and maples, and five other nests were in other kinds of trees. Nine clutches from Arizona were taken between May 2 and June 2, with 6 from May 2–11, and 81 Texas clutches were obtained from March 12 to May 26, with half of these between April 26 and May 8. Millsap (1981) reported on 28 Arizona nests, 16 of which were in cottonwoods and the rest in sycamores, oaks, and pines. Egg dates ranged from April 15 to May 17, with half of them by May 1, and eggs had hatched in half of them by June 10. Five Mexican clutches were obtained between March 13 and May 21. All of these records suggest a rather limited breeding season, with perhaps few if any renesting efforts being made. Clutch sizes are also small in the species; 23 clutches ranged from 1 to 3 eggs, with 20 of them 2 eggs and the average 1.96. Matteson and Riley (1981) reported that all of the 16 nests they studied had two eggs, and Millsap (1981) reported an average clutch of 2.1 eggs in 22 nests.

There is no information on incubation or fledging periods, or indeed on most other aspects of breeding biology. Millsap (1981) stated that in his west-central Arizona study area all eggs had hatched by about June 27, and young were fledged in half the nests by July 28, with the latest fledging date about August 10. Among 22 nests there was a 90 percent hatching success (initial brood size average 1.9 young), and a 97 percent fledging success (final brood size average 1.85 young). Matteson and Riley (1981) reported that, over a two-year period, 19 of 32 eggs in 16 nests hatched (hatching success 59 percent), and that 15 young were fledged (fledging success 79 percent). All told, there were 0.9 young fledged per active nest. All of the nesting territories that were active in one year were reused the following one. All of the montane nests they studied were near igneous rock faces 9–90 meters high, and the two canyon nests were on south-facing cliffs, situated 5–15 meters below the top of the cliff face.

Evolutionary Relationships and Status

Brown and Amadon (1968) described this species as a somewhat aberrant member of the genus *Buteo*, without further speculation as to its possible relationships. Stresemann and Amadon (1979) similarly made no comments on its possible affinities. They placed it in linear sequence between the Andean puna hawk (*B. poecilochrous*) and the Hawaiian hawk (*B. solitarius*), neither of which would seem to be a likely close relative. Palmer (1988) stated that "no useful information" is available as to its affinities.

Oberholser (1974) judged that the Texas population of zone-tailed hawks has perhaps been declining throughout the twentieth century, presumably because of loss of nesting habitat. However, it has nested in recent years in the Edwards Plateau area as well as in Big Bend National Park. In New Mexico the birds still breed in a variety of habitats (Matteson and Riley, 1981). There have also been a few breeding attempts in southern California (Weathers, 1983). As of the mid-1970s perhaps about 20 nests were present in the Big Bend and trans-Pecos areas of Texas, and although only 8 pairs were known from New Mexico the state population was probably several times that number. Similarly, only 12 Arizona nesting sites were known in the early 1970s, but the entire state population was unknown (Porter and White, 1977). Later, Millsap (1981) considered it uncommon but widespread in Arizona, and reported on 28 nests. No zone-tailed hawks appeared on the Audubon Society's 1986 Christmas Counts.

Red-tailed Hawk *Buteo jamaicensis* (Gmelin) 1788

Other Vernacular Names

Eastern red-tailed hawk (*borealis*), Florida red-tailed hawk (*umbrinus*), Fuertes red-tailed hawk (*fuertesi*), Harlan's hawk (*harlani*), Krider's hawk (*kriderii*), western red-tailed hawk (*calurus*); buse à queue rousse (French); aguililla parda (Spanish)

Distribution

Breeds from western and central Alaska, central Yukon, western Mackenzie, northern Saskatchewan, northern Manitoba, central Ontario, southern Quebec, New Brunswick, Prince Edward Island, and Nova Scotia south to southeastern Alaska, Baja California, Sinaloa, Oaxaca, Tamaulipas, southern Texas, the Gulf coast, and Florida, and in the highlands of Central America to Costa Rica and western Panama; also on the Tres Marias and Socorro islands, and on the northern Bahamas, Greater Antilles, and northern Lesser Antilles.

Winters from southern Canada south throughout the remainder of the breeding range. (See Figure 45.)

North and Central American and Associated Insular Subspecies

B. j. alascensis Grinnell: Resident or partial migrant from southeastern Alaska to Vancouver Island.

[*B. j. harlani* (Audubon): Reportedly breeds in Alaska from Norton Sound east to the western Alaska Range, north to tree line, and south to the Alaska Peninsula, and from southwestern Yukon to northern British Columbia, variably merging in most areas with *calurus*. Winters from Kansas and Missouri to the Gulf coast. Now regarded either as a valid subspecies (Mindell, 1983), or as a melanistic morph of *calurus* (Palmer, 1988).]

B. j. calurus (Cassin): Breeds from central Alaska (north to the Yukon, where merging with *harlani*) south to Baja California and Texas. Winters southward, occasionally to Panama, but with more southerly populations residential.

[*B. j. kriderii* Hoopes: Reportedly breeds in the Great Plains from Canada south to Wyoming and Nebraska. Winters south to the Gulf coast. Probably only a leucistic plumage morph of *calurus* or *borealis*, and not recognized by Palmer (1988).]

B. j. borealis (Gmelin): Breeds in eastern North America exclusive of Florida, west to the Great Plains (see *kriderii* description above). Includes *abieticola* Todd. Migratory in north; resident farther south.

B. j. umbrinus Bangs: Resident in peninsular Florida.

B. j. fuertesi Sutton and Van Tyne: Breeds from Texas south to Nuevo Leon and west to Sonora. Probably resident.

B. j. fumosus Nelson: Resident in the Tres Marias Islands, Mexico.

B. j. socorroensis Nelson: Resident on Socorro Island, Mexico.

B. j. hadropus Storer: Resident in the Mexican highlands from Jalisco to Oaxaca.

B. j. kemsiesi Oberholser: Resident from Guatemala to Nicaragua.

B. j. costaricensis Ridgway: Resident from Nicaragua to Panama.

B. j. jamaicensis (Gmelin): Resident in the northern West Indies, excepting the Bahamas and Cuba.

B. j. solitudinus Barbour: Resident of the Bahamas and Cuba.

Description

Adult (sexes alike). *Typical morph* (of *borealis*): Most upperparts fuscous margined with cinnamon-brown to russet, these edges broadest and most conspicuous on the nape and anterior interscapulars; the forehead usually with much whitish; upper wing coverts and secondaries brown, barred more or less distinctly and broadly subterminally banded with fuscous and narrowly tipped with pale brown; the inner webs of the secondaries paling to white internally; the tertials washed with cinnamon-brown and usually with much white; the outermost primaries very broadly tipped with fuscous-black and with the outer webs of this color, the inner webs whitish, unbarred for their basal half or more; other primaries similar except that the dark areas become a little paler and both the dark and the white areas are barred with fuscous-black; upper back dark fuscous; lower back and rump

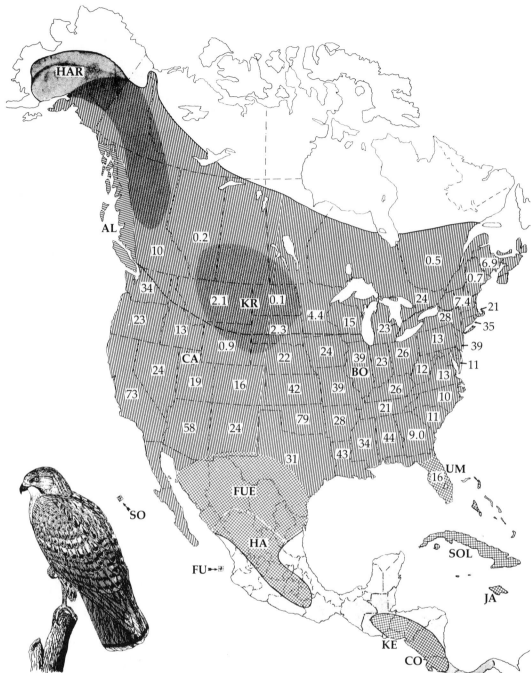

Figure 45. North and Central American breeding range of the red-tailed hawk, with usual northern wintering limits of the migratory populations indicated by dashes and southern wintering areas shown by shading. Relative winter state or provincial density indices (average number seen per Audubon Christmas Count, 1986) are shown for major wintering areas. Approximate breeding or residential ranges are shown for the races alascensis *(AL),* borealis *(BO),* calurus *(CA),* costaricensis *(CO),* fuertesi *(FUE),* fumosus *(FU),* hadropus *(HA),* harlani *(HAR),* jamaicensis *(JA),* kemsiesi *(KE),* kriderii *(KR),* socorroensis *(SO),* solitudinus *(SOL), and* umbrinus *(UM). Since the range limits and taxonomic validity of* kriderii *are still unsettled, its reported Great Plains breeding distribution is shown by a shaded area overlaying the indicated range of* calurus, *as is the apparently extensive zone of intergradation in western Canada between* harlani *and* calurus. *Indicated range limits of these and other contiguous races should not be considered authoritative.*

somewhat paler, upper tail coverts variable, from whitish washed with buff and cinnamomeous to nearly solid cinnamon and more or less barred with brown; rectrices bright hazel to brown narrowly tipped with whitish and crossed by a subterminal band of fuscous-black; lores whitish with black shafts to the feathers; a variably evident malar stripe of dark fuscous; chin and middle of throat usually plain whitish, sometimes streaked with dark olive-brown; sides of throat and breast dark olive-brown to fuscous in the middle of the feathers; rest of underparts whitish with the upper abdomen and sides variably barred and streaked with sepia; the thighs often washed with pale buff and sometimes indistinctly barred with pale brown to pale hazel; under tail coverts whitish; under wing coverts whitish, usually marked with dark brown, especially along anterior edges. Iris yellow (in younger birds) to brown (in older ones); bill bluish horn to black, becoming ashy to lead-colored basally; gape creamy to olive-buff; tarsi and toes yellow to olive-yellow; claws dark.

Dark (melanistic) morph (western North America): Highly variable (see *harlani* below), but typically the head, body, and wing coverts all dark brown to black. Remiges heavily barred with blackish, but the under tail coverts often rufous, and the rectrices distinctly rufous above, often with up to eleven narrow black bands. The tail sometimes with blackish mottling rather than banding, and approaching the typical condition of *harlani*, with almost no rufous evident.

Rufous (erythristic) morph (western North America): Similar to the dark morph, but underparts deeply suffused with rufous, the thighs and under tail coverts barred with dark brown, and the abdominal band wide and chocolate-brown.

Subadult. A complete molt from the juvenal to adult plumage begins early in the spring following hatching, and continues through the summer or into fall, with much individual variation (Bent, 1938). The adult tail is not attained until the second autumn of life; thus some yearlings might breed prior to having their adult tail pattern. The incidence of breeding by such yearlings is believed to be very low (Luttich, Keith, and Stephenson, 1971).

Juvenile (sexes alike). *Typical morph:* Deep brown above, with white and buff streaks and spots, especially on the scapulars; tail pale brownish gray, narrowly crossed with nine or ten fuscous bars, and tipped with whitish; underparts white, with variable brown striping or spotting. *Dark morph:* Similar to adults of the same morph, but more mottled with buff or rufous below, and the tail heavily banded with brown and black, lacking rufous tones. *Rufous morph:* lacking rufous tones, and instead heavily streaked below, as in dark-morph immatures. Iris dull yellow to brownish cream; bill blackish toward tip, lilac gray basally; cere light yellow to olive-buff; tarsi and toes dull yellow to greenish yellow.

Major Plumage Variations (mainly after Lish and Voelker, 1986).

kriderii: Leucistic, with the overall color pattern visible but pale, the tail white to pinkish or pale brown and usually with a subterminal dark band, the head variably white, and the underparts and thighs sometimes immaculate white. White "windows" are visible from above at the base of the primaries in flight (as in ferruginous and rough-legged hawks). Immatures generally resemble adults, with extensive white dorsal spotting and the underparts mostly white, but the rectrices are barred, the paler portions varying from olive-buff to medium brown.

fuertesi: Almost as white below as *kriderii*, but with a rufous tail and a more brownish head. The under wing coverts have dark leading borders. Immatures have heavily barred brownish tails.

calurus: Usually more rufous below than in *borealis*, and with more heavily barred underparts. The under wing coverts are usually dark only anteriorly. Melanistic birds may appear mostly black, but usually have heavily barred reddish tails. Immatures closely resemble those of *harlani*.

harlani: Normally melanistic (blackish) overall, but with the upper tail surface whitish basally, usually banded subterminally with blackish and with varying degrees of rufous tinting, but sometimes with dark gray mottling, or with numerous narrow black bands. Under wing coverts are black, usually mottled with white, and some white mottling is also usually present on upper breast. Immatures are similar, but have banded rather than mottled tail patterns. According to Mindell (1985), black or brown mottling and streaking on the tail is the most distinctive plumage trait of *harlani* (this is lacking in typical *calurus*), but individuals can

Identification

In the hand. This highly variable species can usually be separated from other North American buteos by the combination of having the four outer primaries notched on their inner webs, a wing length of 337–427 mm, and a tail that at least in adults is usually somewhat reddish brown to pinkish, but rarely may be almost entirely white or heavily mottled, streaked, or barred with grayish black, becoming a dirty white basally. Light-morph birds have a characteristic dark patagial stripe along the leading edge of their underwing.

In the field. When perched, typical individuals of this species are best identified by their generally whitish underparts with varying amounts of darker spotting or streaking across the abdomen, and a rusty or red-tinted upper tail surface. The tips of the wings reach nearly to the end of the tail, and usually some white spotting is evident on the scapulars and upper tail coverts, at least in lighter birds. Extremely pale (leucistic) variants may lack any dark markings on the abdomen, and the tail may appear nearly white. Similarly, very dark (melanistic) individuals may have no white spotting on the scapulars and the entire underparts may appear almost blackish (see descriptions of plumage variants above).

In flight, red-tails typically exhibit a conspicuous dark band of feathers along the leading edge of each wing when seen from below, and have variably speckled whitish underwing linings with a ground color that is similar to or at least not notably lighter than the bases of the flight feathers (compare Swainson's hawk). Dark-morph red-tails are extremely variable (as noted above), but generally they too have underwing linings that are not notably lighter than the bases of the adjoining flight feathers. Although the dorsal tail surface typically appears bright reddish when seen from above, this color is not evident from below unless some light is transmitted through the tail by strong back-lighting. When seen from either above or below the primaries appear paler than the secondaries. Leucistic birds often show a white "window" area along the bases of the primaries when viewed from above, potentially causing confusion with ferruginous hawks. Typical *harlani* individuals have no reddish color present on the generally mottled grayish tail. They usually also have some whitish breast streaking amid the dark feathers, and

otherwise range from light to melanistic in general plumage, and intergrade types may have varying amounts of rufous and/or barring in the tail. Intergrades with *calurus* are frequent.

Measurements (of *borealis*, in millimeters)

Wing, males 337–396 (ave. of 35, 369.6), females 370–427 (ave. of 27, 388.8); tail, males 197–240 (ave. of 35, 215.6), females 214.5–254 (ave. of 27, 230.3) (Friedmann, 1950). Average egg size 59 x 47 (Bent, 1937).

Weights (in grams)

Males, ave. of 108, 1028; females, ave. of 100, 1224 (Snyder and Wiley, 1976). Range of weights 710–1550, ave. 1082 (sex and sample size unspecified) (Clark and Wheeler, 1987). Estimated egg weight 71.9, or 5.9% of female.

flight feathers that are much paler at their bases and at the tips, sometimes producing an indistinct windowlike effect when seen from below. Immature individuals of all color morphs lack definite red coloration on the tail, but instead are narrowly barred with brown and black. They sometimes have adultlike white spotting on the scapulars or a darker band of abdominal spotting, and additionally usually show the species's characteristic dark band along the leading edge of the underwing lining, plus lighter "panels" in the primaries. A loud descending scream is the typical alarm call. When soaring, the wings are held slightly uptilted, and they appear to be relatively broader and less swept-back than is typical of the Swainson's hawk. In the Great Plains, soaring red-tails are more likely to be confused with ferruginous hawks, but the darker abdominal band (rather than dark thighs), and the dark leading edges of the underwing linings help separate soaring red-tailed hawks from ferruginous hawks.

Habitats and Ecology

This species has an extremely wide tolerance for habitat variation; Beebe (1974) suggested that it is perhaps second only to the peregrine falcon in its tolerance of diverse habitats among North American raptors, which he attributed to its very broad spectrum of prey. Nevertheless, some clear habitat preferences do exist and have been analyzed by a variety of studies. Petersen (1979) determined habitat preferences by using marked birds; he found that these preferences varied by season, sex, and breeding status. Habitat preferences during winter for both sexes were oriented toward upland pasture, grassland, and hardwood habitats, with females also using lowland hardwoods and males marsh-shrub communities. A similar use of open woods or wooded river bottoms was observed during winter in Iowa by Weller (1964). With spring, males no longer preferred the marsh-shrub areas but were still oriented toward upland hardwoods, pastures, and grasslands, plus lowland pastures. Females continued to use mainly upland and lowland hardwoods, probably as a reflection of their orientation toward a nest site. Summer habitat use of both sexes centered around upland hardwoods and grass-dominated cover types, with hunting the primary summer activity. The members of a pair rarely hunted within 100 meters of

one another, and continued to show significant habitat differences until fall. Then upland hardwoods, pastures, and grasslands were mainly used, with hunting largely occurring in grassy areas, especially those with grasses less than 10 centimeters tall.

Preferred breeding woodland habitats, at least in eastern North America, seem to be those with open rather than closed structures; nests are selectively placed near the edges of dense stands, with relatively few placed in dense, closed-canopy woods or in isolated trees (Gates, 1972; Orians and Kuhlman, 1956).

Howell et al. (1978) compared habitat structure with productivity in red-tailed hawks, and judged that productivity may be related to woodlot structures as well as to percentage of hunting territories in fallow pastures. Sites with high productivity tended to have larger amounts of fallow pasture and smaller amounts of crop pasture, as well as relatively small amounts of the habitat in woodlot. Various specific parameters of woodlot structure, such as canopy height, saplings per unit area, tree basal area, etc., were apparently unrelated to productivity.

Generally the birds seem to favor patches of woodland grass–dominated areas, forest edges, open woodlands, and savanna-like areas with scattered tall trees. Where trees are lacking, as in the arid southwest, they may roost and nest on tall cacti, cliffs, or similar elevated sites, and forage in nearby grassland, semidesert, or desert habitats. Only the treeless arctic habitats have not yet been invaded by the species. Smith and Murphy (1973) found the species to be one of the most diverse in its nesting sites of all twelve raptors nesting in their Utah study area, rivaled only by the great horned owl. Indeed these two highly adaptable predators often coexist in many areas and feed on similar prey, although their differences in activity cycles tend to reduce competition. Although the hawks often provide nest sites for the owls (which nest earlier than the hawks and may usurp their nests), the owls sometimes prey on hawk nestlings, and to some degree the reverse type of predation may also occur. The interactions between these species are complex and the overall effects perhaps vary according to such factors as prey diversity and prey abundance, but in many areas the two species appear to survive and reproduce effectively in the presence of the other (Orians and Kuhlman, 1956; Craighead and Craighead, 1956;

Luttich, Keith, and Stephenson, 1971; Dunstan and Harrell, 1973; McInvaille and Keith, 1974; Houston, 1975; Springer and Kirkley, 1978; Petersen, 1979; Kirkley and Springer, 1980; Gilmer, Konrad, and Stewart, 1983).

Red-tailed hawks seem to maintain rather constant populations in most areas throughout the year; the birds in temperate regions tend to be quite sedentary and most multiyear studies indicate that about 90 percent of resident territorial pairs will breed each year, the breeding population often remaining highly consistent from year to year (Luttich et al., 1970; Petersen, 1979; Janes, 1984a). Obviously different habitats and their associated prey populations are able to support widely differing hawk populations; McGovern and McNurney (1986) summarized literature suggesting average breeding densities of from 1.3 to 24.9 square kilometers per breeding pair in 10 different studies, and Kirkley and Springer (1980) gave ranges of from 1.3 to 50 square kilometers for 14 different studies, with estimated adult nonbreeding rates of from 4.5 to 35 percent (unweighted average of 10 studies, 15.7 percent).

In addition to being widely sympatric with great horned owls, red-tails are also variably sympatric with Swainson's and ferruginous hawks. Schmutz (1977) studied the ecological interactions of these three species in southern Alberta, and found a substantial degree of ecological (especially dietary) overlap among them, although the red-tailed hawk was largely restricted to areas where tall stands of trees were available and thus was rather rare in his study area as compared with the other two species. Gilmer, Konrad, and Stewart (1983) found a lower estimated rate of dietary overlap between the red-tailed hawk and the Swainson's hawk (40 percent) than between the red-tailed hawk and ferruginous hawk (62 percent), and even less between red-tails and great horned owls (20 percent). They considered that nest site differences and prey abundance helped to reduce competition among them. Rothfels and Lein (1983) also studied sympatric populations of red-tailed and Swainson's hawks in Alberta, and determined from measures of nest dispersion that interspecific as well as intraspecific territorial dispersion occurred in these species, the former of which was also indicated by actual observations of interspecific territorial interactions between these species. Similarly, Janes (1984a)

found a strong degree of interspecific territoriality between red-tailed and Swainson's hawks in Oregon, with some pairs of red-tails losing parts of their territories to the later-arriving Swainson's hawks each year of his study, but with no observable effects on the former's reproductive success. Mader (1978) found red-tailed hawks and Harris' hawks breeding sympatrically in southern Arizona, but the red-tails occupied a much greater range of habitats and generally were able to nest in vegetationally simpler and more arid habitats than were the Harris' hawks. The latter species, however, had a much longer nesting cycle, benefited from nest-helping, and was able to nest in dense-canopied or spiny trees such as paloverde (*Cercidium*) and ironwood (*Olneya*) that the red-tailed hawk apparently could not use.

Foods and Feeding

In conjunction with its broad geographic and ecological ranges, the red-tailed hawk has a remarkable capacity for modifying its diet to accommodate local prey sources. Beebe (1974) states that studies made in a particular region, or even a particular time of year, may not indicate the kinds of prey taken elsewhere, or even by the same birds in the same area during a different time of the year. Indeed, two birds living close together may have food preferences and hunting methods as different as two different species.

With this sort of diversity, it is doubtful that any great degree of generalization can be made about the foods of this species. Sherrod (1978) summarized 13 published and unpublished studies of red-tailed hawk foods, analyzing them whenever possible in terms of numerical composition of prey species and categories. In eleven studies where numerical composition was established, mammals comprised from 37 to 88 percent of the total diet, averaging (unweighted average) 68 percent. Birds comprised from 4 to 58 percent, averaging 17.5 percent, reptiles (mostly snakes) and amphibians from nil to 41 percent, averaging 7 percent, and invertebrates from nil to 21 percent, averaging 3.2 percent.

When biomass estimates are used, the relative importance of mammals is made more apparent. Thus, Smith and Murphy (1973) estimated that 92–95 percent of the food biomass on their study area was comprised of jackrabbits (*Lepus californicus*) alone. All told, mammals made up 89 percent of the prey items and 99 percent of the prey biomass, while birds made up 9 percent of the prey items and only 0.2 percent of the prey biomass. Similarly, in Wisconsin, winter foods consisted mainly of cottontails (44 percent biomass) plus mice and voles (28 percent), with birds (mostly pheasants) comprising 19 percent. During spring the importance of pheasants increased and that of rabbits, mice, and voles correspondingly declined. A heavy use of cottontails, *Microtus* voles, and *Peromyscus* mice has also been found in a considerable number of other red-tailed hawk studies from the midwest (Petersen, 1979).

Farther west and north, ground squirrels and hares (jackrabbits or snowshoe) largely replace the cottontail and smaller rodents as primary prey (Fitch, Swenson, and Tillotson, 1946; Craighead and Craighead, 1956; Smith and Murphy, 1973; McInvaille and Keith, 1974). McInvaille and Keith found that the Richardson's ground squirrel (*Spermophilus richardsonii*) was consistently the most important food item brought to nests in Alberta on a biomass basis during five of seven years of study, with snowshoe hare (*Lepus americanus*) most important during the other two years and second in importance in five years. Hares were used increasingly as they became more abundant in the area during that period, but changes in ground squirrel populations were not reflected in the hawks' diet. The Craigheads (1956) observed that breeding red-tails foraged mainly on ground squirrels, while adult wintering birds subsisted largely on jackrabbits and large birds (quail and pheasants), while at the same time immatures were concentrating their hunting efforts on field mice.

Red-tailed hawks are diurnal hunters, typically starting their hunts later than ferruginous hawks but earlier than Swainson's hawks and tending to spend a greater portion of the day in active foraging than any of the other large raptor species found in eastern Utah (Smith and Murphy, 1973). In that area the birds were found to have larger breeding home ranges (averaging 5.7–7.3 square kilometers) than the other two buteos or the great horned owl, but smaller than the home ranges of golden eagles. Petersen (1979) estimated smaller average winter home ranges in wooded areas of Wisconsin (1.64 square kilometers), which in spring remained about the same for males but diminished to only 0.85 square kilometers for females, owing to their increasing involvement with nesting. By summer,

the ranges of both sexes averaged about 1.2 square kilometers, with considerable individual variability. During that period the birds would typically forage from sunrise until midmorning, then rest in cool and shaded tree canopies until hunting resumed in late afternoon and continued until sunset. During fall their home ranges increased to an average of about 2 square kilometers, in part because the birds were spending some of their time defending their territories against migrants.

The commonest hunting pattern of red-tailed hawks is soaring and stooping, followed in frequency by still-hunting from perches (Johnson and Peeters, 1963). Young birds only gradually develop hunting skills, and increasing versatility of hunting techniques (increasing use of aerial searching for prey rather than perching techniques) may improve their hunting success rates (Johnson, 1986).

Social Behavior

It is generally believed that red-tailed hawks establish permanent, perhaps life-long pair bonds, a view supported by the high level of reuse of nest sites and the tendency for population stability within single areas. Petersen (1979) did indeed find the birds to maintain their pair bonds throughout the year, but also observed one case of apparent mate replacement by a female who had been constructing a nest with another male only a short time previously. The other radio-tagged male left the area afterward. However, other pairs maintained their bonds throughout his four-year study, with most territories remaining virtually unchanged during the entire period.

Janes (1984b) estimated that females exhibited a high rate of return to previously occupied territories in Oregon, and the territorial boundaries remained relatively stable from year to year. Plumage variation among the males was too small to estimate their territorial fidelity, although several distinctively marked males were observed to return to previously occupied territories. Similarly, Belyea (1976) found that 41 percent of the nests he studied in Michigan were reoccupied a second year, and 13 percent a third, with prior nesting success apparently not influencing the probability of nest reuse.

Conner (1974) described an observation of aerial courtship involving a pair doing dives, barrel rolls, and ascents and sometimes making contact with their talons while they spiraled downward for about three seconds. As the female began to return to her nest the male approached her from above and grasped her back. This unusual aerial attempt at copulation was brief, and afterward the two returned to the nest tree where normal copulatory behavior commenced. Voelker (1969) reported seeing a pair of birds circling overhead, the two moving in ever smaller circles until they were "face to face"; they then grasped one another's beaks and spiraled to earth, where copulation occurred.

Breeding Biology

In Wisconsin, red-tailed hawks begin nest construction or repair in late February or early March. They typically select sites near woodland edges, with about half the nests in woodland and the remainder in gallery forests or open-growth trees (Petersen, 1979). In a sample of 167 New York nests, only one was in a conifer, while beech (*Fagus grandifolis*) was the most commonly used tree substrate (70 nests), followed by maples (*Acer* spp.) with 45 nests (Bull, 1974). Beech and maple were also found to be the most commonly used nesting trees in a Michigan study (Belyea, 1976). Among 723 Ontario nests, 699 were in trees, 14 were on cliff ledges, and 10 were on tower structures. Deciduous trees were utilized there over conifers at a ratio of about 4:1; elms (153 nests) and pines (100 nests) were the most commonly used nest supports (Peck and James, 1983). Combining the clutch size data for both of these areas, a total of 349 nests had from 1 to 4 eggs, with an average clutch of 2.42 eggs. Kirkley and Springer (1980) summarized average clutch size estimates from their own and 8 other studies; these ranged from 2.0 to 2.9, suggesting that there is little if any apparent geographic variation in this trait. Henny and Wight (1972) provided a similar survey of clutch sizes in this species.

At the northern edge of the species's range in Alberta, where the birds are present for only 5–6 months, within three weeks after arrival mated pairs may select a nest site, produce a clutch, and begin incubation (Luttich, Keith, and Stephenson, 1971). Eggs are laid at approximate two-day intervals, and if the first clutch is lost a second is regularly laid and very rarely a third or possibly even a fourth may be produced (Bent, 1937). Incubation is done mainly by the female, who is

fed as usual by the male. It lasts 28–32 days, with the young hatching at intervals of 1–2 days. The fledging period is normally from 44 to 46 days (Fitch, Swenson, and Tillotson, 1946; Luttich, Keith, and Stephenson, 1971).

In spite of this species's enormous geographic and ecologic range, there seems to be a fair degree of consistency in its reproductive success. Kirkley and Springer (1980) summarized data on average brood size, average number of young fledged per nesting attempt, and average nest failure rate for some 19 studies (not all studies provided estimates of all these parameters). Average brood size in 9 studies ranged from 1.9 to 2.6 young, and average number of fledged young per nesting attempt ranged from 0.9 to 1.64 in 12 studies. Nest failure rates ranged from 16 to 50 percent in 10 studies. Similarly, Mader (1982) summarized productivity data from several studies involving nearly 500 clutches, and Janik and Mosher (1982) estimated that various published studies totaling 930 nests had an overall nest success of 66 percent and an average of 1.35 young produced per nest. Luttich, Keith, and Stephenson (1971) have attempted to fit productivity data such as these to projected survivorship curves for the species, but too many gaps in the data (such as the percentage of yearlings breeding and the incidence of nonbreeding in the adult population) exist to allow for a very convincing "fit." Hubbard, Shipman, and Williams (1988) determined that about 16 percent of 9,388 red-tailed hawks seen in New Mexico were immatures, which compares favorably with an estimated 20.6 percent annual postjuvenal mortality rate calculated by Henny and Wight (1972).

Evolutionary Relationships and Status

This species is morphologically and ecologically quite similar to various South American buteos, especially the red-tailed buzzard (*B. ventralis*) (Voous and de Vries, 1978). The AOU (1983) regards the relationships of *jamaicensis*, *ventralis*, and the ecologically and morphologically similar Eurasian buzzard *B. buteo* as "uncertain."

In contrast to the situation for many North American hawks, the population of the red-tailed hawk has seemingly improved considerably in recent decades. Migration counts in central North America between the late 1940s and early 1970s suggest that an approximate 70 percent population increase may have occurred during that time, and the red-tailed hawk has tended to displace the red-shouldered hawk from some of its upper midwest breeding areas (Petersen, 1979). Additionally, in the northern Great Plains the red-tailed hawk has benefited from increased tree growth in areas that were once pure grasslands, at the corresponding expense of habitat best suited for ferruginous and Swainson's hawks (Houston and Bechard, 1983). Similar trends can be seen in Oregon, where the Swainson's hawk is suffering increasingly from competition for breeding territories with red-tailed hawks (Janes, 1984a).

Illustrating this trend, the Audubon Society's Christmas Bird Count data from the early 1980s as compared to the early 1970s suggest that the red-tailed hawk population wintering in the United States and southern Canada (or probably most of the total North American population) had increased about 33 percent during that interval, to an estimated 350,000 birds (Anonymous, 1986). This makes the species the most abundant wintering hawk and one of the most common breeding hawks in North America, after the American kestrel and broad-winged hawk. Using the estimate of 350,000 birds for the present-day (mid-1980s) wintering population, Christmas Count data indicate that the highest densities of birds occur in California, Arizona, and Oklahoma, with estimated early winter populations of 45,800, 26,200, and 13,500 birds respectively.

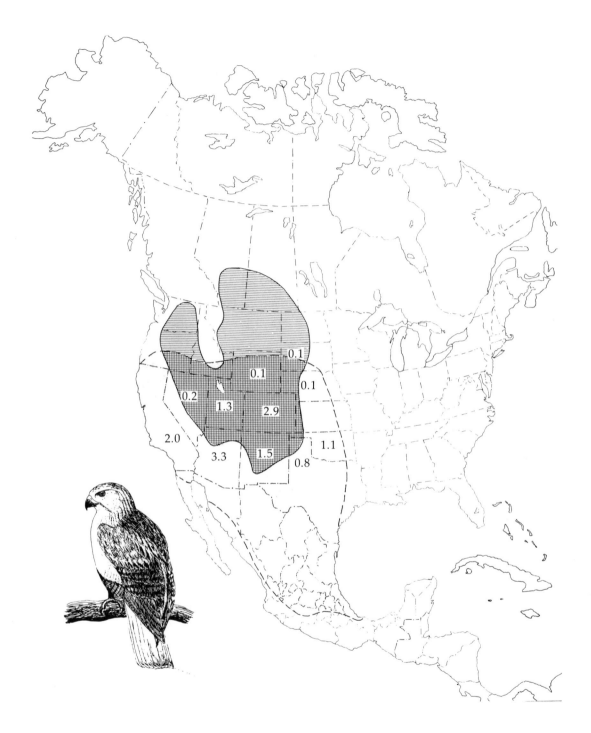

Figure 46. Breeding (hatching), residential (cross-hatching), and wintering (shading) ranges of the ferruginous hawk. Relative winter state or provincial density indices (average number seen per Audubon Christmas Count, 1986) are shown for major wintering areas. The indicated northern limit of wintering is quite subjective and probably highly variable from year to year.

Ferruginous Hawk *Buteo regalis* (Gray) 1844

Other Vernacular Names

Ferruginous rough-leg; buse rouilleuse (French); aguililla patas asperas (Spanish)

Distribution

Breeds in grassland habitats from eastern Washington, southern Alberta, southern Saskatchewan (rarely or formerly also southwestern Manitoba), and western North Dakota south to eastern Oregon, Nevada, northern and southeastern Arizona (where now very rare), northern (formerly also southwestern) New Mexico, northwest Texas (possibly now extirpated), western Oklahoma (where rare), western Kansas (where declining) and western Nebraska (where rare). Recorded in summer (but not known to breed) in northeastern California.

Winters primarily in grassland habitats from the central parts of the breeding range (Nevada, Colorado, and Kansas) south to northern Mexico. (See Figure 46.)

North American Subspecies

None recognized.

Description

Adult (sexes alike). Typically occurring in either of two morphs, with few intermediate individuals. *Light morph:* Feathers of forehead, crown, nape, and sides of face and throat with fuscous shaft streaks and edged with orange-cinnamon, laterally and basally white, producing a distinctly streaked brown and white appearance; most upperparts bright orange-cinnamon to ferruginous (rust-brown), the feathers with large and conspicuous fuscous or fuscous-black centers, these centers largest on the upper back and smallest on the median wing coverts; greater wing coverts deep gray, becoming browner internally; secondaries and inner primaries deep gray, becoming paler, browner and whiter on the inner webs, the secondaries barred with dusky, the inner primaries less distinctly so; the outer primaries largely immaculate white and broadly tipped with blackish on the inner webs, and with the outer webs deep gray; upper tail coverts bright orange-cinnamon to rusty barred with fuscous, and white basally; rectrices varying from whitish to light neutral gray, more or less washed with dull rust and flecked and mottled with rusty and gray; chin, throat, and breast white, the breast feathers with cinnamon shaft streaks; abdomen, sides, and flanks white with wavy cross-bars of rusty and fuscous; thighs deep rusty barred with fuscous; under tail coverts white; axillaries and under wing coverts mixed white and deep rusty. Iris clear light yellow; cere and feet bright yellow to greenish yellow; bill very dark bluish horn, becoming pearl-gray basally; mouth purplish flesh to pale gray; corner of mouth somewhat bluish.

Dark morph: entire head and body, upper and under wing surfaces, and tail coverts deep clove-brown to dark sepia, some of the feathers, especially on the throat, breast, sides, and interscapulars edged with cinnamon; remiges as in light phase except that the deep gray is replaced by deep brown; tail usually grayish, but with the outer webs of the rectrices heavily washed with deep brown. Iris light to medium brown, other softparts as in light morph. A questionably separable "rufous" (erythristic) morph is similar, but is variably reddish throughout, especially on the underparts.

Subadult (sexes alike). By the second fall of life the upperparts are similar to the adult, with very broad cinnamon edgings on the scapulars and wing coverts, but the underparts are still juvenile-like, except that the brown tibiae are attained. Older birds become more extensively white below, the tail progressively whiter, and the upperparts paler (Bent, 1937).

Juvenile (sexes alike). *Light morph:* The entire underparts practically immaculate white (but a pale rufous bloom present on the breast of recently fledged birds); occasional fuscous subterminal spots on the sides and very small paler ones on the flanks; upperparts with less ferruginous or orange-cinnamon; rectrices gray with white basal areas and with an indistinct subterminal fuscous band, the whitish basal portion marked irregularly with fuscous. *Dark morph:* Dark brown over the entire body and wing coverts, the breast sometimes tinged rufous; tail dark brown above, with faint darker bands. Iris brownish yellow; bill black, becoming olive-gray basally; gape and corner of mouth canary-yellow; tarsi and toes dull yellow.

Measurements (in millimeters)

Wing, males 421–440 (ave. of 8, 431), females 427–450 (ave. of 4, 432.3); tail, males 231–246 (ave. of 8, 237.4); females 238.5–252 (ave. of 4, 244) (Friedmann, 1950). Average egg size 61.2 x 48 (Bent, 1937).

Weights (in grams)

Males, ave. of 15, 1059; females, ave. of 4, 1231 (Snyder and Wiley, 1976). Range of weights 980–2030, ave. 1578 (sex and sample size unspecified) (Clark and Wheeler, 1987). Estimated egg weight 73, or 5.9% of female.

Identification

In the hand. The combination of feathered tarsi and a gape at least 42 mm wide at the base distinguishes this species from all other North American raptors except the much larger golden eagle.

In the field. When perched (usually in open grassland or semiarid habitats), this large hawk typically exhibits rufous-cinnamon on the upper wing coverts and scapulars and almost always has extensive amounts of white on the underparts, which sometimes are entirely white except for some cinnamon on the thighs (or "leggings"). The relatively rare dark morph (which probably represents less than 10 percent of the total population and is most common in Saskatchewan and Alberta) may be much more extensively brownish to cinnamon-colored on the underparts, but these birds also show cinnamon or rufous on the upper wing coverts. Both adults and young birds show a conspicuous yellow gape stripe marking the edge of the unusually wide bill. The wingtips nearly reach the tip of the tail. Immatures of the typical light morph resemble adults in having nearly white underparts (with a pale cinnamon blush on the breast of newly fledged juveniles), but have much less rufous on the wing coverts and none on the thighs. Both immatures and adults have pale upper tail surfaces, those of adults often tinted with rufous and those of young obscurely banded with pale gray and black posteriorly, but nearly white basally.

In flight, typical light-morph individuals show almost immaculate underparts save for light brownish speckling on the under wing coverts and a more conspicuous cinnamon to rufous V-mark formed on the posterior abdomen by the thighs and legs. The rare dark-morph birds do not show this contrast, but when birds of either morph are seen from above a distinctive white "window" is typically formed by the bases of the primaries, and the upper tail surface appears relatively unbanded and ranges from almost entirely white to tawny-rufous. From below, very dark birds show white only on their flight feathers and on the tail, the latter lacking distinctly blackish banding posteriorly (except in first-year birds, which are indistinctly barred with blackish and tipped with white) of otherwise very similar dark-morph rough-legged hawks. The strong rufous cast on the upper wing coverts and upper back is also usually apparent in most adult birds. The birds soar on only slightly uptilted wings and also hover at times. They sometimes utter rather harsh alarm calls similar to those of red-tailed hawks, especially when disturbed near the nest, but otherwise are relatively silent.

Habitats and Ecology

This is a species that is primarily adapted to semiarid grasslands of the Great Plains, especially areas of rather unbroken terrain having scattered trees, rock outcrops, and fairly tall trees associated with rivers or streams. Although locally resident year-round in more southern parts of the range, some birds there move into high mountain shortgrass valleys during postbreeding and migration periods (Evans, 1982). At least in Nevada (Herron, 1985) and probably elsewhere the birds occur in habitats that support relatively high densities of rodents and lagomorphs, such as areas having a mixture of grasses, forbs, and white sage (*Ceratoides lanata*), with nesting occurring in scattered junipers overlooking broad, open valleys. Optimum habitat probably consists of unbroken prairie grassland that is at most slightly grazed, with available elevated nesting sites associated with hills and ridge systems separating broad, flat valleys, providing expansive hunting areas and development of thermal air currents (Ensign, 1983).

In conjunction with its large size (being the largest of the North American buteos), its population densities are relatively low. Ensign (1983) estimated a crude density (based on his total southeastern Montana study area) of 5.41 square kilometers per nest for 91 nests, and an ecological density (based on those portions of the study area considered to represent suitable nesting habitat) of 2.22 square kilometers per nest. Similar crude densities were reported by Weston (1969) for Utah, but substantially lower densities were reported by Blair (1978) for a study area in northwestern South Dakota only 60 kilometers from Ensign's study area. Smith and Murphy (1973) estimated that in Utah the annual nesting density of ferruginous hawks averaged about 21 square kilometers per pair over a four-year period. They estimated the average home ranges for nesting pairs to be from about 5 to 6.5 square kilometers, or slightly smaller than those of red-tailed hawks in the same area. A density of a pair per 16.4 square kilometers was estimated for eastern Colorado (Olendorff, 1973), and a pair per 17.4 square kilometers in South Dakota (Lokemoen and Duebbert, 1976). Woffinden (1975) estimated that a reduction in jackrabbit densities had a direct effect on breeding hawk densities in Utah, as well as affecting their breeding success.

Foods and Foraging

To a very large degree this species is specialized for foraging on a restricted food base that consists almost entirely of grassland rodents and lagomorphs. Snyder and Wiley (1976) estimated from available literature sources that mammals comprised 85 percent of the species's prey on a numerical basis.

Lokemoen and Duebbert (1976) estimated from pellet remains and prey brought to nests that mammals comprised over 90 percent of the total biomass taken, with the Richardson's ground squirrel (*Spermophilus richardsonii*) alone totaling 68 percent and the white-tailed jackrabbit (*Lepus townsendii*) another 17 percent. In Smith and Murphy's (1973) Utah study black-tailed jackrabbits (*L. californicus*) comprised over 90 percent of the foods in each of two years, while cottontails and ground squirrels made up most of the remainder. In Ensign's (1983) Montana study white-tailed jackrabbits made up 24 percent of all foods numerically and the greatest single prey biomass, with ground squirrels (*Spermophilus*) and pocket gophers (*Thomomys*) of approximately equal secondary importance. These same three genera were found to be important prey in South Dakota (Blair, 1978; Blair and Schitoskey, 1982) as well as in North Dakota (Gilmer and Stewart, 1983). Ground squirrels and jackrabbits also made

up a preponderance of foods taken in Oklahoma, based on unpublished data of W. G. Voelker that were summarized by Sherrod (1978), and these two groups plus kangaroo rats (*Dipodomys*) comprised about 80 percent of the prey documented by Weston (1969).

It is thus clear that ferruginous hawks tend to prey consistently on fairly large diurnal rodents and lagomorphs of semiarid grasslands, with the relative importance of the specific prey species varying by time and place but rather consistently concentrating on hares, rabbits, and ground squirrels. Although the species soars well, it typically maintains a faster flight, with less stalling and "floating" than is typical of other North American buteos or eagles. These prey are rather seldom attacked from high up; instead the birds prefer to make a low, skimming approach. Hares are pursued in direct low-level chases, as occasionally are fairly large gallinaceous birds such as grouse, partridges, and pheasants. In captures of all of these a determined, goshawklike chase is typical (Beebe, 1974).

Hunting times of the ferruginous hawk tend to be concentrated in early morning and late afternoon hours (Smith and Murphy, 1973; Wakeley, 1974). Four major hunting methods are used, including still-hunting from a perch followed by flights of from under 10 to more than 100 meters to the prey, short-distance (usually under one meter) strikes on prey from the ground, aerial hunting from altitudes of less than about 30 meters, and aerial hunts from altitudes of over 100 meters. Of these, hunting from the ground was judged by Wakeley to be most successful (about 26 percent), followed by high flights (21 percent), low flights (13 percent), and perching hunts (9 percent). Although perching hunts were judged least successful, they comprised over 40 percent of the observed completed strikes. Hunting methods used were apparently not related to variations in cover density (Wakeley, 1974).

Schmutz (1977) estimated an approximate 95–99 percent similarity in the diets of ferruginous hawks and Swainson's hawks where they are sympatric in Alberta, and believed that these two species (as well as the red-tailed hawk) directly compete for food there. He also found the two former species often nesting closely adjacent to one another, suggesting that nest site competition might exist between them. Thurow and White (1983) also found these two species to be significantly associated with respect to nest site placement, which they concluded might be related to a response associated with nest site protection by the Swainson's, a smaller species. These authors calculated a lower index of dietary similarity (56–65 percent), since at least in that region (Idaho) the ferruginous hawk primarily consumed jackrabbits, pocket gophers, and ground squirrels, whereas the Swainson's hawk relied less on these species and more on birds and insects.

Social Behavior

Detailed information on territoriality in the ferruginous hawk is not available, but nesting dispersion patterns suggest that the nests of this species are related to the availability of nesting sites as well as social interactions. Ensign (1983) found no nests of ferruginous hawks closer than 0.8 kilometers from conspecifics, although many Swainson's and red-tailed hawk nests were situated much closer to ferruginous nests than this. Interspecific agonistic territorial encounters were rarely observed by him. Weston (1969) noted that nests facing the same hunting areas were located no closer than 2.1 kilometers apart, but those facing different hunting areas were sometimes as close as 0.6 kilometers apart.

In his studies of 12 nesting pairs, Powers (1981) did not observe any courtship flights in the traditional sense, but believed that pair bonding and pair maintenance was probably achieved by such activities as the gathering and delivery of nest materials, arranging the nest, and food transfer behavior. Prominent territorial advertisement and defense were observed in the forms of perching and flight displays. Perching in conspicuous places was common behavior, especially early in the breeding cycle. The resident male also often soared alone about the nesting territory, sometimes almost out of human sight. Soaring seemed to occur in three contexts, regular soaring of a single bird during hunting and territorial defense, follow-soaring in which a male would escort intruding birds across the nesting territory, and family soaring, in which family groups soared together. A "flutter-glide" flight, consisting of rapid and shallow wingbeats followed by brief glides, was believed by Powers to function largely in territorial patrolling and advertisement, and on several occasions a steep dive, or stoop, was seen performed by the territorial males,

usually in response to intruding raptors of various species. On several occasions a "buoyant flight display" was seen, consisting of flying with deep, heavy wingbeats, accompanied by sharp banking, turning, and some undulations. This also was usually done in the presence of intruding raptors of its own or other species.

Pair bonds in this species may be fairly permanent, given its semisedentary nature, but there is no real information on this point. Schmutz (1987) used indirect evidence (observed mating combinations based on plumage morphs) to conclude that the birds probably remain paired year-round and that mates are probably chosen on the breeding grounds rather than in wintering areas. The question of mate retention and nest site fidelity is made somewhat more difficult to resolve by the fact that this species tends to maintain multiple nests simultaneously. A single pair may tend up to as many as five nest sites in a single year (Weston, 1969), and thus are quite likely to choose and finally use different nests from year to year. As a result, the number of years of reuse of a nest site provides no real clue as to durations of particular pair bonds. However, some favored sites may be used year after year, and there is a record of one nest being successfully used for 28 consecutive years, and fledging 66 young during 19 of these years (Houston, 1987). Seven (out of a maximum of 13) nest territories studied by Ensign (1983) were used in two of three consecutive nesting seasons, and one was used during all three years.

Breeding Biology

Apparently a large number of variables, including nest security, prey availability, and past nesting experience act collectively to influence nest site selection. Several studies of nest site characteristics have been made (Ensign, 1983; Weston, 1969; Lokemoen and Duebbert, 1976; Schmutz, 1977; Powers, 1981; Woffinden and Murphy, 1983), but few consistencies seem to be evident. Thus, in various cases the nest orientation seems to be related to microclimate, direction of prevailing winds (allowing the birds to rise easily from the nest), and visibility around the nest or with respect to prey populations, but in other cases no clear consistent orientational relationships of any obvious kind are evident. Indeed, Olendorff (1973) considered this species the most versatile as to nest sites of all the raptors on his Colorado study area, with 60 percent of the nests in trees but others on erosional remnants, on the ground, on cliffs, and on man-made structures. Loke-moen and Duebbert (1976) found that most nests in their study area were on the ground or in trees, but some were on haystacks. Gilmer and Stewart reported similar findings (1983), while Weston (1969) found the majority of 31 nests in his area to be on the ground, with junipers, cliffs, and cliff rose providing the rest of the substrates.

Typically, nests on the ground are placed on hilltops or other situations affording a panoramic view, while those in trees are typically in lone trees, in open groves, or on the edges of wood-lands. Freedom from human disturbance is prob-ably another factor significant in nest placement (Lokemoen and Duebbert, 1976; Gilmer and Stewart, 1983).

Both members of the pair participate in nest building (Weston, 1969; Angell, 1969; Powers, 1981). However, most of the gathering forays are made by the male, who gathers material from the ground or from shrubs, often tugging at rooted or attached stems. In contrast, the female spends more time at the nest, arranging the materials and molding the nest cup. A loss of the clutch early in the season may cause the pair to move to an alternate nest site and begin over; such may be the adaptive value of the supernumerary nests that are typical of this species (Powers, 1981).

Copulation occurs over a fairly prolonged period, and in 22 instances observed by Powers (1981) only once did the male perform a flight as a preliminary activity; even then the flight was only brief and without obvious elaboration. In 18 percent of the cases copulation followed nest building behavior by the male, in 32 percent the male was perched with the female, and only in 14 percent was the male in flight for more than a minute beforehand. In at least 32 percent of the cases the female was fed prior to copulation, and food transfer either before or shortly after copula-tion occurred in 68 percent of the cases. In nearly all the observed cases the female remained at the copulation site for at least ten minutes, while the male behavior was quite variable. Powers be-lieved that copulation in this species may serve several functions separate from fertilization, in-cluding such things as courtship and territorial display, pair bonding, and reproductive synchronization.

Clutch sizes in this species are seemingly rather variable; Lokemoen and Duebbert (1976)

found a significantly smaller average clutch size in tree nests than in ground nests (4.0 vs. 4.6), and clutch size may also be affected by relative prey abundance in different years. Average an-nual clutch sizes varying from 2.1 to 3.8 eggs have been reported in various years during fluc-tuations in jackrabbit populations, and this factor also affected a variety of other aspects and mea-sures of reproductive success in this species (Smith and Murphy, 1978, 1979; Smith, Murphy, and Woffinden, 1981).

Eggs are apparently laid at approximate two-day intervals, with incubation apparently begin-ning with the first egg (Powers, 1981). Incubation requires about 32 days (Weston, 1969), with near-ly equal numbers of shifts taken by both sexes, but with the males' shifts being of shorter average durations than the females' (Powers, 1981).

Several studies of the nestling and fledgling development and care have been undertaken on ferruginous hawks (Angell, 1969; Howard, 1975; Powers, 1981; Schmutz, 1977; Konrad and Gilmer, 1986), and the birds seem to be fairly typical of buteos in their developmental patterns. Fledging occurs at times varying from 38 to 50 days, with the smaller males leaving the nests as much as ten days before females. Birds as young as 52 days have been observed taking live prey (An-gell, 1969), but generally the fledglings remain dependent on their parents for several weeks after fledging (Blair and Schitoskey, 1982). In one study the postnesting period (the time between fledging and leaving the home range) was esti-mated as 10–40 days (Konrad and Gilmer, 1986).

Numerous estimates of productivity in fer-ruginous hawks are available, which are notable for their high degree of temporal and geographic variability (Gilmer and Stewart, 1983; Lokemoen and Duebbert, 1976; Smith, Murphy, and Wof-finden, 1981). One of the longest records of reproductive success is the study of Smith, Mur-phy, and Woffinden, covering seven years, dur-ing which the total breeding population, number of nesting pairs, total eggs laid, and total young fledged all varied in synchrony with jackrabbit abundance. Average brood sizes did not follow this trend, suggesting that incubation success might decline with larger clutches. Nest abandon-ment, often because of human disturbance, and mammalian predation seem to be common sources of nest failures. High fledging rates of from 2.7 to 3.6 young per nest that have been reported in various areas have been associated with peak jackrabbit populations. Ground nests often seem to be more successful than tree nests, probably because of their greater size and stability and associated higher rates of survival of both eggs and young (Gilmer and Stewart, 1983). Olendorff and Fish (unpublished manuscript cited by Palmer, 1988) summarized productivity data from 20 sources, in which the collective average clutch size (462 clutches) was 3.44 eggs, average brood size (377 broods) was 2.32 young, and the average number of fledged young was 2.10 in 1,531 successful and unsuccessful nests, as compared with 2.93 fledged young in 1,071 successful nests. These productivity estimates averaged higher than comparable ones assembled for red-tailed and Swainson's hawks.

Recent banding data (Schmutz and Fyfe, 1987) suggest that the first-year mortality rate of this species is about 66 percent. Woffinden and Murphy (1988) estimated an adult mortality rate of 25 percent.

Evolutionary Relationships and Status

Probably this species and the rough-legged hawk are fairly close relatives, sharing not only feathered tarsi but also unusually wide gapes and an associated ability to swallow large items. However, the ferruginous hawk takes considerably larger prey on average than does the smaller rough-legged hawk, and they are unlikely to compete much at times when both species might be in contact, such as during winter. Friedmann (1950) evidently did not consider the two species to be very closely related, and placed them in different subgenera.

The ferruginous hawk was on the Audubon Society's Blue List of declining species from 1971 to 1981, and was listed as a species of "special concern" from 1982 to 1986. In 1973 it was placed in a "status undetermined" category by the U.S. Fish and Wildlife Service. It has been categorized as endangered in the state of Oregon, and has been generally believed variously threatened in several other states. Only about six nesting sites have been reported from Arizona (Glinski,

1985a), including one recently active site (Hall, 1985), and in Kansas the known numbers of nest sites and their productivity have declined considerably in recent years (Roth and Marzluff, 1985). Probably Wyoming currently has as good a population as any state; more than 800 nesting pairs are believed present (Oakleaf, 1985a). The Nevada population may be stable at present, and about 240 nesting territories have been identified (Herron, 1985). In southeastern Washington a total of 62 nest territories have been found (Allen, Friesz, and Fitzner, 1985) and in southern Idaho 59 active nests have been found (Bechard and Hague, 1985). In Colorado scattered populations occur on the eastern plains and western desert basins (Lockhart and Craig, 1985), with some very limited nesting extending to western Nebraska. The Oklahoma panhandle also supports a few nesting pairs, and the Texas panhandle had a nesting record at least as recently as 1966 (Oberholser, 1974).

At the northern limits of the species's range in Canada, its Alberta breeding distribution has declined to about 60 percent of the historic range (Schmutz, 1985a), while a similar degree of reduction has occurred in Saskatchewan (Houston and Bechard, 1985). No nesting records exist for British Columbia (Weber, 1980). Some older nesting records exist for Manitoba (Bechard, 1981), and a few nests were found in 1984 and 1985 (*Eyas* 9(2):28–30). Schmutz (1984) estimated a North American population of perhaps 3,000–4,000 breeding pairs in the early 1980s, of which 500–1,000 were in Canada. The species is now considered "threatened" in Canada, and is under consideration for similar listing in the United States (U.S. Dept. of the Interior, 1985).

My analysis of the Audubon Society's 1986 Christmas Bird Count data suggest that about 5,500 ferruginous hawks were then wintering in the species's winter range north of Mexico, with the highest numbers in Arizona (1,250) and Colorado (1,200). A recent study by Warkentin and James (1988) suggested to them that the species may have increased significantly in the past decade, at least in California and locally elsewhere.

Rough-legged Hawk *Buteo lagopus* (Pontoppidan) 1763

Other Vernacular Names

American roughleg, rough-legged buzzard (Britain); buse pattue (French)

Distribution

Breeds in North America from western and northern Alaska, northern Yukon, the Canadian arctic islands (north rarely to Prince Patrick and Bylot islands), and northern Labrador south to northern and southeastern Mackenzie, northern Manitoba, northern Ontario, northern Quebec, and Newfoundland; also widespread in northern Eurasia.

Winters in North America from south-central Alaska (rarely) and southern Canada south to California, southern Arizona, southern New Mexico, southern Texas, Missouri, Kentucky, and Maryland, casually to eastern Texas and the Gulf coast; also winters widely in Eurasia. (See Figure 47.)

North American Subspecies

B. l. sancti-johannis (Gmelin): Breeds in North America as indicated above.

Description (of *sancti-johannis*)

Adult (sexes nearly alike). As in the red-tailed hawk, enormous individual variation in plumage pattern exists. The extremes of this gradient are described here, with intermediate types of great variability present. *Light morph:* Forehead and lores buffy whitish; top of head, cheeks, auriculars, occiput, and nape the same (averaging darker in females), streaked with fuscous and sometimes washed with cinnamon-rufous; upperparts similar but becoming darker toward the rump (especially in females); upper wing coverts varying from dark brown to fuscous-black, the lesser ones usually tipped with partly concealed indistinct bars of darker brownish; secondaries grayish brown, becoming whitish on the inner edge of the inner web, narrowly tipped with whitish, and crossed with about seven rather irregular fuscous or grayish black bars; primaries externally fuscous-black; the inner webs white; upper tail coverts whitish barred with fuscous-black; tail white on basal two-thirds and narrowly but sharply tipped with white; subterminal portion pale mottled gray or grayish

brown, sometimes tinged with umber or cinnamomeous, with a broad zone of black or fuscous-black next to the white tip (averaging broader in females), and anterior to this from three to eight narrow and irregular dark bands (these often lacking in females); chin and throat white, streaked narrowly with fuscous; breast similar but usually more tawny in its wash and with the fuscous streaks broadened out, giving a blackish appearance to the breast; upper abdomen whitish washed with grayish or with pale tawny and sparsely barred or shaft-streaked with fuscous (averaging darker in females); lower abdomen, sides, flanks, and thighs similar but more abundantly barred with fuscous; tarsi feathered to the toes; under tail coverts usually immaculate whitish, occasionally very sparingly marked with fuscous; under wing coverts whitish, marked abundantly with fuscous-black, especially at the wrist. Iris deep light hazel to dark brown; naked eyebrow lead-gray; cere and corner of mouth light yellowish green to orange; bill black, becoming pale blue on the basal half of the lower mandible and below the cere on the base of the upper; toes light lemon-yellow; claws black.

Dark morph (more common eastwardly): General plumage fuscous-black to almost solid black; the occiput and nape often streaked with whitish; remiges as in light morph but the secondaries darker gray; rest of upperparts fuscous-black; rectrices dark brown becoming whitish only on the basal third or less, tipped with whitish, with a very broad subterminal blackish bar; anterior to this up to four narrower blackish bands (best developed in males); chin, throat, and entire underparts fuscous-black, the feathers of the lower breast, abdomen, and thighs sometimes sparsely mottled with brown.

Subadult (sexes nearly alike). By their second fall, immatures are generally similar to adults, but the head and neck are more heavily streaked with dusky, the upperparts are generally deep sepia with tawny edges, and the large blackish abdominal patches of adults are only partially developed. The terminal half of the tail shows indistinct grayish bars, and the barring on the tail's basal half (mainly in males) may be reduced to median spots (Bent, 1938).

Juvenile (sexes alike). *Light morph:* similar to adults, but with the underparts more heavily

Figure 47. North American breeding (hatching) and wintering (shading) ranges of the rough-legged hawk. Relative winter state or provincial density indices (average number seen per Audubon Christmas Count, 1986) are shown for major wintering areas. The breeding (inked) and wintering (shading) Old World ranges are shown on the inset map.

washed with buff, tawny, or cinnamomeous; the dark markings on the abdomen larger and more confluent, forming a practically solid broad fuscous-black band across the abdomen; tail as in adults, but the terminal portion unbarred brown to fuscous, sometimes slightly mottled; throat usually less streaked than in adults. *Dark morph:* averaging slightly more mottled than adults, with brown above as well as below; the under wing coverts dark brown or rufous and the tail dark brown to dull fuscous and unbarred (probably females) or faintly banded (probably males). Iris brownish gray to light gray; bill blackish brown; cere, gape, and toes yellowish to greenish yellow.

Measurements (of *sancti-johannis*, in millimeters)

Wing, males 397–416 (ave. of 22, 407.4), females 395–438 (ave. of 9, 411); tail, males 210–232.2 (ave. of 22, 222.4); females 212.5–234.9 (ave. of 9, 222.1) (Friedmann, 1950). Average egg size 56.6 x 44.9 (Bent, 1937).

Weights (in grams)

Males, ave. of 11, 1027; females, ave. of 17, 1278 (Snyder and Wiley, 1976). Range of weights 745–1380, ave. 1026 (sex and sample size unspecified) (Clark and Wheeler, 1987). Ave. of breeding *s.-johannis*, 5 males 822, 7 females 1080 (Brown and Amadon, 1968). Fourteen adult and juvenile males of nominate *lagopus* (Eurasian) averaged 809, and 10 females averaged 1086 (Cramp and Simmons, 1980). Estimated egg weight 63, or 4.9% of female.

Identification

In the hand. The combination of feathered tarsi and a gape that is no more than 38 mm wide at the base distinguishes this from all other North American raptors.

In the field. When perched (usually on a boulder, cliff, or other elevated site in open country), the wingtips nearly reach the end of the tail, and typical light-morph birds appear to have a rather light brown head, neck, and upper breast, the latter with varying amounts of dark streaking that often blends with a dark abdominal band (this generally much heavier than in red-tailed hawks, and also lacking any white dorsal spotting

as is typical of that species). The whitish base to the dark-tipped tail is often not evident in perched birds, but the lower rump area is also white, and becomes very evident when the birds take flight. Immatures rather closely resemble adults but average somewhat paler and generally more streaked below, and have a broadly banded (two-toned) tail similar to that of adult females. Very dark-morph birds (more common in eastern North America) may appear when perched to be completely black, but often show a whitish nape patch. Dark-morph immatures sometimes also appear almost completely black when perched, and for positive identification such birds may need to be flushed (when the white on the base of the tail and on the undersides of the flight feathers becomes evident).

In flight, birds of all ages and both light and dark plumage morphs have a distinctive underwing pattern, characterized by a black "wrist patch" and white or nearly white bases to the flight feathers that contrast strongly with their darker tips and the usually dark undersides of the body. The tail is strongly banded with black near the tip, but becomes white at the base, providing another strongly contrasting pattern. Immatures typically show a white "window" on the base of the primaries when seen from above (as in ferruginous hawks and some light-morph red-tails), but this is normally reduced and sometimes may even be lacking in very dark adult individuals. Unlike most other buteos, hovering is done rather frequently in this species, and gliding and soaring are done with wings that are slightly uptilted near the body but often bent to a horizontal plane at the wrist. The usual alarm call is a screech that descends in pitch, although this is rarely heard except on the breeding grounds.

Habitats and Ecology

On its breeding grounds, this species is usually associated with areas north of tree line in North America and Eurasia, its southern limits essentially coinciding with tree line, where it nests in trees along this fringe of the boreal forest. Elsewhere in the arctic it mainly depends on riverside cliffs, escarpments of Precambrian rocks, and similar elevated sites (Beebe, 1974). Rough-legged hawks use essentially the same sorts of nesting sites as do other cliff-nesting raptors of the arctic, such as northern ravens, gyrfalcons, and per-

Amadon, 1968). Since much hunting is done from perches, the birds use posts, poles, or other somewhat elevated sites where they are available, but open-ground locations such as rocks or mounds may also be used. Thus, compared with red-tailed hawks, rough-legged hawks tended in one study (Schnell, 1968) to perch in lower trees or perches, less often perched in groves of trees than in lone trees, more clearly preferred utility poles over trees, and spent more time hunting along roads when snow was on the ground. Additionally, rough-legged hawks seem to be somewhat less sedentary (spend more time flying and less perching) than are red-tails, and tend to favor hunting over open areas rather than wooded areas (Weller, 1964). Such differences seem to allow the two species to be present in the same area and feed on much the same foods without an undue degree of contact and competition (Schneli, 1968).

Foods and Foraging

All the evidence indicates that this species is highly adapted to foraging on small mammals, primarily rodents, and especially arvicoline rodents such as voles (*Microtus*) and lemmings (*Dicrostonyx* and *Lemmus*). Fisher (1893), Bent (1937), and more recently Snyder and Wiley (1976), have summarized the earlier literature on the foods of the rough-legged hawk in North America, while Cramp and Simmons (1980) have provided a similar summary for the western Palearctic.

Foods taken on the breeding grounds have been analyzed by White and Cade (1971), Sealy (1966), and Springer (1975), all of whom found that mammals comprised about 80 percent or more of the identified items, with only Springer's study indicating a significant intake of birds (21 percent of the total items counted). Among these are 46 passerines (mostly unidentified fledglings), two lesser golden-plovers (*Pluvialis dominica*), and 14 ptarmigans (*Lagopus* spp.). Rodents made up the great majority of the mammals, including (in descending frequency) the genera *Microtus*, *Dicrostonyx*, *Lemmus*, and *Clethrionomys*.

A similar numerical analysis by White and Cade (1971) had a smaller percentage of birds (13.5 percent) represented, but an otherwise similar spectrum of mammals that was particularly rich in lemmings, while Sealy's (1966) limited analysis indicated that birds were a minor com-

egrines, and they often outnumber all of these other raptors in their nesting density in such locations (White and Cade, 1971). Kuyt (1980) found 18 rough-legged hawk nests as compared with 13 peregrine and 9 gyrfalcon nests during a multiyear survey of the Thelon River areas, Northwest Territories. It is possible that breeding density is limited by available nesting sites and may also fluctuate according to the available food supply. Brown and Amadon (1968) estimated that an average breeding density might be a pair per 7.8–10.4 square kilometers, with a maximum density of a pair per 4 square kilometers and often a density of only a pair per 78 square kilometers. In the Colville River area of Alaska nests averaged 3.8 kilometers apart, with a minimum internest linear river distance of 0.4 kilometer and a maximum of 22 kilometers.

On their wintering areas across much of North America the birds seek out habitats similar to arctic tundra, namely plains, prairies, airports, coastal marshes, and other open-country habitats offering excellent long-distance visibility, including agricultural lands such as pastures and harvested fields where small rodents can be hunted effectively. During such times the birds occupy home ranges or winter territories that may cover about 10–15 square kilometers (Brown and ponent, and that brown lemmings (*Lemmus sibiricus*) were the most important single mammalian food item.

Winter foods are of course variable as to specifics, but similarly are heavily oriented toward microtine rodents, especially cricetid mice and voles of the genus *Microtus*. Schnell (1967) reported that *Peromyscus* mice and *Microtus* voles made up the vast majority of winter and spring foods in Illinois. Birds were represented by a rather small (8 percent) numerical component, although they included prey as large as ring-necked pheasant (*Phasianus colchicus*) and gray partridge (*Perdix perdix*). Similarly, Errington and Breckenridge (1938) found that foods taken in the north-central states from fall through spring included ring-necked pheasants as well as domestic chickens, plus a considerable variety of rodents and lagomorphs, the avian component representing 12 percent of the total numerically.

Social Behavior

These birds have monogamous pair bonds that are maintained at least through the breeding season, although longer durations are uncertain. There is some evidence that pair bonding may occur on wintering areas, as the birds often travel and arrive on the breeding grounds already paired (Bent, 1937). However, after the loss of a mate a new one may be acquired fairly rapidly, sometimes within the same breeding season. Birds arriving on their breeding areas must cope with resident ravens and gyrfalcons for nest sites, as well as with peregrines. Peregrines arrive at about the same time but are able to dislodge roughlegs from nesting sites. Perhaps fortunately, roughlegs are quite versatile in their acceptance of a nesting place, and in the sites chosen they completely overlap these other three species in such variables as distance of the nest site above the river, distance below the cliff brink, and distance above the cliff bottom (White and Cade, 1971).

Pair bonding displays are virtually unstudied in this species, at least in North America. Observations in Europe suggest that the territorial advertisement behavior is like that of typical *Buteo* hawks (Cramp and Simmons, 1980). High circling by a single bird or a pair, presumably centering above the territory, is probably the major advertisement activity, and may be supplemented by calling as the pair fly close together.

Such male soaring may develop into a sky dance in which the male suddenly partially closes his wing, stoops down, climbs back up, finally stalling, and then swings back down again to repeat the maneuver.

Copulatory behavior in wild birds is apparently undescribed but is unlikely to differ from that typical of other North American buteos. Bird and Lague (1976) described this behavior in captive birds, which usually occurred on a log perch. Usually the male would sidle to and fro immediately prior to copulation, and during treading the male would wing-flap for balance and utter high-pitched squeals, and the female would produce a strangled squawk at its termination.

Nest site tenacity is fairly well developed in this species, in spite of its difficulty in retaining its nest sites from year to year in the face of competition from other raptors for similar sites. During annual surveys White and Cade (1971) reported that rough-legged hawks occupied from 69 to 88 percent of the nesting cliffs that were known to have been used previously, the three-year average of 77 percent being higher than any single-year percentage for either the gyrfalcon or the peregrine, and well above that of the northern raven.

Breeding Biology

Where tree-nesting opportunities are lacking, rough-legged hawks typically select cliff-nesting sites that are on ledges or other similar platforms, but where these are not available they may nest on the top of a rock on a steep hillside, on a break in the slope of a high earth bluff, or in similar situations having a broad vista and occasionally even looking directly toward the nesting site of a peregrine or gyrfalcon (White and Cade, 1971). Nests with shelter provided by a caprock or overhangs are preferred, and those that are used repeatedly gradually grow in size from year to year (Beebe, 1974). Other nests may have to be built from scratch, or at least repaired, and this must be done in the three- or four-week period between arrival and the beginning of egg laying.

Clutch sizes are relatively variable, as might be expected from a species that is largely dependent upon a highly fluctuating food supply during the breeding season, and range from about 2 to 7, with 4 being the most typical. In the Palearctic the clutch is typically of 5–7 in good lemming years, and only 2–3 in poor years, the average of 64 clutches from Norway being 3.8 eggs (Cramp and Simmons, 1980).

At least in captivity, eggs are laid at approximate two-day intervals, so that a clutch of five eggs is laid in about ten days. The average incubation period of these eggs (which were initially incubated by the birds and later artificially incubated) was 37 days, or considerably longer than earlier estimates of about 28 days that have been generally accepted (Bird and Lague, 1976). However, Parmelee, Stephens, and Schmidt (1967) estimated a 31-day incubation period for wild birds. They reported a minimum but rather variable fledging period of 31 days, but noted that most young remain at or near the nest until they are at least 40 days old. Independence is apparently not attained for another 20–35 days. Initial breeding probably occurs at two or three years (Cramp and Simmons, 1980).

There is little information on nesting success or productivity for North America, but it is likely that substantial variation in reproductive success occurs from year to year, depending upon rodent populations.

Evolutionary Relationships and Status

As noted in the prior species account, this species and *regalis* are perhaps fairly closely related, although *lagopus* is also clearly a close relative of the Asian upland buzzard (*B. hemilasius*), which is somewhat similar in plumage but has only partially feathered tarsi.

Little quantitative information on the North American population of this species exists, but my analysis of the Audubon Society's 1986 Christmas Counts suggests that about 49,600 birds wintered south of Alaska and the Canadian territories that year. Maximum densities occurred in Montana and Idaho, which had estimated state populations of 5,250 and 3,650 birds respectively. This species is a very widespread and relatively common breeding bird in arctic tundra regions across both North America and Eurasia, making it probably one of the most abundant species of raptors native to North America. Throughout its range its population is probably strongly influenced by local prey population cycles, such as lemmings and *Microtus* voles.

BOOTED EAGLES

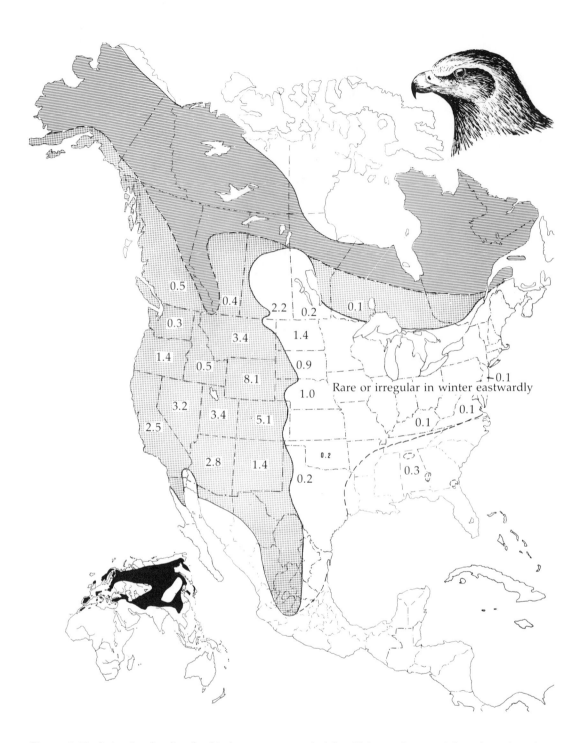

Figure 48. North American breeding (hatching), residential (cross-hatching), and wintering (shading) ranges of the golden eagle. Relative winter state or provincial density indices (average number seen per Audubon Christmas Count, 1986) are shown for major wintering areas. The Old World breeding or residential (inked) and wintering (shaded) ranges are shown on the inset map.

Golden Eagle *Aquila chrysaetos* (Linnaeus) 1758

Other Vernacular Names

Black eagle, brown eagle, calumet eagle, Canadian eagle, mountain eagle, royal eagle; aigle royal (French); águila real (Spanish)

Distribution

Breeds in North America from northern and western Alaska east across Yukon, western and southern Mackenzie, northwestern Manitoba, northern Ontario, and northern Quebec to Labrador, and south to southern Alaska, Baja California, the highlands of northern Mexico, west-central Texas, western portions of Oklahoma, Nebraska, and the Dakotas (very rarely also in western Kansas), and formerly in eastern North American southward in the mountains at least to New York (where last known successful nesting was in 1970); widespread in Eurasia and local in northern Africa.

Winters in North America from south-central Alaska and the southern portions of the Canadian provinces south throughout the western breeding range and more rarely eastwardly; also winters over most of its breeding range in Eurasia. (See Figure 48.)

North American Subspecies

A. c. canadensis (Linnaeus): Resident in North America as indicated above.

Description (of *canadensis*)

Adult (sexes nearly alike). General plumage dark brownish to deep fuscous, generally darkest on the lower surface, under wing coverts, interscapulars, scapulars, primaries, back, and rump; the smaller upper wing coverts, tail coverts, thighs, and feathered tarsi usually paler; occiput and nape pale fulvous at tip, washed with russet to hazel and fuscous basally; secondaries, especially the inner ones, olive-brown to sepia or umber, becoming more whitish basally and more or less mottled with olive-brownish; tail dark fuscous to fuscous-blackish, paling basally to olive-brown or sepia and with two to four irregular, faint zigzag bands of ashy pale umber, the more posterior of these bands usually wider and more distinct in females than

in males. Iris yellowish or clear light hazel to dark brown; cere and gape pale grayish yellow to wax-yellow; bill blackish to bluish slate-black, becoming paler basally; toes grayish yellow to deep chrome-yellow; claws black.

Subadult (ca. 3 years old). Similar to the adult, but with the feathers of the underparts and upperparts basally white; the tail with more white basally than in adults, the darker terminal area forming an indistinct blackish band; the remiges also with whitish mixed with brown basally, this chiefly confined to the secondaries, where indistinct barring is typical. [After the first annual molt the tail has a mixture of juvenal and adult tail feathers, and the outer juvenal primaries are retained for some time. There seem to be two recognizable predefinitive plumage stages, a later "subadult" one (probably representing 2½ to 3½-year-olds) with less white on the tail, wings, and bases of the body feathers than an earlier "immature" plumage (probably of 1½ to 2½-year-olds). These plumages are really only intermediate stages in pattern as well as in chronology between the juvenal and adult plumages. The adult plumage is probably normally acquired at about 3½ years, but because of incomplete molts there always is a mixture present of feathers of several plumage generations (Jollie, 1947; Cramp and Simmons, 1980).]

Juvenile (to 16 months). Similar to older immatures, but the tail clear white for about two-thirds of its length, sharply terminated by a broad blackish band; the bases of the inner primaries and outer secondaries more distinctly whitish than in older birds (most visible from below); the crown and nape less distinctly golden; the body plumage generally darker overall. Iris brown to hazel; bill mostly dull black, but more olive-buff basally; cere, toes, and the edges of the bill deep chrome-yellow.

Measurements (of *canadensis*, in millimeters)

Wing, males 555–610 (ave. of 12, 580.5), females 620–666 (ave. of 17, 633.2); tail, males 320–360 (ave. of 12, 337.4); females 350–390 (ave. of 17, 366.2) (Friedmann, 1950). Average egg size 74.5 × 58 (Bent, 1937).

Weights (in grams)

Males (various races) 3550–4400, ave. of 7, 3924; females 4050–5720, ave. of 4, 4692 (Brown and Amadon, 1968). Average of 31 males, 3477; of 18 females 4913, with 4195.5 the best dividing point for judging sex (Edwards and Kochert, 1986). Fifteen males of nominate *chrysaetos* (Eurasian) averaged 3672, and 19 females averaged 5194 (Cramp and Simmons, 1980). Estimated egg weight 138, or 2.9% of female.

Identification

In the hand. The combination of feathered tarsi and a wing length of at least 550 mm serves to separate this species from all other North American raptors.

In the field. Distinction of golden from bald eagle while perched is simple in adult birds; the tawny-colored (rather than pure white) head and neck of the former immediately separates them. However, juvenile and subadult birds are less simple; golden eagles never show any extensive amount of white on the back or underparts, as is typical of immature bald eagles. Young goldens (especially first-year birds) are more likely to exhibit white at the base of the tail than are those bald eagles that still lack any white on the head, and additionally younger bald eagles (to three years) retain dark outer edges on their outermost tail feathers that are lacking in golden eagles.

In flight, golden eagles of all ages exhibit dark wing linings, and most birds show at least lighter (immatures and subadults) if not actually whitish (juveniles) bases to their primaries; subadults and adults also may show a variable degree of lighter and darker mottling or barring on the undersides of their flight feathers. First-year birds also have a white band at the base of the tail (immature bald eagles may also have white at the base of the tail but it is typically much more diffusely sprinkled rather than organized into a definite basal band and thus does not produce such a distinctly half-white, half-black tail). Additionally a diffuse white "window" is sometimes visible at the base of the spread primaries when seen from above, in a pattern similar to that typical of rough-legged hawks. Thus, any flying eagle with dark wing linings and no white on the head can safely be identified as a golden eagle, and any eagle with extensive whitish spotting on the wing linings

and/or underparts, regardless of whether it has any white on the head, can similarly be recognized as a bald eagle. Golden eagles typically but not invariably soar with their wings slightly uptilted (rather than held rigidly horizontal as bald eagles normally do), and they have relatively longer tails than do balds. Golden eagles are generally to be found farther from water than are balds. A buteo-like leisurely soaring over arid rimrock or mountainous country is a typical hunting method. A steep dive to capture prey such as rabbits on the ground is fairly common, but occasionally birds up to the size of geese and cranes are attacked in the air. The wing shape of golden eagles, together with their slightly uptilted flight profile, remind one more of large buteo hawks than do bald eagles, which have seemingly larger heads (owing to their larger and longer bills), although these differences are not easily appreciated unless both are visible simultaneously. Both species are typically silent in flight, but the golden eagle sometimes utters a series of repeated yelping notes when near the nest, especially when carrying prey.

Habitats and Ecology

This species occurs in nearly all habitats of the western states, from desert grasslands to above timberline, perhaps avoiding only densely forested areas where hunting is impossible. Generally it occurs in grass-shrub, shrub-sapling, and young woodland growth stages of forested areas, or in forests with open lands nearby for hunting. Essentially, it needs only a favorable nest site (usually in a large tree or cliff), a dependable food supply (mainly of medium to large mammals and birds), and broad expanses of open country for foraging. It especially favors hilly or mountain country over flat habitats, where takeoff and soaring are facilitated by updrafts; deeply cut canyons rising to open or sparsely treed mountain slopes and crags represent ideal habitat (Beebe, 1974).

Wintering habitats of golden eagles in western states tend to be those having available perches plus native shrub-steppe vegetation types (mainly *Artemisia* and similar shrubs), which are habitats that usually have good populations of black-tailed jackrabbits (*Lepus californicus*). Such habitat selection tendencies have been observed in Utah (Fischer, Ellis, and Meese,

1984). In the eastern United States wintering habitats are quite different, and in one survey (Millsap and Vana, 1984), 82 percent of winter records were associated with riverine or wetland systems, mainly steep river valleys or associated reservoirs or wetlands. Estuarine marshlands, barrier islands and associated sounds, and the mouths of major rivers represented the primary coastal habitats.

Breeding densities of golden eagles are relatively low everywhere, as a reflection of territorial spacing and foraging requirements for the species. In Utah, six breeding pairs had home ranges of from 17 to 31 square kilometers, averaging 23 square kilometers, and nesting population densities of from 42 to 52 square kilometers per pair (Smith and Murphy, 1973). There the birds concentrate along high, north-south-oriented ridges and hills, but seem to avoid the intervening broad and flat valleys (Smith and Murphy, 1982). Caminzind (1969) estimated a density of 99–156 square kilometers per pair in his Utah study in the same area. In southern California, 27 year-round home ranges averaged 93 square kilometers (Dixon, 1937), while Spofford (1971) estimated probable winter home ranges of golden eagles ranging from approximately 130 to 260 square kilometers in the Appalachians. One of the highest known breeding densities in North America or anywhere else must be that of the Snake River canyon in southern Idaho, where annual breeding densities in two years ranged from a pair per 5 to 8 kilometers of river, along a 240-kilometer stretch of river that supported up to 56 breeding pairs. There the maximum density is probably a collective function of the availability of suitable nest sites, an adequate prey base, and minimum nesting territory size (Beecham and Kochert, 1975). Differences in population densities in two areas of Sweden (average of 10.2 kilometers between nests in mountains versus 17 kilometers in forests) was attributed to higher hare and grouse prey densities and possibly also topographic and habitat diversity differences in these two areas (Tjernberg, 1985). Average breeding densities in Wyoming have been estimated as 60 square kilometers per pair, and 5.3 kilometers between active nests. The range of nine previous studies from various areas was from 41 to 251 square kilometers per pair (Phillips, McEneaney, and Beske, 1984).

Territoriality probably is an important aspect of spacing and associated limitations to popula-

tion density; it has been studied by Bergo (1987) in a sparse breeding population in Norway. He judged that the regular spacing of nests was probably facilitated and maintained by territorial defense throughout the year, especially by use of aerial activities such as undulating flight displays as well as by soaring and "hanging on the wind." There each pair had a "core area" of their home range that varied from 10 to 35 square kilometers, but this was only a small part of the pair's total hunting range.

Foods and Foraging

A very large amount of information is available on golden eagle foods; Sherrod (1978) listed ten original sources or summaries for the species, seven of which provided a percentage analysis of major food types numerically. In these the mammalian component ranged from 77 to 97 percent of the identified items (unweighted average 82.6 percent), birds ranged from 3 to 28 percent (unweighted average 12.6 percent), and the remainder was mostly comprised of reptiles and fish. Brown and Amadon (1968) suggested that, depending upon locality and prey availability, from 70 to 98 percent of the species's foods on a biomass basis consists of small mammals, mainly

lagomorphs and rodents. This estimate agrees with studies done in the Snake River canyon of southern Idaho, where mammals comprised an estimated 82 percent of prey biomass (and jackrabbits alone about 60 percent), birds comprised 16 percent, and reptiles and fishes occurred in only trace quantities (U.S. Dept. of Interior, 1979).

A comparably large amount of information is available for the Eurasian population of golden eagles, which has been summarized by Cramp and Simmons (1980). These authors commented on the wide range of prey taken, but noted that it consisted mainly of medium-sized mammals, especially lagomorphs (rabbits and hares), and birds, mostly tetraonids (grouse and ptarmigan), with a greater reliance on rodents toward the south, particularly marmots. The most comprehensive of the studies on golden eagle foods in North America was that of Olendorff (1976), who reviewed the entire North American literature

available at that time, representing more than 7,000 identified food items. Of these, there were 52 species of mammals, 48 birds, 5 reptiles, and 2 fish. Lagomorphs comprised 54 percent of the total, marmots, ground squirrels, and prairie dogs 22 percent, game birds 8 percent, wild ungulates 6 percent, passerine birds 3 percent, domestic livestock 1 percent, and mammalian predators 1 percent. In descending frequency of occurrence, the most common prey species were black-tailed jackrabbit (*Lepus californicus*), arctic ground squirrel (*Spermophilus parryii*), white-tailed jackrabbit (*L. townsendii*), desert cottontail (*Sylvilagus audubonii*), and yellow-bellied marmot (*Marmota flaviventris*).

Among the studies that have analyzed prey on a biomass basis is that of Bloom and Hawks (1982), who examined over 1,100 prey items from nests and found that four species (black-tailed jackrabbit, Nuttall's cottontail *S. nuttallii*, yellow-bellied marmot, and chukar *Alectoris chukar*) were the most frequently encountered, accounting for 90 percent of all items, with livestock accounting for less than 1 percent (and some of this probably being carrion). Lagomorphs comprised 91 percent to the total estimated prey biomass. Similarly, six species of jackrabbits, cottontails, ground squirrels, and prairie dogs comprised nearly 90 percent of the prey items found at 41 nests in Texas and New Mexico (Mollhagen, Wiley, and Packard, 1972).

Because of traditional concern by ranchers, especially sheep ranchers, over golden eagle depredation, many papers have been devoted to this controversial topic. The evidence for lamb-killing by golden eagles and ways of reducing such behavior were reviewed by Matchett and O'Gara (1987), who stated that depredation levels vary greatly, depending on natural prey densities, availability of carrion, ranching practices, weather, and a variety of other factors. Probably most lambs are killed by young eagles, and lambs are more often taken during periods of decline in jackrabbit populations or during cold periods when their usual rodent prey are relatively inactive. The use of conspicuous scarecrows, especially in combination with harassment and increased human activity, offers the best means of protecting lambs from eagles according to these authors. Brown and Watson (1964) similarly found that sheep management practices had a great effect on eagle depredation levels, and losses to eagles were infinitesimally small when compared to those associated with poor management practices.

Eagles hunt largely by soaring-searching techniques, and need either thermals or orographic winds (topography-related updrafts) to help them gain sufficient altitude to hunt and cover the large areas of their foraging ranges. Thus, the birds seek out areas with such updrafts, and avoid areas of downdrafts or dead air. They are unable to take off with any significant amount of prey, possibly even as little as two kilograms, when they are in the center of such downdraft areas (Snow, 1973a). It is unlikely that even an adult eagle can normally take off with more than about three kilograms of prey, although an eagle has been observed in flight carrying a jackrabbit weighing more than 3,170 grams (Kalmbach, Imler, and Arnold, 1964). However, adult black-tailed jackrabbits average only about 2.3 kilograms (Smith and Murphy, 1973). Huey (1962) reported that an adult male eagle is apparently able to carry a load of only about 900 grams easily, or less than 25 percent of its body weight, assuming an average male eagle weight of 4,000 grams. Certainly the amount of wind and the opportunity to take off downslope must greatly influence carrying ability, and even eagles carrying small prey will use wind to assist in this.

Eagles typically hunt by using favorite perches located near an area having regular updrafts that allow them to rise to soaring height, from which they can scan their hunting areas, which are usually upwind. Eagles also prefer to make attacks upwind, in order to have the maximum degree of aerodynamic control and maneuverability, and perhaps also to force the prey to face into the wind while trying to escape. Hatch (1968) has described the cooperative hunting tactics of two golden eagles that worked together in successfully killing a red fox (*Vulpes vulpes*), in which an immature eagle distracted the fox while an adult attacked it from behind. Dekker (1985) likewise observed hunting behavior of golden eagles, reporting that surprise was a basic ingredient in their attacks on ground squirrels. Seven ground squirrels were captured during low flapping flights and eight taken during low, high-speed glides following soaring flights, the low final approach probably helping to screen the attack.

Collopy (1983) found that golden eagles use both solo and tandem-hunting techniques, the male typically flying in front of the female during tandem hunts and apparently leading the hunt. Overall, 20 percent of 115 hunts he observed were successful, with the smaller and more agile males nearly twice as successful as females during solo hunts. Males also nearly always initiated the attacks during tandem hunts, the female attacking only when the male was unsuccessful. Surprisingly, solo hunts had a higher success rate (29 percent vs. 5 percent) than tandem hunts. However, this statistic is perhaps misleading in that tandem hunting might be mainly directed toward larger and/or more elusive prey than can be caught easily by single birds, as has also been observed in the aplomado falcon.

Social Behavior

More perhaps than to any other bird of prey, people have been inclined to attribute lifelong monogamous pair bonds to eagles, although inadequate evidence is available to support this assertion. Brown (1976a) stated that a bond lasting for the life of any one individual is probably the usual eagle mating pattern, but that if a mate dies the other is likely to try to obtain a new one immediately, even if it is an immature. Such a mate replacement may occur within 10 weeks in a single breeding season (Dixon, 1937), or it may occur by the following season. An unpaired bird may occupy an area for up to two years before obtaining a mate, and it is probable that most pairs are formed only after a territory has been established. Although monogamy is typical, casual bigyny by the male has also occurred (Cramp and Simmons, 1980).

The territories of eagles are extremely large, and perhaps for that reason it is impossible to defend them efficiently. Brown (1976a) suggested that the term "home range" is a better term for the area occupied year-round by eagles, since the birds do not overtly defend this entire area. However, Bergo (1987) studied territoriality in golden eagles in Norway, concluding that territoriality was expressed there in a variety of ways that facilitated spacing of nests. Perhaps the most spectacular of these is the "undulating flight," which has often been interpreted as courtship behavior (Bent, 1937). However, Harmata (1982) found that over 70 percent of these flights were associated with the appearance of intruders, and Bergo's observations were similar. Bergo did, however, also see undulating flights at low altitudes when no intruders were apparent, and judged that such flights might have a sexual function as well. Most of the examples of undulating flight that he saw occurred when the performer's mate was in the core area (the males displayed nearly three times as frequently as females, but with fewer undulating "waves" per sequence), and most were performed within a kilometer of the nest. A few undulating flights were intermediate in form between "undulating" and "pendulum" (or figure-eight) in form, but true pendulum flights were also seen twice. Mock attack and evasion between mates was observed several times, and fledged young were also observed to make mock attacks on their own parents. Bergo believed these mock attacks to have sexual significance when performed between pairs.

Other aerial activities commonly observed by Bergo that might constitute territorial advertisement included slow glides (such as "hanging on the wind") and soaring, plus much less frequent fast glides, flapping flights, and diving. Conspicuous and inconspicuous perching were also frequently seen but of uncertain significance in territorial advertisement. "Exposed perching" (use of the highest and most exposed available perch) and calling were probably not a part of advertisement behavior, at least in this population.

Additional data on territoriality were provided by Collopy and Edwards (1989) and by Tjernberg (1985); Tjernberg concluded that the distributional pattern of nests in Sweden was affected more by spacing behavior than by a shortage of suitable nest sites. There was a high stability in size and spacing of pairs, with territories typically occupied annually although not all pairs bred every year. Further, no new territories were formed in between existing territories, although in five cases where one member of a pair disappeared a new mate was present the following breeding season. Smith and Murphy (1973) found that three of five pairs occupied territories during all four years of their study, while the other two occupied territories for three years, although the majority of specific nest sites were used only a single year (perhaps because of human disturbance). Brown (1976a) stated that the average number of golden eagle nest sites per territory in his experience is 2.6, but the known range is 1–11.

Ellis and Powers (1982) described the occurrence of copulatory behavior in golden eagles at times when fertilization was clearly not a functional part of the behavior, such as during the incubation and fledging periods, and judged that such behavior might serve as a territorial display, help maintain pair bonds or synchronize reproductive readiness, or serve as displacement behavior in conflict situations.

Breeding Biology

The first phase of nesting biology probably involves the choice of nest site, which is likely to be either the nest used the previous year or a fairly nearby location. Nest building can be done at almost any time of year; indeed Brown (1976a) suggested that the only times it is not likely to be done are when the sites are snow-covered or the pair have a brood in the nest or just out of it, and are thus occupied by hunting. In Scotland perhaps 95 percent of the sites are on cliffs, and some of the pairs with three nests also have rock sites that they sometimes use. On the other hand, all but one of 21 nests in a western Washington study were in Douglas fir (*Pseudotsuga menziesii*) trees, with the other on a cliff, and most were at forest edges or in small stands near clear-cuts or open fields (Bruce, Anderson, and Allen, 1982).

Caminzind (1969) noted that 87 percent of 31 nests he observed in Utah were on cliffs, while the rest were on the ground or some artificial structure. More than half of this total faced west, and only 4 percent faced east, although Caminzind did not believe this to provide any evidence for or against the sun's possible effect in influencing nest site selection. McGahan (1968) noted that 62 percent of 92 nests in Montana were on cliffs, and most of the rest were in large-limbed trees such as Douglas fir. Of the cliff nests, nearly half faced south, while only about 10 percent faced north, which McGahan attributed to possibly advantageous warming effects in early spring. Similarly, 37 Alaska nests were mostly oriented toward the south or southeast, and all but one were on cliff faces (Ritchie and Curatolo, 1982). Data from these and other areas suggest that directional exposure of nests may be related to a strong sensitivity to thermal stress in young eagles (Mosher and White, 1976).

In the area of plains, hills, and buttes in

northeastern Wyoming nearly 90 percent of the eagles nested in trees, with a preference shown for larger trees. Ground nests there tended to be on butte tops, while tree nests were found close to watercourses. Both deciduous trees and ponderosa pines were used, with the nests usually placed in the upper one-third of the tree in either case (Menkens and Anderson, 1987).

Brown (1976a) judged that from 10 to 25 percent of eagle pairs do not lay during any given year in Scotland, but only scant evidence exists that nonbreeding is more common in years of poor prey supply. Caminzind (1969) reported that 13 percent of 31 nests in his study area during two years had no eggs present. While nest abandonment for various reasons may account for some such cases, probably some of these nonlaying pairs involve immature birds that may nonetheless form pairs, defend territories, and construct nests. However, Teresa (1980) reported a case of a nesting pair in which one bird was in subadult plumage but which nevertheless fledged two young. She stated that about 2 percent of 564 observed nesting attempts had one pair member (probably the male) a subadult, and that such pairs had an estimated breeding success rate of 80 percent. Similarly, Steenhof, Kochert, and Doremus (1983) found that about 5 percent (17) of 340 territorial pairs of eagles in Idaho included one subadult bird, three of which failed to lay eggs. Of six subadults for which the sex was known, four were males.

Eggs are laid at intervals of about 90–120 hours, with the clutch size typically two or three, but sometimes only one and rarely four (Brown, 1976a). Of 82 Scotland clutches, the average size was 1.91, with 72 percent of the nests having two eggs (Cramp and Simmons, 1980). Similar or slightly higher clutch sizes seem typical for western North America, with Caminzind (1969) reporting an identical average of 1.91 eggs in 23 clutches in Utah, Beecham and Kochert (1975) an average of 2.04 for 89 clutches in Idaho, and McGahan (1968) an average of 2.1 eggs for 20 clutches in Montana. Three-egg clutches rarely comprise more than about 10 percent of the total in any of the available samples; Jenkins and Joseph (1984) estimated that of 267 clutches from North America and Scotland a collective average of 9.7 percent of the nests had three-egg clutches. Furthermore, for 1,309 successful nestings, only 16 of the pairs (1.01 percent) successfully fledged

three young, with unusually high numbers in 1981 (9 of the 16 cases) apparently being associated with extremely high jackrabbit densities that year.

Incubation is done by the female for the most part, but cases have been known of males taking an almost equal share (Brown, 1976a). Although some early literature suggests incubation periods of as short as 35 days (summarized by Brown and Amadon, 1968), most recent observations indicate a minimum 41-day (Caminzind, 1969; Olendorff, 1973) or 43- to 45-day incubation period (Brown, 1976a; Cramp and Simmons, 1980; Hobbie and Cade, 1962). Because of the fairly long egg-laying interval, the young are often substantially different in size and age. This may frequently lead to the older nestling killing the younger (Brown and Amadon, 1968), although the incidence of this is probably quite low, and the rare fledging of "triplets" in nests would suggest that the younger birds sometimes can hold their own in defending themselves and in competition for food. Collopy (1981, 1984) suggested that intense sibling competition for food during periods of low jackrabbit abundance may produce a differential mortality among the sexes (females surviving better than males). He found no differences in the food-provisioning rates of one-chick broods and two-chick broods, although different pairs provisioned their broods at significantly differing rates. Most of the food for chicks was captured by the male, while the females typically fed and tended the offspring.

The fledging period is about 72–84 days, but the young remain largely dependent upon their parents for as much as about 11 weeks afterward, may remain within their parents' territory for several more weeks thereafter, and at least in one case wintered on the edge of the parents' home range (Walker, 1987; Hobbie and Cade, 1962).

Breeding success in golden eagles tends to be fairly variable, often following fluctuations in primary prey populations such as jackrabbits. Thompson, Johnstone, and Littlefield (1982) found annual reproductive success rates to parallel jackrabbit abundance closely in Oregon, with a 15-year mean of 1.08 young fledged per breeding territory, 1.7 young fledged per successful nest, and 51 percent of the nests successful. Beecham and Kochert (1975) estimated a similar average of 1.1 young fledged per nesting attempt, 1.8 young fledged per successful nest, and 65 percent of the nests successful over a three-year study. McGahan (1968) found a slightly higher average number of young fledged per nest (1.4) in a three-year study, while Caminzind (1969) estimated a two-year average of 1.13 young hatched per nest and 0.84 fledged per nesting attempt. Over a six-year period, productivity in New Mexico, Colorado, and Wyoming remained fairly constant, with annual averages of from 1.2 to 1.5 birds fledged per nest (of 264 total nests), and with 82–95 percent of the hatched young fledging annually (Boeker and Ray, 1971). In the Snake River Birds of Prey Natural Area of Idaho the average clutch size over eight years was 1.98 eggs for 365 nests, and the estimated hatching success was 69 percent. The average number of hatched young was 1.4 for all nests, and 1.77 for all successful nests, while the average number of fledged young was 1.03 for all nests and 1.62 for all successful nests (U.S. Dept. of Interior, 1979). Breeding success in Scotland has been found to depend largely on the degree of human interference as well as relative food supplies and extent of pesticide usage (Cramp and Simmons, 1980).

Evolutionary Relationships and Status

The golden eagle is one of a number of similar "booted eagles" that all have feathered tarsi and are placed in the genus *Aquila*. Brown and Amadon (1968) suggested that the golden eagle along with *verreauxi* (the Verreaux's eagle of Africa), *audax* (the wedge-tailed eagle of Australia) and *gurneyi* (Gurney's eagle of New Guinea) may all have been derived from the same ancestral

stock. Stresemann and Amadon (1979) suggested that *chrysaetos, audax,* and possibly *gurneyi* form a superspecies.

In a general review, Snow (1973a) presented unpublished data of L. G. Huegly suggesting that at least 16,000 golden eagles were resident in eight western states, including Wyoming (3,063), Colorado (2,600), Montana (2,433), New Mexico (2,050), Nevada (1,833), Oregon (1,600), Idaho (1,383) and Utah (1,367). My very similar estimates of 1986 winter populations, based on the Audubon Society's Christmas Bird Count data, are of 18,520 birds for the whole of North America except Alaska and the Canadian territories, with the largest populations occurring in Wyoming (3,140), Colorado (2,120), Montana (1,985), Nevada (1,400) and Utah (1,115). Oakleaf (1985a) believed that Wyoming alone may support over 4,000 breeding pairs. There do not appear to be any good estimates for Canadian or Alaskan breeding populations. Snow (1973a) suggested that the total North American population may be at least 50,000 birds, and Braun, Hamerstrom, and White (1975) judged that as many as 100,000 may be present, but it is difficult to imagine where all of these birds might be breeding. Palmer (1988) recently estimated a North American population of about 70,000 birds, which seems to me to represent the upper numerical limits of probability.

A recent and extensive bibliography of the golden eagle and other members of the genus *Aquila* is available (LeFranc and Clark, 1983). Palmer (1988) noted that he consulted over 700 sources on this species, and cited 242 of these.

Family Falconidae
(Caracaras and Falcons)
Tribe Polyborini
(Caracaras)

Figure 49. North and Central American residential range of the crested caracara, including forms cheriway (**CH**), pallidus (**PA**), *and* lutosus (**LU**), *the last-named form often being considered a distinct insular species. The South American residential range is shown on the inset map.*

Crested Caracara *Polyborus plancus* (Miller) 1777

Other Vernacular Names

Audubon's caracara, common caracara, Guadalupe caracara (*lutosus*), Mexican buzzard, Mexican caracara, Mexican eagle; caracara huppé (French); carancho, quebrantahuesos (Spanish)

Distribution

Resident (*plancus* group) in south-central Florida (centering around Lake Okeechobee, where declining), Cuba and the Isle of Pines, and from northern Baja California (where rare), (rarely also southern Arizona and formerly southern New Mexico), Sonora, Sinaloa, Zacatecas, Nuevo Leon, central and southern Texas (where declining), and rarely also southwestern Louisiana (Cameron Parish), south locally through Central America and throughout most of South America and the Falkland Islands; also (*lutosus* group) formerly on Guadalupe Island, off Baja California (extinct since about 1900). (The southern hemisphere form *plancus* is larger and sometimes considered specifically distinct, in which case *cheriway* is the appropriate specific name for the more northerly continental populations; additionally the generic name *Caracara* is believed by some authorities to have priority over *Polyborus*.) (See Figure 49.)

North American and Associated Insular Subspecies

P. p. cheriway (Jacquin): Breeds from the Gulf coast of the southern U.S., Cuba, and Mexico south to Panama. Includes *audubonii* Cassin, which was accepted by Palmer (1988).

P. p. pallidus (Nelson): Resident on the Tres Marias Islands, Gulf of California.

P. (p.) lutosus (Ridgway): Extinct, formerly on Guadalupe Island, Mexico. Often considered a distinct species.

Description (of *cheriway*)

Adult (sexes alike). Most upperparts, sides, and abdomen very dark fuscous, darkest on the top of the head; interscapulars buff to pale buffy white, usually whiter toward the tip, crossed by broad transverse spots of deep fuscous-black, the pale interspaces narrow, and the tips whitish; outermost five or six primaries tipped with fuscous-black and variably whitish inwardly, the outer webs mottled with gray, the inner webs crossed by four to seven incomplete gray bands; basal two-thirds of tail white, crossed by 11 to 14 narrow transverse bands of pale grayish black and a broad terminal very dark fuscous-black band; lores, orbital area, and cheeks bare; auriculars, chin, throat, and sides of neck whitish, washed with ivory-yellow; breast and sides of neck buff to pinkish buff, the area speckled to barred, with most feathers crossed by several transverse spots of deep fuscous-black (this color varying from grayish black to brown in various races), the spots larger on the midventral parts of the breast (often more so in females), merging with the solid blackish abdomen; rear abdomen and under tail coverts buff to dirty pinkish buff. Iris raw sienna; cere, bare eye ring, and facial skin usually dull carmine-red or orange (the color varying from pale yellow to vivid red under varying degrees of excitement); bill mostly whitish, becoming bluish at base; tarsi and toes chrome-yellow to very pale yellow; claws black.

Subadult (sexes alike). Similar to adult, but the breast buff-tinted, and the back and abdomen darker than in adults, with blackish cross-barring on the feathers of the abdomen, lower breast, and median upper wing coverts. This stage is probably typical of one- to two-year-old birds; the age of initial breeding is uncertain. In the closely related genus *Phalcoboenus* attainment of the definitive plumage perhaps requires about four or five years (Brown and Amadon, 1968), but in this species the definitive plumage (and presumably also sexual maturity) is apparently reached in two years (Palmer, 1988).

Juvenile (sexes alike). Crown brown, narrowly streaked with black; nape buff, somewhat streaked with brown; rest of upper surface dark sepia, the feather tips of the upperparts and wing coverts dull buff to buffy white, producing a spotted effect; tail broadly tipped with brown to sepia, the rest white to buff, narrowly barred with brown or gray; middle of most primaries barred with white and pinkish buff; chin and throat like nape but unstreaked; breast and abdomen brownish, streaked anteriorly with dull cinnamon, the thighs the same but the feathers tipped lighter; under tail coverts buffy white; wing lining dull olive-brown. Iris brown; bill pale green to nearly white, the bare portion of the face dull pinkish to dull grayish white; tarsi and toes yellowish white to pale grayish.

Measurements (of *cheriway*, in millimeters)

Wing, males 370–418 (ave. of 10, 393.5), females 373–408 (ave. of 10, 391); tail, males 228.6–254 (ave. of 10, 240.5); females 223.5–254 (ave. of 10, 242) (Friedmann, 1950). Eggs average 59.4 × 46.5 (Bent, 1938).

Weights (in grams)

Males, ave. of 14, 834; females, ave. of 10, 953 (Hartman, 1961). Range of weights 800–1300, ave. 1006 (sex and sample size unspecified) (Clark and Wheeler, 1987). Captives of the larger South American race *plancus* weigh up to 1585 g (Brown and Amadon, 1968), and a male from Tierra del Fuego weighed about 1460 g (Humphrey et al., 1970). Estimated egg weight 70.9, or 7.4% of female.

Identification

In the hand. This is the only North American hawk in which the nostrils are slitlike, with the posterior end the lower one.

In the field. The distinctive bicolored head, with a black crown and white cheeks, plus bright reddish skin around and in front of the eyes, easily serves to identify perched birds, which spend a good deal of their time scavenging or hunting for prey on the ground. Immatures appear more streaked and brownish on the undersides than do adults, and have less bright facial skin and legs. Unlike eagles, the birds are quite vocal, and utter a variety of loud, often rattling or croaking calls, often with head-tossing movements. Caracaras sometimes call from perches in a high, cackling call that is the basis for such vernacular names as "caracara," "caraira," and "carancho."

In flight, the undersides and upper surfaces of the wings are mostly blackish, except for contrasting whitish "windows" crossing the barred primaries, and the tail is also white with contrasting narrow cross-bars and a broader terminal black band. Although related to falcons, the caracara has a rounded, distinctly eaglelike wing outline, and it sometimes sails for extended periods with flattened (when soaring) or slightly down-tilted (when gliding) wings. It has a loud, grating alarm call.

Habitats and Ecology

The caracara is associated with open-country habitats, which in Florida are typically dry prairie areas with scattered cabbage palms (*Sabal palmetto*), wetter prairies, and to some extent also improved pastures and sometimes even rather wooded areas having associated limited stretches of open grasslands (Kale, 1978). The center of its Florida range is the Kissimmee Prairie, which consists of shallow ponds and sloughs with scattered hummocks of live oaks and cabbage palms (Quincy, 1976).

In the southwest, such as in Texas, the birds occur in coastal grasslands and lowland brush country, typically nesting in brush or woodlands on prairies or hill slopes (Oberholser, 1974). This same general habitat association of arid prairies, grasslands, and desert brushlands at lower altitudes is typical of the species in general (Brown and Amadon, 1968). No good information on population densities exist, but Hector (1981) reported an estimated caracara density of 4.78 birds per 40 hectares on his aplomado falcon study area in eastern Mexico. The home range of a Florida pair has been estimated at about 8.3 kilometers in diameter, or roughly 55 square kilometers (Kale, 1978).

Foods and Foraging

To a greater degree than any other species in this book, this is a scavenger as well as a predator, often feeding at carrion, but perhaps being more attracted to the associated insect larvae than to the carrion as such (Brown and Amadon, 1968; Oberholser, 1974). Bent (1938) regarded the caracara as mainly a carrion eater, but mentioned among its foods a considerable array of mammals, birds, reptiles, frogs, and fish, plus a variety of invertebrates including crustaceans, insects, and worms.

Glazener (1964) observed that in Texas caracaras regularly cruise along highways at daybreak, finding and feeding on recently killed animals, including rodents, rabbits, armadillos, opossums, skunks, raccoons, coyotes, snakes, and a variety of birds. He also saw caracaras robbing turkey vultures (*Cathartes aura*) that had been feeding on such carrion, by chasing them and causing them to regurgitate their food in flight, then often catching it in midair.

Whitacre, Ukrain, and Falxa (1982) observed caracaras hunting cattle egrets (*Bubulcus ibis*) and a brown jay (*Cyanocorax morio*), either hunting in pairs or individually. Although none of these hunts was observed to end in kills, the remains of four cattle egrets at a caracara nest suggest that caracaras are probably occasionally successful predators. Caracaras have also been observed attacking cattle egrets in Florida (Layne, Lohrer, and Winegarner, 1977; Quincy, 1976).

Richmond (1976) observed a nest in Costa Rica for about 23 hours, and during that period food was brought in at a rate of about one item per hour. Most of this material appeared to be freshly killed, and included items unlikely to have been carrion. These included such birds as three northern jacanas (*Jacana spinosa*), an unidentified small bird, a cotton rat (*Sigmodon hispidus*), the tails of various lizards (*Iguana* and *Ameiva*), a large frog, an eleotrid fish, and a large insect or tarantula spider. She believed that, apart from a peccary jaw, all of the other identified foods represented probable prey rather than carrion. Kilham (1979) estimated that at least 11 and probably all of 13 prey items brought to a nest in Costa Rica were birds, and two of these were probably blue-winged teal (*Anas discors*). However, all had been partially plucked or dismembered when brought to the nest, and Kilham was unable to tell if they had been captured alive or found dead.

Social Behavior

It is assumed that pair bonds in caracaras are monogamous and long-term, but very few observations have been made on this species's social behavior. Tinkham (1948) observed a bird throwing back its head until its crest rested on its back, fluffing out its throat feathers, and uttering a croaking call, which Tinkham believed to be a mating call, since a possible female was nearby. Similar rattling calls accompanied by head-tossing behavior have been seen by other observers during probable agonistic encounters, and thus its display significance is still very uncertain. Similarly it is unknown whether one or both sexes perform the display. Kilham (1979) observed a nesting pair performing allopreening on six occasions. In one typical case the male initially

preened the female's crown and later the female nibbled the side of the male's neck. Generally the male did more preening of the female than vice versa, presumably as a pair maintenance activity. Kilham saw four incomplete copulations, two of which were on a tree limb where the birds fed, one on a small tree in a marsh, and one on the nest. In three of the four cases the male alighted on the back of the female without obvious preliminaries, staying on for about four seconds. In the fourth case the male first presented food to the female, then mounted as she stood with head lowered and while holding the food in her beak. Obvious display flights comparable to territorial flights in other falcons or hawks have not yet been reported.

Breeding Biology

In contrast to typical falcons (tribe Falconini), caracaras actually construct their nests, which are usually large, crudely built structures made mostly of sticks. These nests are often reused from year to year, and indeed abandoned caracara nests are frequently used by other raptors, such as the aplomado falcon. Unlike typical hawk nests, they are not lined with finer materials nor is greenery added.

Caracaras have a fairly prolonged breeding season. A total of 238 Texas clutches in the Western Foundation of Vertebrate Zoology extend from February 1 to June 26, with the largest number (124) from March, followed by April (57). A total of 61 Florida clutches are from January 1 to April 7, with half of these obtained between January 21 and February 22, and the largest number (24) obtained during February. There is even a report of a late December nest holding a well-grown young (Bent, 1938), suggestive of possible double-brooding.

In Florida, the commonest nest substrate (46 nests) was the cabbage palm (*Sabal palmetto*), with saw palmettos (*Serenoa repens*), pines, and unidentified "palms" and "bushes" also represented. In Texas, the most frequent substrate was of oaks (*Quercus* spp.) (93 nests), followed by Spanish dagger (*Yucca treculeana*) (33), mesquite (*Prosopis glandulosa*) (16), huisache (*Acacia farnesiana*) (11), hackberries (*Celtis* spp.) (6), ebony (*Pithecellobium flexicaule*) (6), yaupon (*Ilex vomitoria*) (6), elms (*Ulmus* spp.) (5), and various other supports (11). Considering the entire sample from Texas and Florida, the range of 292 reported nest heights was from about 1.5 to 22 meters, but 32 percent of the total were between 3 and 4 meters, and more than 70 percent between 1.5 and 5.3 meters. Clutch sizes ranged from 1 to 4 eggs, with 3 eggs the modal number (in 53 percent of the nests), and 377 clutches averaged 2.58 eggs.

The egg-laying interval is still uncertain, but field notes of T. C. Meitzen in the Western Foundation of Vertebrate Zoology indicate that 2 or 3 days seems to be typical. The incubation period is about 30–32 days, and incubation is shared by both sexes. If the first clutch is lost, a second and even a third may be laid. It has been suggested (Bent, 1938) that in Florida the birds may even be double-brooded, judging from the presence of eggs in a nest that had held a well-developed chick about three months earlier. Judging from the lack of proven (albeit suspected) double-brooding elsewhere, the case more likely involved successive use of the same nest by two different females.

The young were said by Bent (1938) to remain in the nest for two or three months following hatching, but Newton (1979) reported a fledging period of more than 42 days. Kale (1978) indicated that about eight weeks are needed for fledging, which also seems to be an unusually long nestling period. A closely related but smaller caracara, the chimango (*Milvango chimango*), has been reported to have a nestling period of only 32–34 days (Fraga and Salvador, 1986).

There are apparently no estimates of nesting success and productivity for this species. Apparently many of the pairs normally raise two young, with families remaining near the nest for some time.

Evolutionary Relationships and Status

Brown and Amadon (1968) believed that *Polyborus* is very closely related to the South American genus *Phalcoboenus*, and less closely related to *Milvango*. Vuilleumier (1970) merged *Milvango*, *Phalcoboenus*, and *Polyborus* into the single genus

Polyborus, leaving only the forest caracaras in a separate genus (*Daptrius*). He also included *lutosus* within the species *plancus*, recognizing it as a "strong subspecies."

The status of the crested caracara in the United States is distinctly precarious. In 1973 the U.S. Fish and Wildlife Service listed its status as "undetermined," but this classification was changed in 1977 to "threatened." It was on the Audubon Society's Blue List of apparently declining species from 1971 to 1979, and appeared again in 1981. It has since been listed as a species of "special concern." In Florida it is classified as "threatened," and habitat changes such as prairie destruction, increasing traffic-related roadside mortality, and illegal killing have collectively diminished the population. Kale (1978) judged that during the period 1973–75 Florida supported a minimum of 350 birds, and Layne (1982) stated that about 150 active territories were present in 17 counties of Florida during the period 1973–78, with the best populations in Glade, Highlands, Okeechobee, and Osceola counties. He estimated (in Palmer, 1988) that about 150 breeding pairs and 100–200 prebreeders were present in 1984.

Habitat loss appears to be the major cause of population declines in the southwest. The last definite nesting record for Arizona was in 1960, and the species is now only a rare vagrant in New Mexico. One pair nested in Kaufman County, Oklahoma, in 1980 (*American Birds* 34:909). It probably decreased in southern Texas at least through the 1970s, where for example between the 1950s and mid-1970s the breeding population dropped from 35 pairs to one or two pairs on the Welder Wildlife Refuge (Porter and White, 1977). However, there were 23 caracaras on the Welder Wildlife Refuge during the Christmas Count in 1986, and 11 in 1987. Fairly large numbers were seen at Palmetto, Texas, and at Attwater Prairie Chicken National Wildlife Refuge during the 1987 counts. Based on the 1986 Christmas Counts, I estimate the U.S. population as 2,280 birds, with nearly all of these in Texas.

Tribe Falconini

(Falcons)

Figure 50. North and Central American breeding (hatching), residential (cross-hatching), and wintering (shading) ranges of the American kestrel, including races dominicensis (**DO**), guadalupensis (**GU**), nicaraguensis (**NI**), paulus (**PA**), peninsularius (**PE**), sparverius (**SPA**), sparverioides (**SP**), and tropicalis (**TR**). Relative winter state or provincial density indices (average number seen per Audubon Christmas Count, 1986) are shown for major wintering areas. The South American breeding or residential range is shown on the inset map. Indicated range limits of contiguous races should not be considered authoritative.

American Kestrel *Falco sparverius* (Linnaeus) 1 7 5 8

Other Vernacular Names

American sparrow hawk, desert sparrow hawk (*phalaena*), eastern sparrow hawk (*sparverius*), little kestrel, little sparrow hawk (*paulus*), San Lucas sparrow hawk (*peninsularis*), sparrow hawk; crecerelle d'Amérique (French); cernicala chitero, gavilan chitero (Spanish)

Distribution

Breeds from western and central Alaska, southern Yukon, western Mackenzie, northern Alberta, northern Saskatchewan, northern Manitoba, northern Ontario, southern Quebec, New Brunswick, Prince Edward Island, Nova Scotia, and Newfoundland south to southern Baja California, Sinaloa, the Gulf coast, and southern Florida; also resident on the Bahamas and the Antilles (excepting the Cayman Islands); locally in the highlands of Central America including eastern Honduras and northeastern Nicaragua. Also widespread as a resident through most of South America.

Winters from south-central Alaska, southern British Columbia, the northern United States, southern Ontario, southwestern Quebec, and Nova Scotia south throughout the breeding range, including the northern Bahamas and virtually all of Central America, the northern populations migrating as far south as Panama. (See Figure 50.)

North American and Associated Insular Subspecies

F. s. sparverius (Linnaeus): Breeds in North America as indicated above excepting the areas occupied by the following races. Includes *phalaena*.

F. s. paulus (Howe and King): Breeds from northeastern Texas or southern Louisiana to South Carolina, probably resident.

F. s. sparverioides Vigors: Resident of Cuba and the Isle of Pines; recent colonizer of southern Bahama Islands (north to Long Island, Rum Cay, and San Salvador).

F. s. dominicensis Gmelin: Resident of Hispaniola and Jamaica.

F. s. peninsularis Mearns: Resident of southern Baja California and the lowlands of Sonora and Sinaloa.

F. s. guadalupensis Bond: Resident on Guadalupe Island, Mexico.

F. s. tropicalis Griscom: Resident from Chiapas, Mexico, south to northern Honduras.

Description (of *sparverius*)

Adult male. Forehead and lores white; crown and occiput slate-gray to slate with a somewhat variable but usually fairly extensive patch of brown to cinnamon-rufous or hazel; most upperparts pale cinnamon-rufous to deep hazel, the nape usually slightly tinged with apricot-buff and with a small concealed blackish spot, the scapulars and upper back crossed by a variable number of black bars; upper wing coverts slate-gray to slate marked with rather large and conspicuous squarish black spots, primaries black with six to nine broad transverse bars of white on the inner webs, the white areas broader than the black ones and widening toward the edge of the feather, where they are confluent; primaries, especially the inner ones, narrowly tipped with white; secondaries basally black for two-thirds of their length, then broadly slate and narrowly tipped with white, the inner webs crossed by four to seven white bars similar to those of the primaries; tail rich hazel with a broad subterminal band of black, and tipped fairly broadly with white; superciliaries, chin, throat, cheeks, and auriculars white; a black mustachial stripe beginning in front of the bare preorbital space and extending downward across the malar region; another black stripe crossing the posterior edge of the auricular area; a third and very much smaller black transverse mark at the lateral termination of the nape; breast, upper abdomen, sides, and flanks varying from light pinkish cinnamon to cinnamon, sometimes practically immaculate; lower abdomen, thighs, and under tail coverts white to very pale buff, spotted with black. (Birds from Cuba, the Isle of Pines, and the southern Bahamas differ in that two distinctive but intergrading morphs exist, one with rich rufous below in both sexes, the male washed with slate above, and the other morph white below, spotted [in females] with rufous on the sides of the breast.) Iris dark brown; cere, orbital skin, tarsi, and toes ranging from deep cadmium to reddish orange; claws black.

Adult female. Entire head as in male, but the chestnut feathers of the crown usually with slate shafts; scapulars and most upperparts brown to pale brown barred with blackish; primaries as in the male, but the whitish areas heavily tinged with brown; secondaries as in the male, but with the slate replaced by brown; tail like the upper back, but slightly more hazel, crossed by a broad subterminal and nine to eleven narrower black bands, the outer web of the outermost pair irregularly edged with whitish; throat white; rest of underparts white to pale buff, the breast, sides, flanks, and upper abdomen streaked with fairly pale tawny-olive to pale umber; thighs, lower abdomen, and under tail coverts immaculate; under wing coverts white to pale buff transversely spotted and incompletely and irregularly barred with pale tawny-olive to pale umber. Softparts as in male, but the cere, orbits, tarsi, and toes usually slightly paler and generally more yellowish, less orange.

Subadult: Unique among North American hawks in that a nearly adultlike, sexually dimorphic body plumage is attained prior to fledging, and molts again into a nearly adultlike plumage during the first winter. However, both sexes retain their juvenal wing and tail feathers until their second summer of life (molting them after their first breeding season). Males up to a year old can sometimes be recognized by their generally heavier black upperpart barring, and females can sometimes be recognized by their juvenal tail feathers, which have a more poorly defined subterminal band than do adults.

Juvenile. Male: similar to the adult male, but with the black bars on the scapulars and upper back averaging broader and more numerous, the breast and upper midabdominal area with short, narrow blackish streaks, and the feathers of the crown and occiput usually with dark shafts. *Female:* similar to the adult female, but the ventral streaks darker, umber to pale sepia, and with the under surface usually washed with pale ochraceous. The crown has heavier shaft streaks and the tail a less well-defined subterminal bar. *Both sexes:* iris brown; bill pale horn; cere pale greenish yellow; tarsi and toes pale yellowish white.

Measurements (of *sparverius,* in millimeters)

Wing, males 174–198 (ave. of 64, 183.1), females 178–207 (ave. of 68, 195); tail, males 116–142 (ave.

of 64, 129.4), females 119–142 (ave. of 68, 129.5) (Friedman, 1950). Eggs average 35 × 29 (Bent, 1938).

Weights (in grams)

Males, ave. of 69, 111; females, ave. of 111, 120 (Dunning, 1984). Males 94–126 (ave. of 20, 114.4), females 132–160 (ave. of 20, 147.1) (Bowman, 1987). Estimated egg weight 16.2, or 13.5% of females.

Identification

In the hand. This is the only North American raptor with circular nostrils, a wing length of no more than 207 mm, and toes that are entirely scutellated from their bases to their tips.

In the field. When perched, this tiny falcon appears mostly rusty brown above, with both sexes exhibiting a conspicuous black nape and ear patches, and a black "mustache." Males additionally exhibit distinctive light bluish wing coverts, and sometimes their bright reddish brown tail color is visible too, while females appear rather uniformly barred with rufous and darker brown tones. This hawk regularly perches on telephone or similar small wires; the larger hawks typically perch on posts or poles.

In flight, its very small (175–200 mm long) body, rusty brown color, pointed wings, and deep and rapid wingbeats help identify this falcon, which at considerable distance is perhaps more likely to be misidentified as a killdeer (*Charadrius vociferus*) or even a mourning dove than as one of the other variably larger falcons. However, female kestrels are about as large as male merlins, and thus size alone does not serve to identify the species. It far more often hovers while searching for prey below than do merlins or indeed perhaps any of the other North American hawks, when its rusty brown tail is somewhat spread and its rusty brown color (heavily barred in females, singly barred in adult males) is easily seen. Most of the kestrel's prey (often large insects) is captured by pinning it to the ground rather than during extended aerial chases. When alarmed, the birds utter rapidly repeated and high-pitched *klee* or *killy* notes.

Habitats and Ecology

This is a very widely distributed, ecologically versatile species, ranging from alpine zones down into desert habitat in various parts of the west, but generally favoring open savannalike areas with a few trees, forests edges near open areas, farmsteads, suburbs or parks in cities, and the like. The presence of nest cavities in trees (or sometimes birdhouses, earthen banks, or other nontree cavities), elevated perches, and open terrain for hunting insects and small vertebrates probably constitutes minimum habitat needs. Balgooyen (1976) listed perches, a food supply, a territory with available nest sites, and the presence of open vegetation as limiting factors for kestrels. Winter habitat needs are similar (though of course breeding sites are not needed), and probably the availability and location of perch sites strongly influence distribution patterns (Macrander, 1983). At least in some areas the northern flicker (*Colaptes auritus*) is a critical element in the ecology of kestrels because of its excavation of nest cavities (Balgooyen, 1976), but similar-sized woodpeckers also are able to provide nest sites where flickers are uncommon or lacking.

Hunting is done over open areas that typically offer a considerable amount of space between plants, which appears to be vital to kestrel predation because the birds obtain nearly all their prey from or just above the soil substrate. Since over 95 percent of their attacks are typically initiated from perches at ranges of up to 275 meters, a convenient perching site offering excellent long-distance visibility is also clearly important. Most perches used are under 10 meters high, and there is no clear relationship between size of perch and striking distance (Balgooyen, 1976).

Millsap (1984) analyzed habitat characteristics used by hunting kestrels in Arizona, noting that although this species exhibits a high level of habitat diversity during winter (sharing with redtailed hawks and Cooper's hawks the highest indices of distributional diversity), some sexual dimorphism in habitat use occurs, with males associated with more densely vegetated areas than females (areas having stands of arborescent vegetation larger than two hectares). Similar habitat segregation was observed by Koplin (1973) in California, which he attributed to intersexual competition and ecological segregation. Mills (1976) later confirmed a habitat separation by sex during winter, apparently as a means of reducing winter competition, with the larger females possibly forcing males into less suitable (more wooded) habitats.

Perhaps because of its small size and ability to use insects as well as vertebrates as food, the kestrel differs from larger North American falcons in being able to maintain fairly high population densities. Millsap (1981) found it to be the most common falconiform in six of ten vegetational communities in his Arizona study area. Similarly, Smith and Murphy (1973) found it to have the smallest average home range size (0.68 and 0.81 square kilometers in two years) of any of the hawks in their Utah study area, and about the same as that of the similar-sized and similarly insectivorous burrowing owl (*Athene cunicularia*). Wintering territories of birds in Alabama were estimated by Macrander (1983) as averaging only 0.15 square kilometers, while in Pennsylvania six pairs occupied nesting boxes within an area of 1.3 square kilometers (Nagy, 1963). Craighead and Craighead (1956) estimated an average territory size of 130 hectares (but much larger winter home ranges), while Balgooyen (1976) made a similar estimate of 109 hectares for 32 kestrel territories. He found the birds to defend their entire home range; thus the home range and territory were identical. Sixty-one percent of the 18 territories that were present in one year were also active the following season, suggesting a fairly high level of territorial fidelity. The minimum distance between active nests that he observed was only 42 meters. Millsap (1981) reported an average nearest-nest distance of 1.1 kilometers in Arizona. A population density of 1.2–1.7 pairs per square kilometer was estimated by Smith, Wilson, and Frost (1972) in Utah; Enderson (1960) estimated a lower breeding density of 0.75 pairs per square kilometer in Illinois.

Foods and Foraging

A large amount of information on kestrel foods and foraging behavior is now available; Heintzelman (1964) made an early effort at surveying spring and summer foods of kestrels, and Sherrod (1978) listed ten studies or summaries of dietary analyses of this species. Six of these studies provided a numerical composition of food items, in which the mammalian component ranged from 1 to 54 percent (unweighted average 16 percent), the bird component ranged from nil to 20.5 percent (unweighted average 6 percent), and invertebrates ranged from 32 to 99 percent (unweighted average 74 percent). Much more useful is a relative biomass analysis, given the weight diversity of such prey, and on that basis Smith and Murphy (1973) estimated the kestrel's diet in Utah to consist of 38 percent mammals (compared to 26 percent numerically), 57 percent birds (compared to 16 percent numerically), and 2 percent invertebrates (compared to 52 percent numerically). The most important vertebrate prey were starlings (*Sturnus vulgaris*) and deer mice (*Peromyscus maniculatus*), these two species contributing over 60 percent of the total biomass intake. Smith, Wilson, and Frost (1972) found that vertebrates comprised only 20 percent of the species's foods in Utah numerically, but made up 96 percent of the total prey biomass, with voles (*Microtus* spp.) contributing 62 percent alone.

Millsap (1984) compared frequency and biomass estimates of kestrel foods in Arizona, finding that birds contributed 35 percent of biomass and 24 percent of the food items, with horned larks (*Eremophila alpestris*) and various sparrows or other small passerines most important, while mammals comprised 32 percent of the biomass and 20 percent of the items, with deer mice the most important single species. Reptiles contributed 28 percent of the biomass and 17 percent of the items, with lizards (mostly *Uta* and *Sceloporus*) the most important components. The insects taken were mostly beetles, grasshoppers, and cicadas, which contributed less than 5 percent of the estimated biomass. The average prey weight was 14.8 grams.

Balgooyen (1976) estimated that birds, mammals, reptiles, and insects all contributed important components to the diet of kestrels he studied in California, with birds and mammals taken through the early part of the breeding season, lizards and insects becoming more important later in breeding, and insects eventually constituting a high proportion of the diet from midseason to fall departure. He estimated that prey weights ranged from 0.05 to 89 grams, the latter about 80 percent of the bird's body weight and more than they can actually carry, which is about half of their body weight. Apparently kestrels are able to capture insects more readily than other prey, and thus concentrate on them when they are readily available, thereby selecting for a continuous flow of energy input. Prey is typically captured with the feet and killed with the beak, although some prey are captured with the beak. The birds hunted nearly all day long on Balgooyen's study area, although other studies have indicated hunting peaks in morning and evening. Probably prey abundance and activity pattern variations influence the periodicity of kestrel foraging in various areas. Similarly hunting methods may vary by place and according to prey taken; Balgooyen found that nearly all hunts were initiated from perches, with hovering, hawking flights, and foraging on foot comprising less than 3 percent of observed hunts. Hovering was always associated with wind, and occurred in areas lacking perches. When capturing mammals from a hovering start, the bird would dive head-first, compared with feet-first when capturing insects, thereby gaining additional speed and force.

Social Behavior

The social and sexual behavior of kestrels has been very well documented by a variety of studies (Willoughby and Cade, 1964; Mueller, 1971; Porter and Wiemeyer, 1972; Balgooyen, 1976). Pair bonding in kestrels is strong and tends to be permanent. Craighead and Craighead (1956) found a pair nesting in the same tree cavity for six consecutive years; 26 of 31 pairs studied in Utah (Smith, Wilson, and Frost, 1972) maintained their territories during two consecutive years, and 21 of these did so for three consecutive years, even though only eight pairs used the same nest site all three years. Balgooyen (1976) found that 11 of 18 territories were active a second successive year. These figures do not exclude the possibility of mate exchanges or pair replacements in a territory, however. Indeed, studies suggest that early in the pair-forming season promiscuous matings are fairly common as females move around among the territories of two or more males before forming monogamous pair bonds with one of them (Cade, 1955; Balgooyen, 1976). Males may also leave their initial mate to form a new pair bond with another female (Fast and Barnes, 1950).

The general process of kestrel pair formation is initiated by the male establishing a territory that often is the same as that used the previous year, which probably promotes re-pairing with the same female. Concurrently with or slightly before copulation behavior is initiated, females begin to associate and hunt with a territorial male. Aerial displays, the search for a nest site, and courtship feeding of the female by the male follow and are important components of pair bonding. Pairing is probably completed when the female associates exclusively with a single male on his territory (Balgooyen, 1976).

Willoughby and Cade (1964) studied pair-forming behavior in captive kestrels and suggested that in addition to its role in egg fertilization, copulation may serve to bring and hold potential mates together early in the season, especially inasmuch as it may begin prior to courtship feeding. They recognized three vocalizations, the "klee" (or "killy"), the whine, and the chitter. The klee is the most common vocalization and is uttered by both sexes at any time of year, usually under conditions of generalized excitement. The whine varied from a simple to a treble (three-part) call; all of these variations are associated with food or copulation, and are used by adults of both sexes as well as by fledglings when food-begging. The chitter is the most frequent call uttered between males and females, and occurs during courtship feeding, copulation, nest site inspection, and the feeding of nestlings; apparently it indicates sociable, nonaggressive tendencies. Mueller (1971) added two vocalizations to this list, the "klee-chatter," which was uttered by either sex when very hungry and food was removed by the investigator, and the "whine-chitter," also uttered by birds that were extremely hungry, especially females.

Willoughby and Cade (1964) recognized courtship feeding as an important pair-bonding activity. In the complete sequence a male whines at the sight of food, whine-chitters as he picks it up, chitters as he flies to the female, and holds it to her beak while still chittering (Figure 51A). In the Eurasian common kestrel and also in this species the male may try to entice a female to a nest site by calling to her, whether or not he is holding prey (Figure 51C,D). Males may "flutter-glide" toward the female with shallow and quick wingbeats while carrying foods, and females at times also beg for food by similar flutter-gliding. Copulations are initiated by the female bowing deeply with her back sloping slightly and her tail held in line with the body or slightly angled up. During copulation the male flaps his wings to maintain balance (Figure 51B), and holds his toes balled into fists. Besides the flutter-glide, kestrels perform a flight display called the "dive display" by Willoughby and Cade. It is mainly performed by the male during the early phases of courtship, but extends into the nestling period. It consists of a series of climbs and dives, with 3–5 klee notes uttered near the peak of each ascent. The vertical depth of the dive is about 10–20 meters, and at times the male may swoop above the perched female at the bottom of the dive.

Mueller (1971) recognized several additional displays not mentioned by Willoughby and Cade. One of these is the "curtsey," in which the body is held horizontal, with the back feathers erected, the legs flexed, and the head and tail touching or nearly touching the substrate. The bird may relax slowly, then quickly resume the curtsey posture. Typically, two birds assume this posture in parallel orientation, head to tail and about 10 centimeters apart, as agonistic display, after one of them initially displays frontally toward the other. The "bow" is a frontal display rather similar to the curtsey, but all the body feathers are raised, and the tail is usually spread. This display perhaps indicates a higher level of aggression, and more often is followed by an attack. In both displays the black and white markings on the sides of the head are exhibited, as are the markings of the back and tail, suggesting that they may be important territorial signals.

After copulation has become frequent and the sexes begin to hunt together through the male's territory, nest site searches begin. The male begins this search, with the female following. Sometimes the male will stimulate following by uttering chitters or whine-chitters, or by presenting her with food, often after flutter-glides to her. As the female follows the food-holding male he flies to the lip of the nest cavity and loudly chitters (see Figure 51D). Before she arrives he flies to another perch where the food is transferred. At other times he may chitter from the lip of the nest cavity without food (see Figure 51C), then will enter the cavity and remain there for up to 300 seconds as the female perches nearby. As the male emerges the female flies to the cavity and also enters it. Frequently copulation follows immediately after she emerges from the cavity. Apparently females do not explore nest cavities on their own, but will enter established cavities without the male. They may enter it periodically during the day and roost in it at night, indicating the pivotal importance of the nest cavity in the breeding cycle (Balgooyen, 1976).

Breeding Biology

The final choice and occupation of a nest site is the first stage of the nesting phase. Apparently kestrels are dominant over all the species of

Figure 51. Social behavior of kestrels, including prey presentation by a male (A) and copulation (B) of the American kestrel (after Willoughby and Cade, 1964). Also shown is the calling posture of male Eurasian common kestrel when enticing female to nest scrape, without prey (C) and while holding prey (D) (after sketches in Cramp and Simmons, 1980).

woodpeckers, as well as over most other cavity-nesting birds using similar-sized cavities and cavity-dwelling mammals such as chipmunks and squirrels, and kestrels may even evict flickers from active nests. Competition with screech-owls (*Otus* spp.) may be severe and prolonged, and sometimes the eggs or young of both screech-owls and kestrels have been found in the same nest (Balgooyen, 1976).

In Balgooyen's study area kestrels favored nest sites with tight-fitting entrances, those with entrances facing east, and sites located on east-facing slopes (most storms in that area coming mainly from the south and southwest during the breeding season). The birds also tended to favor sites on troughs near ridges, which provide both weather protection and easy return flights to the nest when carrying prey.

In Utah, Smith, Wilson, and Frost (1972) found that 19 of 41 nests were in woodpecker holes, 7 in natural cavities, and 13 in niches in rocky cliffs or in buildings. In Ontario, natural sites are more often in deciduous than coniferous trees, in dead rather than in living trees, and in woodpecker cavities than natural cavities. However, nest boxes tend to be used where they are available, nests are sometimes in buildings, and one of 233 nests was in an earthen burrow. Half of 121 nests were at heights of 5–12 meters, and cavity openings were usually 5–20.5 centimeters in diameter (Peck and James, 1983).

Eggs are laid at the rate of one every 24–72 hours, or normally on alternate days (Balgooyen, 1976; Heintzelman and Nagy, 1968). Clutch sizes tend to average between 3.6 and 4.7 eggs (Heintzelman and Nagy); Balgooyen (1976) found an average of 4.0 eggs in 40 California clutches. Peck and James reported that 135 of 151 Ontario nests had from 3 to 5 eggs present, and the average clutch of this sample was 4.3 eggs.

Both sexes regularly incubate, with the male often incubating in the morning and evening hours but sometimes during the night as well. Balgooyen estimated that perhaps 15–20 percent of the total incubation was performed by the male. This pattern differs from that of most larger falconiforms, in which males typically only incubate while the female is eating food brought her by the male. Perhaps this behavior reflects the relative ease with which the female can capture her own food as compared with the inefficiency of the male carrying many such small items to the nest. The male sometimes brings no food when he relieves the female on the nest, and they may then feed at a food cache. After the female has been gone for up to two hours she will return to a position near the nesting site and call. The male will then emerge from the nest and soon resume hunting.

The incubation period lasts 29–31 days, averaging in one study 30.9 days (Heintzelman and Nagy, 1968) and 28.4 days in another (Willoughby and Cade, 1964). Since hatching often occurs over a period of about 3–4 days, it is clear that incubation must not normally begin until the clutch is complete or nearly so. The young grow very rapidly, gaining their adult weight in only 16–17 days. Fledging occurs at 29–31 days, and thereafter a close flock of siblings develops that persists until the young disperse (Balgooyen, 1976).

Because of efficient parental defense, there is usually a high survival rate for the fledglings. Balgooyen reported a high hatching rate of 89 percent, followed by an even more remarkable 98 percent fledging success, or an overall reproductive success of 87.5 percent. A somewhat smaller but still impressive 78 percent hatching success was reported by Heintzelman and Nagy (1968), while Smith, Wilson, and Frost (1972) estimated a 67 percent hatching success and a 72 percent fledging success in Utah. Renesting following initial clutch losses is known to be frequent, and such efforts may increase the overall productivity somewhat (Bowman and Bird, 1985). Similarly, the production of second broods in a single season has been reported to perhaps occur in some areas such as Colorado and Oklahoma (Black, 1979; Stahlecker and Griese, 1977); more convincing evidence exists for such southern locations as Florida, Tennessee, and southern California (Cade, 1982). Recently Toland (1985c) reported on double-brooding in 14 of 45 nests studied during two successive years; 12 of the same pairs were studied both years, and two of these were double-brooded both years. The average clutch size was not significantly smaller for second clutches, but the hatching success of second broods was significantly lower. The overall average nesting success was 85 percent for first nestings and 73 percent for second nests. In more than half of the cases the same nesting site was used for both broods, and the postfledging dependency period for young of the first brood averaged less than for those of second broods or for single-brooded nests.

Evolutionary Relationships and Status

There is no doubt that this is a very close relative of the considerably larger Eurasian common kestrel (*F. tinnunculus*), and both species are part of a widely ranging group of 9 or 10 species of small falcons that are widely scattered around the world in a mostly allopatric manner (Cade, 1982). Palmer (1988) suggested that the lesser kestrel (*F. naumanni*) of Eurasia may be the nearest relative of the American kestrel.

There is also no doubt that this is the most abundant of the North American hawks; Cade

(1982) estimated that it might number about 1.2 million pairs in North America alone, plus a possibly equally large Neotropical component, which would make it perhaps the most abundant of the world's kestrels. However, this story is not totally lacking in problems; the southeastern race *paulus* has been listed as "threatened" in Florida (Kale, 1978). The causes of this population decline are uncertain but may involve habitat loss. A recent study (Hoffman and Collopy, 1988) suggests that in north-central Florida this form's populations have declined more than 80 percent since the early 1940s, with the loss of suitable nest sites an apparently contributing factor.

Based on the 1986 Audubon Christmas Counts, I estimate the wintering Florida population at about 12,000 birds, but this number includes substantial numbers of migrants from farther north. My estimate of total wintering populations in the U.S.A. and southern Canada is 236,000 birds, or only a small fraction of Cade's (1982) estimate of the total North American breeding population. This estimate of course excludes all the birds wintering from Mexico to Panama, which if Cade's estimate (1.2 million pairs) is accepted must represent about 90 percent of the total North American breeding population.

Merlin *Falco columbarius* Linnaeus 1758

Other Vernacular Names

American merlin, black merlin (*suckleyi*), eastern pigeon hawk (*columbarius*), pigeon hawk, prairie merlin (*richardsonii*), Richardson's pigeon hawk (*richardsonii*), taiga merlin (*columbarius*), western pigeon hawk (*bendirei*); faucon emerillon (French); halcon palomero (Spanish)

Distribution

Breeds in North America from northwestern Alaska, northern Yukon, northwestern and central Mackenzie, southern Keewatin, northern Manitoba, northern Ontario, northern Quebec, Labrador, and Newfoundland south to southern Alaska, southwestern British Columbia, central Washington, eastern Oregon, Idaho, northern Montana, western South Dakota, northwestern Nebraska, northern Minnesota, southern Ontario, southern Quebec, New Brunswick, and Nova Scotia. Formerly or rarely also in eastern Montana, western North Dakota, eastern Iowa, northern Wisconsin, northern Ohio, and northern Michigan. Also widespread in Eurasia.

Winters in western North America from south-central Alaska, southern British Columbia, Wyoming, and Colorado southward to Mexico, locally or occasionally across southern Canada in Alberta, Saskatchewan, Manitoba, southern Ontario, southwestern Quebec, New Brunswick, Nova Scotia, and Newfoundland, and in the eastern United States regularly from southern Texas, the Gulf coast, and South Carolina south through Middle America and the West Indies to northern South America. Also widespread in Eurasia. (See Figure 52.)

North American Subspecies

F. c. columbarius Linnaeus: Breeds widely in northern North America as indicated above, excepting the areas occupied by the following races. Highly migratory, wintering south to Peru. Includes *bendirei* Swann (Stresemann and Amadon, 1979).

F. c. suckleyi Ridgway: Breeds from western Oregon (Puyallup River) north to Sitka, Alaska. Somewhat migratory, occasionally wintering south to California and New Mexico.

F. c. richardsonii Ridgway: Breeds from the grovelands of southern Canada south to Montana and western South Dakota, occasionally or rarely to western Nebraska and western North Dakota. Usually winters south to northern Mexico, but some birds are resident in the cities of southern Canada.

Description (of *columbarius*)

Adult male. Most upperparts slate to blackish slate, becoming lighter to slate-gray on the greater upper wing coverts, outer webs of the inner secondaries, lower back, rump, and upper tail coverts; feathers of the nape whitish washed with pinkish buff to pale cinnamon, producing a mottled nuchal band; feathers of the forehead and of the lateral edges of the crown grayish white to pale cinnamon-buff, forming an indistinct frontal line and a fairly definite superciliary stripe on either side; primaries and outer secondaries black crossed by fairly broad white bars or spots on the inner web; rectrices colored like the lower back and rump, but tipped with white and very broadly subterminally banded with black, anterior to which are several (three or four in *columbarius, bendirei,* and *suckleyi,* up to six in *richardsonii*) somewhat irregular but fairly complete narrower lighter gray bands; lores whitish with fine blackish hairlike plumes; cheeks and auriculars whitish; chin and upper throat white to pinkish buff; other underparts similar, but more heavily tinged with cinnamon and with broad fuscous shaft streaks, widest on the flanks; thighs, lower abdomen, and under tail coverts more intensely tinged with pinkish cinnamon, especially the thighs; under wing coverts whitish and somewhat washed with tawny, with narrow blackish shaft streaks and two or three broad, deep grayish black transverse bars, breaking up the white into spots. Iris very dark brown; eyelids and cere greenish yellow to yellow; bill bluish, tinged with greenish at base and blackish at tip; tarsi and toes wax-yellow to deep cadmium-yellow; gape pale blue; claws black.

Adult female. Upperparts washed with brownish black, sandy, or fulvous tones, which almost obliterate the slate color; lower back, rump, and upper tail coverts more like that of the male, only very slightly washed with grayish black; the pale

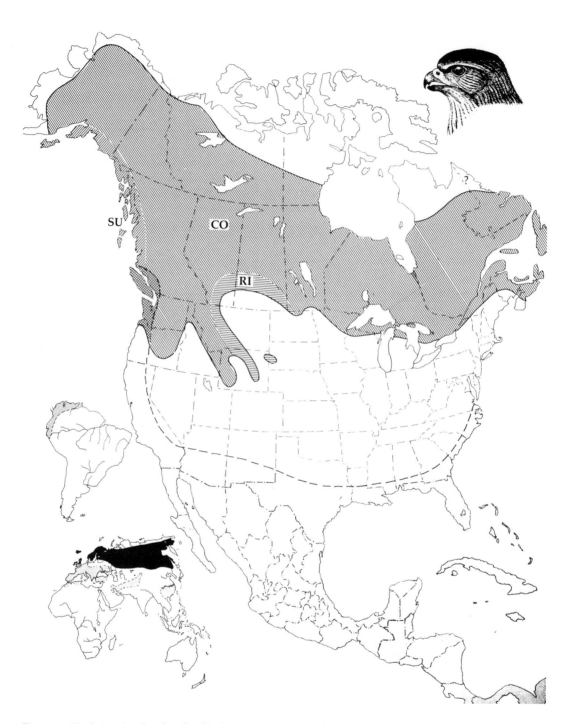

Figure 52. North American breeding (hatching), residential (cross-hatching), and wintering (shading) ranges of the merlin, including races columbarius (**CO**), richardsonii (**RI**), and suckleyi (**SU**). The South

American and Old World breeding (inked) and wintering (shading) ranges are shown on the inset maps. Indicated range limits of contiguous races should not be considered authoritative.

bars or transverse spots on the inner webs of the remiges washed with pinkish buff to very pale pinkish cinnamon, and generally larger than those in the male; pale bands on the rectrices whitish and more or less tinged with pale pinkish buff on all but the median pair, which are washed with pale gray; the thighs less washed with pinkish cinnamon, not sharply different in tone from the abdomen and breast; sides of head averaging more heavily washed with pinkish buff than in males.

Subadult (sexes nearly alike). The juvenal plumage is held until the spring after hatching, when a prolonged molt into adult plumage occurs, the body feathers molting first and the wings and tail later, the molt lasting until as late as November (Bent, 1938). Initial breeding occurs at one year of age.

Juvenile (sexes nearly alike). Similar to the adult female, but browner and darker above, no trace of slate color anywhere, the entire upperparts being dark sepia to fuscous; scapulars, interscapulars, and upper back with narrow tawny-chestnut edges to the feathers; nuchal band generally broader and more pinkish buffy than in adult females; the pale markings on the remiges deeper, ochraceous-buff to ochraceous-salmon; entire underparts washed with pale cinnamon-buff; pale tail banding sexually dimorphic as in adults. Iris grayish brown to dark brown; cere, eyelids, and most of bill bluish black, the bill darkening to black at the tip; tarsi and toes pale greenish yellow.

Measurements (of *columbarius*, in millimeters)

Wing, males 182–200 (ave. of 28, 188.9), females 192.5–215 (ave. of 32, 207.8); tail, males 114–128 (ave. of 28, 121), females 120–140 (ave. of 32, 133.6) (Friedmann, 1950). Eggs average 40.2 × 31.3 (Bent, 1938).

Weights (in grams)

Males, ave. of 14, 158; females, ave. of 14, 213 (Snyder and Wiley, 1976). Males 129–187, ave. 155; females 182–236, ave. 210 (sample sizes unspecified) (Clark and Wheeler, 1987). Forty adult males of *columbarius* in fall averaged 159, and 72 adult females averaged 218, with imma-

tures averaging about 6 g lighter (Clark, 1985c). Sixteen adult males of the Eurasian race *aesalon* averaged 162 (range 125–234), and 14 females averaged 212 (range 164–300) (Cramp and Simmons, 1980). Estimated egg weight 21.7, or 10.2% of females.

Identification

In the hand. The combination of rounded nostrils, a wing length of no more than 215 mm, and toes that are somewhat reticulated rather than scutellated at their bases (see Figure 66E) serves to identify this species.

In the field. When perched, this species appears dark brown (females and immatures) to almost slate-gray (males) above, and variably barred or streaked below with buff and dark brown; the dark crown is set off by a narrow buffy streak above the dark brown eye, and the face lacks the black-and-white, almost clownlike patterning typical of American kestrels. Merlins are about the same size (250–300 mm long) as sharp-shinned hawks, but lack the rufous underparts and red eyes typical of adult sharpshins.

In flight, merlins have the same general size and outline as do American kestrels, but lack any brown tones above or extensive buffy to white underparts, and instead show the same kind of dark underpart patterning seen in immature peregrines, with which they might be easily confused. However, besides being considerably smaller they lack the broad dark malar patch of that species, and instead have only a weakly developed "mustache." When alarmed, merlins utter a rapid series of high-pitched *ki* notes very much like those of other falcons. Hunting is usually done by chases and shallow dives on prey, but occasionally they will stoop on prey from greater heights.

Habitats and Ecology

The habitats of merlins in North America are rather variable regionally, with the northernmost populations breeding in coniferous forests that are often near open areas such as forest edges, bogs, or lakes, with nests often near water, such as on lake shores or islands in lakes. At the extreme northern edge of their range the birds may nest on the ground under tree branches, but otherwise normally depend on old corvid nests,

especially those of crows or ravens (*Corvus* spp.). The northwestern race *suckleyi* probably has a similar ecological range, favoring open rather than extremely dense forests, although sometimes nesting high in dense stands of conifers, and also often near lakes. The interior race *richardsonii* favors areas of mixed grasslands and deciduous trees, these often being quaking aspens (*Populus tremuloides*), and frequently uses old nests of crows or of black-billed magpies (*Pica pica*) (Trimble, 1975; Evans, 1982).

Wintering habitats of the northern and northwestern races are extremely diverse, ranging from deserts to tropical forests, but the race *richardsonii* typically prefers prairie habitats having scattered trees, utility poles, or other open areas with nearby hunting perches, including open farmland with fence posts. Urban areas are also sometimes used, especially by immature birds (Trimble, 1975).

Merlin populations are probably everywhere limited by an adequate food base, a source of available and suitable nest sites, and locally perhaps also by the presence of pesticide residues. The food base is normally provided by a variety of small to medium-sized birds that are usually associated with grasslands, wetlands, or forest edges. Nest sites are typically old nests, although

tree cavities, building roofs, cliff ledges, and ground nesting sites all may be used on occasion, perhaps in the absence of suitable stick nests or as an alternative to them (Trimble, 1975).

As for most bird-dependent hawks, population densities of merlins are not high, and the birds have correspondingly large home ranges and territories and are widely dispersed through available habitats. Three pairs of merlins have been found nesting along a 4.8-kilometer stretch of river in Saskatchewan (observation of L. W. Oliphant, cited by Trimble, 1975), and Craighead and Craighead (1940) found merlin nests to be separated by at least 3.2 kilometers in the boundary waters area of Minnesota. Similarly, Oliphant (1974) observed that a male had a hunting range of up to 3.2 kilometers from its probable nest site, while Lawrence (1949) found a nesting pair hunting in a circular area radiating out about 1.6 kilometers from their nest. Becker (1984) reported that three males had home ranges of from 13 to 28 square kilometers (average 21.3), and ranged a maximum of 8–9 kilometers from the nest during hunting. Preferred hunting habitats consisted of a patchy mixture of grassland and shrubs such as sagebrush (Becker and Sieg, 1987).

As an exception to this general pattern, the recently expanding merlin population in the city of Saskatoon, Saskatchewan, merits attention. In about 12 years that population has increased from 1 to 16 pairs in an area of only 29 square kilometers, and thus the density as of 1982 was approximately two square kilometers per pair. This represents the highest recorded density for this species, including the European population, which has a maximum recorded density of 3.9 square kilometers per pair (Oliphant and Haug, 1985). In 1982 the average nearest internest distance in Saskatoon was 1.2 kilometers, and varied from 0.7 to 2.3 kilometers. The probable reasons for this astonishing density include an abundance of American crow (*Corvus brachyrhynchus*) and black-billed magpie nests, a large urban population of house sparrows (*Passer domesticus*) as a prey base, and the proximity of the South Saskatchewan River, which apparently provided an original population source for the nesting population. Nonurban population densities of merlins in southern Canada are far lower, and in southern Alberta one such population was estimated at only 1.9 pairs per 100 square kilometers (unpublished report of J. K. Schmutz cited by Oliphant and Haug, 1985).

Foods and Foraging

This is a strongly bird-adapted predator, with only a secondary use of other prey, including a few small mammals (bats, squirrels, pocket gophers, mice, and voles) and occasionally lizards, snakes, toads, frogs, dragonflies, and other larger insects. Sherrod (1978) summarized the then-available information on the species's foods in North America, including ten published summaries. Six of these were numerically analyzed, and in these the incidence of birds in the prey ranged from 25 to 100 percent (unweighted average 82 percent). Mammal remains never exceeded 7 percent numerically, and the incidence of insects was quite variable, but mainly consisted of dragonflies, which are sometimes eaten in large numbers by juveniles (Oliphant, 1974) or migrating birds (Allen and Peterson, 1936). Birds predominantly included small to medium-sized passerines, including especially such open-country birds as horned larks (*Eremophila alpestris*), various longspurs or sparrows, and locally also house sparrows in some areas. Some woodland passerines such as thrushes, warblers, kinglets, vireos, and the like also were part of the prey list summarized by Sherrod.

A pair of merlins nesting on the Saskatchewan prairie subsisted almost entirely on open-country passerines that were abundant within their territory, including such genera as *Eremophila, Molothrus, Pooecetes, Melospiza, Ammodramus,* and *Calcarius* (Fox, 1964). A similar dependence on grassland passerines was established in southern Alberta by Hodson (1975, 1978). In a study of merlins in southeastern Montana (Becker, 1984), birds comprised over 90 percent of the total prey items, and most of the species represented were open-habitat forms such as horned larks, lark buntings (*Calamospiza melanocorys*), and vesper sparrows (*Pooecetes gramineus*). However, urban-breeding merlins in Saskatoon depended largely (76 percent of total identified prey) on house sparrows (*Passer domesticus*) for their major prey, and especially on newly fledged young (James and Smith, 1987). Merlins at the edge of tundra near Churchill, Manitoba, were found to prey on nestlings or fledglings of various tundra-nesting species, such as the savannah sparrow (*Passerculus sandwichensis*). Likewise, in an Alaskan study (Laing, 1985) birds were the only items found among prey remains at four nests; these included 22 species

but these were mainly various sparrowlike forms (*Spizella, Junco, Calcarius, Passerella, Zonotrichia,* and *Plectrophenax*) that varied in habitat associations from tundra to spruce forest. The average biomass of 113 items was calculated at about 25 grams (exclusive of some unidentified remains and a ptarmigan of unknown species). Temple (1972a) noted that shortly after hatching most of the prey items brought to a Newfoundland nest were of species weighing under 50 grams, while later (when the larger female was actively helping the male feed the young) most of the prey were of species that averaged more than 50 grams. It is thus possible that the birds adaptively adjust the size of their prey to accommodate nestling food requirements. The two sexes appear to take different-sized prey during both summer and winter, the female taking birds weighing up to about 200 grams (Dickson, 1988).

The typical hunting behavior of merlins is rather accipiterlike, the birds often making surprise attacks on prey from inconspicuous perches or searching flights to flush prey ahead of them. Stooping from considerable heights is also done (Brown and Amadon, 1968; Beebe, 1974). Boyce (1985b) observed merlins attacking by diving almost vertically on flocks of shorebirds, and selectively chasing any trailing individuals. If unsuccessful in singling out a flock member the bird would ascend and repeat the attack. Thus, experienced shorebirds are likely to learn to remain in the middle of the flock, probably exposing inexperienced juveniles to higher risk as peripheral or trailing individuals. Dickson (1988) reported that of some 230 observations of hunting methods, 44 percent consisted of low flight, 32 percent were of still-hunting from perches, and 19 percent were searches from higher elevations.

Merlins are almost entirely daytime hunters, with apparent peaks in hunting effort in early morning and from late afternoon to dusk, occasionally extending their hunting time to twilight, when bats probably are sometimes hunted (Lawrence, 1949; Johnson and Coble, 1967). During the winter as well as during the breeding season the birds sometimes engage in food-caching behavior (Oliphant, 1974; Warkentin and Oliphant, 1985). Page and Whitacre (1975) observed a high level of exploitation by an adult female merlin of wintering shorebirds in coastal California, especially of small sandpipers (*Calidris minutilla, C. alpina,* and *C. mauri*). She killed an average of 2.2

birds per day, which weighed an estimated total of 71 grams, or an average prey size of 32 grams. Most (81 percent) of the hunts were directed toward birds initially on the ground, but some were directed toward flying birds and some toward birds initially on buildings or in trees. The usual attack method was a low, fast, and horizontal flight. Stoops were often seen, but no observed kills were obtained in this manner. Prey was captured at a 13 percent success rate among 343 hunts, and most prey were taken out of flocks rather than from chases of individual birds. Buchanan et al. (1988) observed a somewhat higher success rate per hunt (22.5 percent), with the most common techniques being stooping at a flock and chasing individual birds separated from the flock, but the highest success rate occurring during low stealthy attacks.

Social Behavior

Merlin pairs typically winter separately, the males normally arriving back on the breeding grounds about a month in advance of females. In some cases the female may remain on the nesting territory all year, and the male joins her on his return. Thus, pair bonds must be reestablished or newly established each year. The birds use the same general area and perhaps the same territory each year, but often the nest site is varied (Brown and Amadon, 1968; Cade, 1982). There are records of nest sites in Britain being used for as long as a century (Newton, 1979), but certainly this is simply a matter of a site being highly desirable, and being consistently chosen by the available birds. One site on the side of a cliff in Newfoundland was used continuously for 23 years (observation of S. Temple, cited by Trimble, 1975). On the other hand, pairs in Montana were never observed to use the same nest in successive years (Ellis, 1976), and it is more common for males to show fidelity to a particular territory than is the case for females (Hodson, 1975).

The pair-forming displays of merlins have been well described by Feldsine and Oliphant (1985), and seem to be remarkably similar to those of larger falcons such as the peregrine, prairie falcon, and gyrfalcon. These authors recognized 4 vocalizations and 14 courtship displays. The vocalizations include a rapidly repeated *ki* or *kek,* which varies in intensity, speed, rhythm, and number of syllables according to the situation. Both sexes utter it, with the male's call higher in pitch. Both sexes also utter repeated *tic* or *chip* sounds, often alternately when the pair is not in visual contact. A chuttering or bleating call is used by the male ready to copulate, and sometimes females utter it in encounters involving food. Females also have a distinct food-begging whine.

Aerial courtship displays include power flying (Figure 53A), a strong flapping flight accompanied by rolls, which is a major display of territorial males toward rival males or to attract females into their territory. A more intense version of this is power diving (Figure 53D), and a less intense version is the rocking glide (Figure 53B), the latter sometimes also being used by females on seeing a female intruder. Flutter flying (Figure 53E) is usually a circular or figure-eight flight with shallow wingbeats, performed by either sex, perhaps as a territorial or sexual display. High circling and soaring are also sometimes performed by birds of both sexes, perhaps as a mild territorial display. The male's slow-landing display (or "hitched wing" display of peregrines) is commonly used as a terminal part of aerial displays as the bird lands very slowly and alights with the legs stiffly outstretched and the head bowed (Figure 53I). When used as a precopulatory display it may be accompanied by the copulation chutter call. Females usually perform food begging (Figure 53G) from a perch, while sitting in a fluffed plumage and uttering a series of monotonous whines and wails. Food transfers may occur air-to-air (Figure 53F), air-to-perch (with either sex perched), or with both birds perched, usually accompanied by calling. Supplanting of one bird by another on a perch may occur between rival males, by a resident female supplanting her mate, or by a mated pair supplanting an intruder. On a few occasions tail chasing by a female chasing her mate (Figure 53C) has been seen; such chases may end in copulation or in the female attacking her mate. A female may solicit copulation by bowing deeply and fanning her tail (Figure 53H), often after an exchange of vocalizations between the pair. Both sexes will enter potential nests, and in one case a male in this situation was observed to extend and droop his wings while trembling them and fan his tail nearly vertically as he extended his neck forward and then withdrew it while arching his back and facing the female (Figure 53J). Com-

Figure 53. Social behavior of the merlin, including power flying (A), rocking glide (B), tail chasing (C), power diving (D), slow flutter flying (E), aerial food transfer (F), food begging (G), copulation solicitation (H), slow landing (I), and male nest display (J). After sketches in Feldsine and Oliphant (1985).

pared with the larger falcons, tail exhibition is seemingly better developed in merlins, whereas bowing is apparently much less well developed.

Copulations are often associated with food exchanges and are quite common early in the breeding season, such as during the period of nest site selection (Laing, 1985). Although the majority of nest sites in Britain are on the ground (Cramp and Simmons, 1980), in North America ground sites are common only in extreme northern areas where no tall trees exist. In North America, old corvid nests are most commonly used, with crow nests perhaps preferred over magpie nests and conifers over deciduous trees (Palmer, 1988). However, in one small sample of nests from various parts of North America, more were located in deciduous forests and in deciduous trees than in coniferous forests or coniferous trees (Apfelbaum and Seelbach, 1983). In the urban-nesting Saskatoon population nearly all nests are placed in coniferous trees, and crow nests are preferred over magpie nests, perhaps because the latter tend to be placed lower and have reduced visibility from the nest as compared with crow nests (Warkentin and James, 1988).

Breeding Biology

In contrast to other North American falcons, merlins at least at times "decorate" their nest site with greenery or other materials, which are typically gathered from immediately around the nest site, and sometimes produce a rim around the nest scrape. Over most of their North American range merlins lay clutches of 4 or 5 eggs, although replacement clutches typically are smaller (Bent, 1938). Fairly large samples from Britain and Norway indicate clutch sizes there averaging from 3.96 to 4.25 eggs (Cramp and Simmons, 1980). A total of 21 Ontario nests had 3–5 eggs, the average clutch size being 4.2 (Peck and James, 1983). Temple (1972b) reported an average clutch of 4.3 eggs for 15 Newfoundland clutches, and similarly a Canadian study of *richardsonii* indicated an average clutch of 4.5 eggs in 156 nests (Hodson, 1975), while in Montana the same race had an average of 4.3 eggs in 48 nests (Becker and Sieg, 1985).

Eggs are laid at two-day intervals, although the final egg of a clutch is sometimes deposited after a longer interval. Partial incubation begins while the clutch is being laid, with the female doing the incubation except during periods when the male has brought in food. Males apparently do all the hunting during the incubation period (Laing, 1985). Incubation normally lasts 28–32 days, or about 26 days per egg (Cramp and Simmons, 1980). With hatching, males begin to bring food more frequently to the nest, and the females begin to help with hunting after only about a week of brooding. The nestlings develop rapidly and may fly at 25–35 days of age, perhaps most often at about 27–30 days. By the second week after fledging they are starting to catch insects successfully, but they are not independent of their parents until they have been out of the nest for about five weeks (Fox, 1964; Oliphant, 1974; Trimble, 1975).

A considerable amount of data are now available on merlin productivity in North America, which have been summarized by Oliphant (1985b) for the entire continent (but mainly from the Canadian prairies), and by Becker and Sieg (1985) for the race *richardsonii*. Generally, reproductive success since 1950 has been favorable, with 3–4 young raised per successful nest, tending to allay fears by Fox (1971) and others concerning the impact of pesticides on merlin reproduction in North America (although in Europe pesticide effects are still apparently depressing reproduction). Apparently reproductive rates in more northerly populations are somewhat lower, with 2–3 young per successful nest typical of most areas, even though counts on east coast migration points indicate a population improvement in recent years. Inadequate information is available on productivity in the race *suckleyi* (Oliphant, 1985b). Becker and Sieg (1985) mentioned the rather consistent year-to-year productivity statistics in their study area over four years, and found that losses were highest just prior to hatching and secondly during the period between hatching and fledging. Oliphant and Haug (1985) reported that in an expanding merlin population the average number of fledged young per nesting attempt was 3.7 during a period when the breeding population was increasing at a rate of 26 percent per year. A more recent estimate of productivity in that population (Warkentin and James, 1988) is of 4.2 young reared per successful nest, the highest such statistic among six studies of prairie-breeding North American merlins.

In a study of the Saskatoon merlin population, James (1985) determined that pairs in which one member was a yearling produced broods averaging fewer chicks (3.1) than did adult pairs

Merlin and ruddy turnstone

(4.1). The mean initial breeding age was two years for females and 1.5 years for males, and the oldest known breeder was six years old. An annual maximum mortality rate of 22 percent was estimated, and territorial turnover rates of 27 and 64 percent were estimated for males and females respectively.

Evolutionary Relationships and Status

Apparently this is a rather isolated species of falcon within the genus *Falco*, based on a phylogenetic dendrogram proposed by Fox (1977) and reproduced with some modifications in Cade (1982).

The continental population of merlins is somewhat uncertain. Its status was listed as "undetermined" by the U.S. Fish and Wildlife Service during the 1970s, based on the fact that the numbers of birds migrating through and wintering in Florida had seemingly declined (Evans, 1982). Similar evidence of declines was also noted elsewhere during this period, such as a decrease in merlin counts in Minnesota of 83 percent between 1960 and 1967 as compared with 1935–42 whereas total raptor counts increased 87 percent during this period. Both of these areas are probably on the migratory route of *columbarius*, whereas Alaskan populations of this same race were apparently not declining in the 1960s (Trimble, 1975). During the same general period there was a very substantial decline in productivity and range contraction of *richardsonii*, which has been attributed to eggshell thinning from

pesticide contamination as well as to loss of
native grassland habitats (Fox, 1971). As noted
above, more recent productivity estimates support the view that this population is now reproducing quite well, and so the dangers from
pesticides may for the present at least have
subsided. Population data on the northwestern
race *suckleyi* are still only very poor, but that race
may have declined only slightly during the period of major population decline of *richardsonii*
(Trimble, 1975).

A recent analysis of Audubon Christmas
Counts for the period 1980–83 as compared with
the period 1969–72 suggests a 43 percent increase
in merlin populations wintering in southern Canada and the United States, and an estimated
population there of about 8,000 birds (Anonymous, 1986). The merlin was on the Audubon
Society's Blue List of declining species from 1972
to 1981, and was listed as a species of "special
concern" from 1982 to 1986.

Aplomado Falcon *Falco femoralis* Temminck 1822

Other Vernacular Names

Northern aplomado falcon (*septentrionalis*); halcon azul plomizo (Spanish)

Distribution

Resident in open country and savanna habitats from northeastern Mexico (currently north to about Veracruz; formerly to southeastern Arizona, southern New Mexico, and west-central and southern Texas), south locally to Chiapas and casually to Belize, Guatemala, eastern Honduras, and northeastern Nicaragua; also resident from western Panama south generally throughout South America. (See Figure 54.)

North American Subspecies

F. f. septentrionalis Todd: Formerly bred from the southwestern U.S. south casually to Guatemala and Belize, rarely farther. (The last documented nesting record for Arizona was in the late 1800s, but a few unverified reports of breeding in southeastern Arizona have occurred since the late 1960s. The last documented breeding occurred in New Mexico in 1952 and in Texas in 1941. In recent years some captive-raised birds have been released in Texas.)

Description (of *septentrionalis*)

Adult (sexes alike). Considerable individual variation occurs among adults as to degree of cinnamon or tawny pigmentation on underparts; extreme morph types are described here. *White-breasted morph*: a very narrow frontal line on forehead whitish to buff; crown, occiput, and nape deep neutral gray with a slight ashy bloom; a white superciliary stripe that as it crosses the nape becomes a collar of pinkish buff to very pale pinkish cinnamon; most upperparts neutral gray slightly washed with drab, the interscapulars usually somewhat darker than the back or wings; primaries grayish black darkening to fuscous terminally, the inner webs crossed by numerous (up to 13) white bars, these bars not quite reaching the edge of each feather; secondaries gray, broadly tipped with white and mottled and barred with white on the inner webs; upper tail coverts gray barred with white and tipped with grayish white, the four white bars narrower than

the gray ones; rectrices black with a grayish cast, crossed by 8–13 narrow white bands; lores buffy white; a broad, conspicuous whitish to buffy superciliary stripe from the eye to the occiput, which becomes more cinnamomeous posteriorly; a broad deep gray stripe running from the posterior angle to the eye back over the upper edge of the auriculars and extending broadly down the side of the neck; another smaller one of similar color starting at lower border of bare suborbital skin, passing downward across the cheeks, forming a dark "mustache," leaving the middle area of the ear coverts, the chin, throat, and whole breast white, these areas variably suffused with light buff and usually narrowly streaked with black; sides of the breast, sides, flanks, and upper abdomen black with narrow indistinct transverse whitish bars; rest of abdomen, thighs, posterior flanks, and under tail coverts very pale cinnamon; under wing coverts very pale cinnamon to white, those forming a broad border to the bend of the wing immaculate, the others barred and serrated with black. Iris deep yellowish brown to dark hazel or burnt umber; cere and eye ring light yellow; bill yellowish basally, blue-gray to slate terminally; tarsi and toes cadmium-yellow; claws black.

Cinnamon morph: Similar to the above, but the forehead narrowly pinkish buff, the superciliary stripe, cheeks, auriculars, and nuchal collar all pinkish cinnamon, and the chin, throat, and breast all buffy. Variation between these extremes is continuous, with the cinnamon morph typical of more southern populations.

Subadult. Not well known, but year-old birds are more brownish gray dorsally (rather than bluish gray) than adults and more plumbeous than juveniles. The breast still heavily streaked with black, but light-colored barring or spotting is also present, and breast streaking decreases while the barring increases with each subsequent molt. The tail still juvenile-like and the wing linings still black.

Juvenile (sexes alike). Darker and more blackish and brownish rather than grayish above, the feather margins brownish, and the upper breast heavily streaked with black; the black underpart feathers rarely margined with whitish, and the ventral area streaked with darker feathers, especially on the breast. Tail with eight or nine brown bars or spots of dull white, the wing

Figure 54. *North and Central American breeding (hatching) and residential (cross-hatching) ranges of the aplomado falcon. The dashed line indicates maximum northern historic range limits. The South American breeding or residential range is shown on the inset map.*

linings entirely black. Cinnamon-breasted extreme more deeply colored than the white-breasted morph, with cinnamon, cinnamon-buff, tawny, or russet tones prevalent. Iris brown; cere and eye ring initially bluish gray, gradually becoming brown; tarsi and toes pale yellow.

Measurements (of *septentrionalis*, in millimeters)

Wing, males 248–267 (ave. of 8, 257), females 272–302 (ave. of 7, 290.4); tail, males 172–193 (ave. of 8, 182), females 192–207 (ave. of 7, 198.7) (Friedmann, 1950). Eggs average 44.5 × 34.5 (Bent, 1938).

Weights (in grams)

Single male, 235 (Russell, 1964); females (sample size unspecified) 271–305 (Haverschmidt, 1968); two males 220 and 223, one female 332 (Howell, 1972). Males 208–305, ave. 265; females 310–460, ave. 391 (sample sizes unspecified) (Clark and Wheeler, 1987). Average weight, 6 males 260.5, 7 females 406.7 (Snyder and Wiley, 1976). Estimated egg weight 29.2, or 7.5% of female.

Identification

In the hand. The combination of oval nostrils, a wing length of 248–302 mm, and toes that are scutellated to their bases serves to separate this species from other North American raptors.

In the field. When perched (usually in open, semiarid country), this falcon exhibits a dark brown to blackish color above, with an unusually long tail and a buff-and-black facial pattern characterized by a broad buffy postocular stripe and a similarly light area behind the dark "mustache." The birds are also mostly buffy-white to cinnamon-colored below, but with blackish flanks (and some black breast-streaking in young birds).

In flight, the general shape and size is very similar to that of a prairie falcon, but instead of blackish axillaries the flanks and midabdomen region are blackish, and the birds otherwise usually appear buffy to cinnamon rather than pure white on the undersides of the body. The wings appear generally dark when seen from above, with a whitish trailing edge, and are heavily patterned with black and white barring below; the tail is strongly banded with black and

white. The species has a relatively long tail, and long narrow wings as compared with other North American falcons. Much of their prey is captured on the wing via direct pursuit rather than by stooping. When alarmed near the nest the birds utter repeated *kek* calls similar to those of other falcons, but faster and higher in pitch than those of prairie falcons.

Habitats and Ecology

In the United States the aplomado falcon inhabited open grasslands having scattered arborescent yuccas (*Yucca elata* and *Y. treculeana*) and mesquites (*Prosopis*) that were scattered about a half kilometer apart. Elsewhere in its range it also is known to inhabit oak and pine savannas, open grasslands, coastal deserts, riparian woodlands in desert grassland, and even marshes. However, the encroachment of heavy grasses or of brush of up to about a meter tall into grassland cover causes the birds to abandon an area, probably because of the difficulties of capturing prey in such thick cover (Hector, 1981, 1987).

In eastern Mexico Hector (1981) found aplomado falcons in palm savanna, oak savanna, acacia savanna, and cut-over rain forest habitat. All of these habitats contain an open overstory of woody vegetation and a shallow understory of herbs and forbs, suggesting that the species responds more to habitat lifeform structure than to the exact species composition of vegetation. Habitat preferences of observed prey species were mostly (80 percent) of "variable" type, but brushy-, savanna-, and woods-adapted species were all represented at rates of 2 percent or more. Studies of habitats near nest sites as compared with nonnesting sites in the same area suggested that the birds favor trees that are larger and more sparsely distributed. Estimated bird (prey) densities were also about one-third higher in used areas, and insect densities were lower, but in both cases not significantly so. Two bird species that were found to have unusually high densities in nesting habitats and that were frequently taken as prey were the great-tailed grackle (*Quiscalus mexicanus*) and groove-billed ani (*Crotophaga sulcirostris*). Trees in the area are used as observation stations, preening and roosting sites, for providing protection from wind and sun, as plucking perches, as cache sites, and as nest sites. Probably the falcons depend on nest-building raptors such as caracaras or corvids such as ravens for many of their nest platforms, but arboreal bromeliads serve in some areas with little or no modification (Hector, 1981).

Foods and Foraging

Until the studies of Hector (1981, 1985), almost no good information existed on the foods of this falcon. Hector collected prey remains at 18 plucking or nesting sites and also made direct observations of foraging. In all, 240 prey remains and 82 prey observations were obtained. Of these, bird prey comprised 94 percent of prey items found (representing 99.8 percent of total prey biomass), and 35 percent of prey observed being captured, eaten, or cached (representing 97.3 percent of total prey biomass). Insects, especially moths, beetles, cicadas, and orthopterans made up most of the remainder; the only mammalian prey record was of a single bat.

Among the birds, 16 species were represented, with doves, icterids, and cuckoos the most important groups. White-winged doves (*Zenaida asiatica*), mourning doves (*Z. macroura*), great-tailed grackles, common ground-doves (*Columbina passerina*), yellow-billed cuckoos (*Coccyzus americanus*), and groove-billed anis made up over half of the estimated total prey biomass, while small birds in the 20- and 60-gram categories accounted for 30 percent of the biomass. In all, over 90 percent of the prey weighed 100 grams or less, and 77 percent of the bird prey was in this weight category. The overall average weight of prey in Hector's sample of prey remains was 82.2 grams. The largest prey observed was of a plain chachalaca (*Ortalis vetula*), a species with an adult weight of at least 470 grams, but birds weighing as little as 8 grams were also consumed. The foods of nestling birds were estimated to contain a higher proportion of insects (25 percent numerically) than adults.

Prey is usually captured by direct flight from observation posts, and either captured in full flight or forced to the ground and pursued on foot. The birds will readily enter thick cover when chasing prey, and often hunt in pairs when chasing birds. Swifts and swallows, although common in the areas, were not observed prey, although some slower aerial foragers such as caprimulgids were taken. Similarly, slower insects such as beetles, cicadas, and moths were selected over such swift fliers as dragonflies. The birds were observed to make most of their bird captures early in the morning, and to spend time later in the day hawking insects. In drier areas where avian prey is scarce, they may also tend to concentrate more on reptiles and rodents. They are also known to steal small mammals taken by kestrels and black-shouldered kites, and have even been observed stealing crayfish from herons (Hector, 1985).

Most of the hunting is done within 500 meters of the nest, which is a much smaller foraging range than has been reported for other bird-eating raptors (Newton, 1979). Large birds are typically caught by the pair working cooperatively, the first falcon (usually the male) climbing and hovering above the trapped animal until the second arrives and tries to flush it from its refuge. The first one may then attack in a slanting dive or swoop, while the second gives chase in a low, horizontal path. After each dive the male rebounds and then stoops again, thus following an undulating path as the prey attempts to escape. If it should reach ground cover the falcons attempt to extract it or force it from cover again, but may abandon the chase if the cover is too thick (Hector, 1981).

Social Behavior

Very little is known of the social and pair-bonding behavior of this species. The birds are usually found in pairs throughout the entire year, and Hector (1981) noted that 29 percent of the hunts he observed involved two birds simultaneously chasing the prey, suggesting but not proving that cooperative hunting by a pair is fairly common. Of the 106 observed pair hunts, 32 percent were successful, but of 254 solo hunts 59 percent were successful, suggesting that cooperative hunting is not invariably more efficient. However, if only chases of birds are considered (thus excluding insect hawking), pair hunts were successful 44 percent of the time while solo hunts were successful only 19 percent of the time, indicating that for large and elusive prey such as birds hunting by the pair may be a distinct advantage.

Nesting territories are apparently fairly small, and usually the birds are to be found within a kilometer of the nest. Normally during the entire nesting season the female remains within view of the nest and participates in most of the territorial encounters that occur. Her larger size probably helps to intimidate intruders. However, males may continue to attack conspecifics long after the female has returned to the nest, indicating that the female's role may be mostly nest defense and that of the male mostly territorial defense. For this reason females may be largely limited to hunting fairly near the nesting site. Both sexes are apparently partially committed to feeding the other adult and the young, and females are better able to carry heavy prey back to the nest than are males (Hector, 1981). Apparent courtship-feeding of the female by the male has been reported (Snelling and Leuck, 1976).

Breeding Biology

Nesting platforms used by these falcons in the American southwest included sites built by Chihuahuan ravens (*Corvus cryptoleucus*), Swainson's hawks, and crested caracaras, and perhaps also black-shouldered kites and white-tailed hawks. Of 15 sites studied by Hector (1981) in eastern Mexico, three had been made by caracaras, three by black-shouldered kites, eight by roadside hawks (*Buteo magnirostris*) or brown jays (*Psilorhinus morio*), and at least one by a gray hawk.

Bent reported 29 egg dates from Texas as extending from March 14 to May 26, with 15 between April 12 and 26. Data from 80 Texas clutches in the Western Foundation of Vertebrate Zoology extend from March 2 to May 18, with 40 of these between April 3 and 22. A large percentage of these were in Spanish dagger (*Yucca treculeana*), with some in mesquites, and the platforms presumably mostly represented old raven or hawk nests situated from about 2.5 to 8 meters above ground. Trees at the Mexican nesting sites studied by Hector (1981) averaged 46 meters apart and 9.5 meters high; the nests themselves ranged in outside diameter from 28 to 100 centimeters. Sixteen Mexican clutches ranged from March 15 to April 23, with 7 between April 7 and 12. Of 96 total clutches, the range was 2–4 eggs, with most (52) clutches having 4 eggs, and the mean clutch size was 3.52 eggs. However, most of these were eggs collected by a professional egg collector, F. B. Armstrong, and Hector (1987) believes that many of these may actually represent artificially contrived clutches. Among 28 sets obtained by other collectors the average is only 2.64 eggs, which Hector considers a more reliable estimate of normal clutch size. Brown and Amadon (1968) reported that two or three eggs are the normal clutch, with 4 rarely present. Three eggs appear to be the usual clutch in southern South America (Johnson, 1965).

The incubation period is 31–32 days, and both sexes are said to incubate (D. Hector, in Palmer, 1988). Hector (1981) found that, of 15 nesting attempts in which young were successfully fledged, the average brood size was 2.2 young. Larger nests (at least 50 centimeters across) had associated slightly larger average brood sizes (2.4 vs. 2.1) than did smaller nests, but this difference was not found to be statistically significant. Twelve of the 17 pairs that he studied in Mexico actually nested. The fledging period is 4–5 weeks, with fledglings remaining in their natal areas for at least a month (D. Hector, in Palmer, 1988).

Evolutionary Relationships and Status

Although a typical *Falco*, there are no obvious close North American relatives of this species. Cade (1982) suggests that it is most closely related to the orange-breasted (*F. deiroleucus*) and bat (*F. rufigularis*) falcons, both of which are Neotropical falcons with similar ranges to the aplomado's, the former preying mostly on birds and the latter taking a variety of small birds, bats, and insects on the wing.

The sad history of the aplomado falcon in the United States has been recounted by Hector (1987). He stated that the reasons for its early historic (pre–pesticide era) decline are uncertain, but most probably resulted from habitat change that made hunting more difficult; these causes probably have been recently supplemented by pesticide contamination that has produced egg-thinning effects at least as great or greater than those observed in peregrines. The aplomado falcon has recently (Anonymous, 1985) been proposed as an addition to the federal list of endangered species, and its North American status has been reviewed by Millsap (1984). Some limited attempts at reestablishing the species in the American southwest have been made by hacking young birds, but predation by great horned owls (*Bubo virginianus*) has interfered with their success (Glinski, 1985c; *American Birds* 40:1226). The last validated nesting in the U.S. was in southern New Mexico during 1952, and there was a possible but unconfirmed nesting in Arizona during the late 1960s. The last known Texas nesting was in 1941, and the last documented Arizona nest was found in 1887. However, recent sightings have been made in western Texas, southeastern Arizona, and southwestern New Mexico (Hector, 1980). Between 1985 and 1988 a total of 17 birds had been released in Texas, with more releases planned in 1989 and releases in Arizona to begin in 1991.

Peregrine Falcon *Falco peregrinus* Tunstall 1771

Other Vernacular Names

Duck hawk, Peale's peregrine (*pealei*), peregrine, tundra peregrine (*tundrius*); faucon perelin (French); halcon pollero (Spanish)

Distribution

Breeds or formerly bred in North America from northern Alaska, northern Mackenzie, Banks, Victoria, southern Melville, Somerset, and northern Baffin islands, and Labrador south locally along the Pacific coast to southern Baja California, and interiorly in the Rocky Mountains to southern Arizona, New Mexico, and western Texas, occasionally to the coast of Sonora and northern interior Mexico; formerly also to Kansas, Arkansas, Louisiana, Tennessee, Alabama, and Georgia; also breeds in western Greenland, southern South America, and over much of the Old World including Africa and Australia. (Eliminated as a breeding bird through much of continental North America since the 1950s, especially in areas east of the Rockies and south of the Canadian Arctic, but currently being locally reestablished through reintroductions in various parts of the historic range.)

Winters in the Americas from southern Alaska, the Queen Charlotte Islands, coastal British Columbia, and the eastern United States south through Middle America, the West Indies, and South America; in the Old World generally winters through the breeding range, the northernmost populations usually migrating to tropical regions. (See Figure 55.)

North American Subspecies

F. p. anatum Bonaparte: Breeds in North America as indicated above, with the exception of areas occupied by the following races. Northern populations relatively migratory, wintering at least to the Gulf coast, the southern ones generally resident.

F. p. pealei Ridgway: Breeds from the Aleutian Islands to the Queen Charlotte Islands; relatively sedentary. This form's Alaskan range is sometimes limited to southeastern Alaska.

F. p. tundrius White: Breeds in tundra areas of northern North America and Greenland. Migratory, wintering well to the south of the breeding range from Baja California and the Gulf coast south to Chile and Argentina.

Description (of *anatum*)

Adult male. Forehead buff; crown, occiput, nape, and anterior interscapulars deep neutral gray to blackish plumbeous; the interscapulars tipped and crossed by a single band of slate-gray; posterior upperparts deep gray, generally palest on the rump and upper tail coverts; upper wing coverts slate narrowly tipped with paler gray, the greater coverts obscurely banded with slate-gray; primaries slate-black externally and terminally, the inner webs whitish to gray barred with slate-black for most of their length; the primaries very narrowly tipped with whitish; secondaries gray, paling to white on the inner webs, and barred with dark slate; rectrices gray, tipped with dirty whitish and crossed by eleven or twelve bars of blackish slate; lores like forehead, but separated from the eyes by a narrow blackish or lead-black circumocular area that merges dorsally into the dark crown, and is continuous with the cheeks and auriculars and a broad malar stripe (narrow and sometimes broken in *tundrius*) on either side of the upper throat; chin, throat, and breast white, often tinged with ashy or buff; the lower breast with a few somewhat rounded spots of blackish; rest of underparts white to buffy, transversely spotted or narrowly barred with blackish, the bars most pronounced on the sides, flanks, and thighs (this underpart spotting and barring heaviest in *pealei* and weakest in *tundrius*); under wing coverts white barred narrowly with black, the white interspaces much wider than the dark bars. Iris very dark brown; cere, bare orbital ring, tarsi, and toes bright yellow; bill pale bluish, the culmen darker and the tip blackish; claws black.

Adult female. Similar to the adult male, but usually somewhat more tinged with light ochraceous-buff below; the buffy whitish bases of the nape feathers more often visible; the breast more spotted, reaching the anterior part of the breast, the abdomen, sides, flanks, thighs, and under tail coverts more heavily barred with black, the bars deeper, wider, and more numerous; and the light area on the inner webs of the primaries lightly tinged with very pale ochraceous-buff.

Figure 55. North American range of the peregrine falcon, including known historic range (narrow hatching) and probable current breeding (wide hatching) or residential (cross-hatching) ranges of races anatum *(AN),* pealei *(PE), and* tundrius *(TU). Major wintering areas of migratory populations are shown by shading. The breeding (inked) and nonbreeding (shading) ranges in South America and the Old World are shown on the inset maps. Indicated range limits of contiguous races should not be considered authoritative.*

Subadult. Apparently identical to adults by the end of the first year, although initial breeding probably does not occur until the second year of life (Cramp and Simmons, 1980).

Juvenile (sexes nearly alike). Deep fuscous-brown above, with buffy feather edgings (these poorly developed in *pealei* and quite conspicuous in *tundrius*); ochraceous or deep pinkish cinnamon below, very heavily marked with blackish streaks or spots (females more heavily marked than males); and tail coverts barred. Iris dark brown; bare orbital skin bluish white to grayish blue; cere grayish to greenish yellow; bill grayish to bluish, becoming darker at the tip; tarsi and toes pale yellow to greenish yellow.

Measurements (of *anatum*, in millimeters)

Wing, males 301–327 (ave. of 20, 314.2), females 340–376 (ave. of 22, 356.3); tail, males 138–154 (ave. of 20, 145.1), females 167–192 (ave. of 22, 178.9) (Friedmann, 1950). Eggs average 52 × 41 (Bent, 1938).

Weights (in grams)

Males (of *tundrius*) 550–647 (ave. of 12, 611); females 825–1094 (ave. of 19, 952) (White, 1968). Males 453–685, ave. 581; females 719–952, ave. 817 (sample sizes unspecified) (Clark and Wheeler, 1987). Ranges of adults of various Eurasian races, males 445–750, females 925–1333 (Cramp and Simmons, 1980). Estimated egg weight 48.2, or 5.1% of female.

Identification

In the hand. The combination of circular nostrils, a wing length of at least 300 mm, and the tenth (outermost) primary longer than the seventh and only slightly shorter than the longest (ninth) serves to separate this species from other North American raptors.

In the field. Perched birds of most races and age classes have a more distinctly two-toned or "hooded" head than do gyrfalcons, the darker crown, nape, and a well-developed blackish malar stripe generally contrasting with a more whitish throat and ear patch (this is least developed in the tundra race, which are generally paler, and in *pealei,* which are more sooty gray). The overall upperpart coloration ranges from bluish black or slate-gray to rich brown, and at

least in adults the underparts are only weakly spotted, the breast often nearly immaculate. A well-developed yellow eye ring is present in adults, but in younger birds this appears to be grayish or pale bluish and is less conspicuous.

In flight, the rather large size and rather uniformly spotted and strongly barred underwing surface (the wing linings not noticeably darker than the undersides of the primaries) help to separate this species from the prairie falcon and dark-phase gyrfalcons, both of which have rather two-toned underwing patterns that are produced by wing linings that are generally darker than the flight feathers. However, gyrfalcons have broader wings and more robust bodies than peregrines, and their primaries are distinctly paler rather than heavily barred near their bases. Peregrines fly with extreme power and speed, and often stoop in vertical dive on their prey from great heights, as well as sometimes engaging in direct pursuit, the typical mode of hunting in gyrfalcons.

Habitats and Ecology

Probably the most common habitat characteristic of this species is the presence of tall cliffs (typically over 50 meters), which serve both as nesting and perching sites, providing an unobstructed view of the surrounding area. The nest site component requires the presence of ledges, potholes, or small caves that are relatively inaccessible to mammalian predators and also provide protection from the elements (rain, excessive heat or cold, etc.). Rarely, man-made structures (such as tall buildings) may serve as substitute nesting sites, or steep cutbanks and gravel slopes may likewise serve in the absence of cliffs, and in the Mississippi Valley a small population utilized the holes and stubs of very large trees for nesting. A source of water (river, coast, lake, marsh, etc.) is almost always close to the nest site, probably in conjunction with a localized and adequate prey base of small to medium-sized birds (or sometimes mammals), which is the other major habitat need.

Climatically, the peregrine is extremely tolerant; indeed it has one of the largest of breeding ranges of all species of birds, and in North America it breeds or has historically bred from desertlike to temperate coniferous forest and arctic-tundra habitats. Beebe (1974) states that the only environment that the peregrine seems to be

incapable of tolerating is that of Antarctica. Wintering habitats are correspondingly diverse, and probably only require an adequate prey base and suitable perching sites. Beebe noted that the number of species of appropriately sized birds that regularly fly high in a way that exposes them to the specialized kind of attack used by peregrines is not large, and these are mostly of highly mobile, flocking and colonial-nesting species, such as shorebirds and waterfowl. Thus, the distribution of peregrines is distinctly limited,

even where cliffs or other ideal nesting sites happen to be available.

As extremely fast-flying birds, it is not surprising that peregrines have very large home ranges and territories and a consequent low population density throughout their range. In Great Britain, home ranges have been estimated at 44–65 square kilometers, averaging about 52 square kilometers (Brown and Amadon, 1968). This is close to the average local density of a pair per about 51 square kilometers estimated for

Britain, or with eyries spaced on average from 5.2 to 11.2 kilometers apart in various parts of Scotland. In northeastern Scotland most hunting is done within about two kilometers of the nest cliff, but up to six kilometers away or more when the female also begins to hunt. However, in continental Europe, hunting may occur up to 15 kilometers or more from the eyrie (Cramp and Simmons, 1980).

A great deal of information on home ranges and densities is available from North America, and considerable variability is evident. In areas of dense populations the home ranges tend to be small; Cade (1960) reported that no nests were located closer than 0.4 kilometers apart, with a minimum territorial radius of about 90 meters around the nest, but some birds attacking intruders as far as 1.6 kilometers from the nest (Cade, 1960). In a similarly dense island population of British Columbia, the average distance between territorial pairs was about 1.6 kilometers (Beebe, 1960). The densest known nesting population from the arctic is at Rankin Inlet, Northwest Territories, where an average internest distance of 3.3 kilometers has been found, or a pair per 17 square kilometers (Court, Gates, and Boag, 1988). Beebe (1974) stated that the densest populations occur where a lake or stream provides a conveniently located corridor (up to about 4 kilometers away) for nesting birds, so that pairs can be organized in linear or peripheral associations. In such cases the adjacent pairs sometimes are separated by as little as a kilometer, and local populations probably are comprised of an inbred tribe or clan.

On the other hand, over many areas the population densities are far lower. Thus, Burnham (1975) estimated a density of a pair per 500 square kilometers in western Greenland. Hickey (1942) estimated that about 19 pairs historically occurred in an area of about 26,000 square kilometers around New York City, or a pair per 1,370 square kilometers. Bond (1946) similarly estimated that a pair per 5,180 square kilometers was probably typical of western areas of North America where the birds were considered fairly common, but a pair per 51,800 square kilometers or more was typical where the species was rare. It is possible that the entire pre-DDT population of peregrines in the eastern United States numbered only about 350 pairs (Barclay, 1985), and even in the western parts of North America large areas of mountainous terrain amounting to thousands of

kilometers of cliffs are apparently completely devoid of peregrines, presumably for lack of a prey base (Beebe, 1974).

Foods and Foraging

Wherever the peregrine occurs throughout the world it is a species adapted to living almost exclusively on living birds, the great majority of which are taken in full flight, either being struck dead after a nearly vertical stoop or being seized from various angles while fleeing. Of 1,240 prey items recorded in Britain, at least 117 species of birds were recorded (and 145 in Europe as a whole), including 123 gamebirds (mainly grouse), 168 shorebirds, 294 doves and pigeons, and 366 medium-sized passerines. Most prey items were in the 100–500 gram category, but birds up to the size of geese have been killed (Brown, 1976a). Cramp and Simmons (1980) report a prey weight range of 10–1,800 grams, with females averaging heavier prey than males, and average prey weights of from about 217 to 398 grams recorded for various parts of the European range.

Sherrod (1978) has summarized the North American data on peregrine foods and listed 14 sources of original or secondary information. Of these, nine included a percentage analysis of food types, in which the bird component ranged from 70 to 100 percent (unweighted average 95.8 percent), the rest mainly consisting of mammals and insects. Species represented in large numbers include rock doves (*Columba livia*), mourning doves (*Zenaida macroura*), crested auklets (*Aethia cristatella*), least auklets (*A. pusilla*), ancient murrelets (*Synthliboramphus antiquus*), common snipes (*Gallinago gallinago*), gray jays (*Perisoreus canadensis*), and Lapland longspurs (*Calcarius lapponicus*). Relatively few of the North American studies have provided relative biomass data, but Porter, White, and Erwin (1973) judged that in Utah shorebirds comprised the largest relative biomass (60 percent) as well as the largest numerical component (44 percent) of the birds' diet, with doves (mourning dove and rock dove) of secondary importance (19 and 17 percent respectively). In their Rankin Inlet study, Court, Gates, and Boag (1988) found that 19 species of birds and three species of mammals were used, with considerable prey flexibility evident. Indeed, during a year of high microtine (lemming and vole) abundance there was a dramatic increase in territorial peregrines and high reproductive success, suggesting that, like the local population of rough-legged hawks, the birds were able to exploit this prey base effectively.

Besides the fact that female peregrines are likely to take larger prey than males, there are also age-related differences in food preferences. Thus, Ward and Laybourne (1985) reported that during autumn migration on Assateague Island, immature peregrines primarily preyed on solitary migrant birds, especially northern flickers (*Colaptes auritus*), whereas adults took approximately equal numbers of flocking shorebirds and other species. Suggested reasons for these differences included differences in diurnal hunting times (immatures foraging earlier in the day), aerodynamic differences (adults having heavier wing loading), differences in hunting experience, and the development of specific search images for shorebirds by adult peregrines.

Adult and immature peregrines also apparently differ significantly in hunting efficiency, as do breeding adults during the breeding season versus outside it, perhaps because of the usual

abundance of young, inexperienced prey usually available during the breeding season. There may also be individual differences in hunting efficiency, and some "playful hunting" that reduces the apparent hunting efficiency estimates. Because of all these variables it is hard to provide a single statistic for hunting efficiency, but the overall range of estimates varies from 7 to 83 percent success, with the highest success rates typical of breeding adult males during the nesting period, and the lowest rates typical of juveniles (Roalkvam, 1985).

During the nesting period, males tend to range farther from the eyrie than do females, in one study the female usually remaining within a kilometer of the site and the male typically more than a kilometer away (Enderson and Kirven, 1983). Toward the end of the nestling period and during the early postfledging period females may begin to range more widely, at times as far as 14 or even 18 kilometers from the eyrie (Mearns, 1985). It is possible that at least wintering females tend to take heavier prey near their plucking site and lighter prey (under 100 grams) when far away, implying an energy cost minimization in aerial transportation of prey (Albuquerque, 1986).

Peregrine hunting behavior mainly involves still-hunting (attacks on passing birds made from a perch) and aerial "waiting-on" (circling or maneuvering for position) followed by a stoop at varied angles of attack (Cade, 1982). At times special techniques may also be employed, such as the "solitary flushing" of ground-dwelling birds such as quail, or the aerial "hawking" of insects (Czechura, 1984).

The spectacular aerial stooping of peregrines has received a great deal of attention from writers, especially with regard to the possible maximum speed attained during diving. Cochran and Applegate (1986) used radio tracking to estimate the peregrine's normal flapping-flight ground speed as only about 49 kilometers per hour. Brown (1976a) reviewed the evidence on flight speeds, suggesting that although the birds can probably not fly faster than 128 kilometers per hour in level flight, in a full stoop they might easily attain and exceed the maximum flight speed possible in a falling object in the absence of air resistance (185 kph), perhaps moving as fast as 288 kph, assuming that the dive is vertical and the wings are closed. During a stoop the wings are perhaps not actually fully closed, but their leading edges are held back in such a way as to be nearly parallel to one another, probably reducing drag to the minimum while retaining maximum control of direction. Using radar data, Alerstam (1987) calculated far slower speeds during stooping than these estimated maxima. However, as Brown pointed out, there is no need to attain a higher speed than necessary when attacking slow-flying birds; indeed too high a speed might be counterproductive. As he also noted, the birds apparently select for subnormal or diseased birds as prey when given a choice, perhaps because their chances of successfully making a kill are considerably higher with such birds than with normal ones.

Social Behavior

In areas where peregrines are resident all year the pairs frequently roost near their nesting ledge through much of the winter, indicating that the pair bond must be retained throughout the nonbreeding period. The same may be true of migratory birds; Albuquerque (1986) observed a pair of birds of *tundrius* phenotype wintering in southern Brazil that exhibited aerial displays and courtship flight during the winter months. However, Beebe (1974) stated that in some areas of high latitude the male may arrive on the eyrie site a week or more before the female and begin some calling and subdued display flights, which are followed by great excitement upon the arrival of the female. He noted that isolated pairs, or pairs reproducing in regions supporting only very low populations, are likely to form permanent pair bonds prior to their first breeding attempt, and thereafter remain mated for life. Furthermore, they are likely to try to reoccupy the same breeding territory year after year thereafter.

Evidence generally indicates that indefinite monogamy is the rule, although bigamous males have been reported twice, once with the extra female at the same nest and once with the females having separate nests. The time required to replace a lost mate is highly variable, ranging from a minimum of a few hours to the following year. Most cases of known mate replacement have involved new females, perhaps since females may be prone to abandon the territory on the death of their mates (Cramp and Simmons, 1980). Court (1986) estimated a maximum turnover rate of adult males as 16 percent, and 30 percent for females, as well as a maximum annual mortality rate of 16 percent for males and 27

percent for females, with the youngest territorial females three years old and the youngest territorial male two. Newton (1985) judged that in Scotland the average age of first reproduction was 2.5 years in males (range 2–4) and 2.0 years (range 1–3) in females, based on a small sample, with an average annual survival rate (both sexes) of 89 percent for breeding birds. Bull (1974) reported that one pair of banded adults was mated for at least 14 years.

Courtship display in the peregrine is perhaps the best-studied of any falconiform bird, and has been well summarized in recent references (Cade, 1982; Cramp and Simmons, 1980). As with other falcons, it is closely associated with and perhaps impossible to separate from territorial establishment and advertisement activities that are largely performed in the air, mostly but not entirely by the male. These flights (similar to those illustrated for the merlin and described for the gyrfalcon) include single birds performing high circling, undulating flights, and figure-eight flights, while those done by the pair together include high circling and "flight play," during which one bird dives toward the other, passing closely and often executing brief "flight rolls" when near, with or without talon presentation. Talon grasping may also occur, as well as symbolic food passing, and even "aerial kissing," in which the birds touch one another's bills in midair, perhaps as a ritualized form of courtship feeding. Undulating flights typically take the form of "Z-flights," in which the bird makes a series of steep descents followed by short periods of level flight of 50–300 meters before descending again. A "V-flight" display consisting of a very steep stoop followed by a rapid and similarly steep bouncelike ascent has also been observed (Czechura, 1984). These display flights, although primarily sexual in function, also serve as territorial markers, and are supplemented by far-carrying calls (Cramp and Simmons, 1980). Intruders may be dealt with by high circling followed by stooping, the two birds often talon-locking and falling nearly to the ground. Such agonistic encounters are most frequent between males, but sometimes the resident female will attack the intruder from below as the male is attacking from above. Either sex may perform an eyrie-flyby-and-landing display at the appearance of an intruder, accompanied by repeated double-syllable "creaky calls" that sound like the noise made by a rusty hinge, *Wi-chew.*

More specifically courtship-associated behavior are a variety of perching displays that are mostly oriented toward the nesting site, plus a few other aerial displays. One of the latter is the slow-landing or hitched-wing display. In this display the male performs a "slow flight" with the wings held high, the wingbeats slow and mostly from the wrist, the legs extended well forward, and the tail depressed as he approaches a perch. He then lands with a final upward bound and vertical drop in a deliberate fashion ("slow landing"), and immediately assumes a distinctive "hitched-wing" posture (Figure 57C) before performing ledge displays. Females also perform an aerial display called the "begging flight" (Figure 57D), in which they fly with the tail fanned and depressed and with wingbeats kept below the horizontal while uttering a harsh wailing call. This same aerial begging display of the female has been called the "cuckoo flight" (Figure 56A), but both sexes sometimes perform this type of flight after apparent copulation attempts in the air.

Perched displays are very similar to those of the gyrfalcon. A pair-greeting ceremony typically begins early in the season, when a solitary bird calling on the ledge is joined by its mate, after which both birds begin to call together while holding their heads low (Figure 57A). This "head-low bow" display is the most frequently performed behavior throughout the breeding cycle; it is performed by both sexes but mainly the male, especially when the mate approaches or shows an intention of doing so. The bill is usually directed toward the ground, and is often accompanied by vigorous bowing movements and by calls (the wailing and creaky calls). This posture along with site-showing behavior is among the first shown at the nest ledge; it gradually increases interest by the female in the site, and she then performs her own ledge displays, the two thereby forming a mutual ledge ceremony. In the male's display he utters continual creaky calls while approaching the scrape in a horizontal, head-low posture, making high-stepping, swaggering steps ("tiptoe walk"). He then settles on the scrape and performs turning and body-bowing motions, along with repeated creaky calls, sometimes pausing to look at the female. Her display is a simpler version of the male's. The female may also make scraping movements at the site, but only for about five days prior to actual egg laying. Courtship-feeding of the female (Figures 56C, 57B) begins about the time of the ledge ceremonies. When it occurs on the ground it is usually a bill-to-bill transfer, during which the female normally approaches and takes the food from the male. For a period of about 10–15 days prior to egg laying the female may perform a juvenilelike begging posture while squatting and calling (Figure 57E), as well as the aerial begging flights mentioned above.

Copulation sequences are usually incomplete at first, with full ceremonies beginning about two weeks before egg laying. Female soliciting (Figure 56B) may be preceded by aerial display, courtship feeding, or ledge ceremonies. It is accompanied by the whining call, usually while the male is still at some distance, and often by the body-bow or moving the spread tail up and down. The male typically performs a head-low appeasing display or, more commonly, a hitched-wing posture. He also performs body-bows and utters chittering calls. He then approaches the female in a slow flight (Figure 57F), and while mounted beats his wings while holding his head with the bill pointing down somewhat in a curved-neck posture and with his talons bunched into a fist (Figure 57G). Following treading he departs in a slow flight.

Breeding Biology

Nest sites of peregrines are very similar to those used by prairie falcons, but peregrines are apparently more selective in their site selections. They favor more inaccessible sites at greater distances below the brink of a cliff, and predominantly choose open ledges or shelves that usually are protected from above by cliff overhangs. The sites also tend to be on higher cliffs, and the birds apparently avoid using small potholes or crevices that are often used by prairie falcons. At least in Utah, cliff sites facing north or east are favored, probably to avoid thermal stress to the young, whereas the earlier-nesting prairie falcon tends to select sites facing south and west (Porter, White, and Erwin, 1973).

The eggs are laid at intervals of 2–3 days, with the usual clutch numbering 3 or 4 eggs. Of 98 British clutches the average was 3.4 eggs, and of 43 French clutches the average was 3.04 eggs (Cramp and Simmons, 1980). North American data are similar; Bull (1974) reported on 39 New York nests that ranged from 2 to 5 eggs (average 3.23), and said that up to two replacement

Figure 56. Social behavior of the peregrine falcon, including aerial begging ("cuckoo flight") (A) and solicitation posture (B) of female, and prey presentation by male (C). After sketches in Glutz, Bauer, and Bezzel (1971).

Figure 57. Social behavior of the peregrine falcon, including head-low bowing of pair (A), courtship feeding (B), hitched-wing display of male (C), aerial begging ("flutter glide") of female (D), terrestrial food begging by female (E), and copulation sequence (F–G). A–C after sketches in Cade (1982), D–G after Cramp and Simmons (1980).

clutches have been reported for that population. Similar but perhaps slightly higher average clutch sizes are typical of arctic breeders; Court (1986) reported an average clutch size of 3.62 eggs for a Canadian population, but with ten percent of the territorial pairs failing to lay in a sample of 29 nesting territories. Females nesting for the first time laid smaller clutches and appeared to lay later than did experienced females, but their overall reproductive success was no lower. Similarly, birds nesting in areas of high nesting density did not show either a lowered clutch size or lowered reproductive success than did those nesting in other areas. Court, Gates, and Boag (1988) discounted some suggestions that arctic-breeding peregrines lay smaller clutches on average than more southerly ones, and further noted that arctic breeders are as productive as some of the most productive populations from more temperate latitudes.

Although the incubation period of the peregrine has often been listed as 28–29 days, there is good evidence that at least for the race *pealei* the usual interval between the laying of the last egg and the nearly synchronous hatching of the young is 32–34 days, and perhaps as long as 35 days (Nelson, 1972). Porter, White, and Erwin (1973) similarly reported that a 32–34 day incubation period is likely in birds of the Utah population. Incubation begins with the laying of the last or penultimate egg and is by both sexes, with the female doing the majority of the incubation (Tarboton, 1984). Similarly the female does most of the brooding of young, which is nearly continuous for the first three days or so, but gradually diminishes, so that little brooding occurs after about 17 days beyond hatching. The fledging period is 35–42 days (Cramp and Simmons, 1980).

Court (1986) reported asynchronous hatching in an arctic population, and noted that about 7 percent of all chicks hatched died as a result of brood-reduction effects of asynchronous hatching, usually through starvation of the last-hatched chick during the first five days following hatching. Mortality was seemingly related to the inability of last-hatched chicks to compete with older ones during feeding, rather than to the amount of total food available to the parents during that period.

The breeding success of various populations in western Greenland and Britain has been evaluated by Moore (1987), who estimated breeding

success rates of 73–92 percent for three Greenland populations and a 56 percent success rate for Great Britain between 1976 and 1980. The average brood size ranged from 2.3 to 3.2 young, and the average number of young fledged per pair ranged from 1.3 to 2.3. This compares with a Canadian (1988) estimate of 73 percent of all territorial pairs successful in fledging young, and a fledging success of 2.03 young per successful pair between 1981 and 1985 (Court, Gates, and Boag, 1988). In Baja California and the Gulf of California there has recently been a site occupancy rate approaching 80 percent (of about 100 sites), with good average clutch sizes (3.3–3.5) and about two young produced per pair (Porter et al., 1987). In Ungava, the number of young per pair has increased from 2.7 to 3.05 between 1980 and 1985, and the number of young per successful pair from 2.9 to 3.2 (Bird and Weaver, 1988).

Evolutionary Relationships and Status

This species is apparently part of a group (the subgenus *Rhynchodon* of Fox, 1977) of peripheral members of the genus *Falco* that also includes *peregrinoides* (the Barbary falcon of North Africa) and *kreyenborgi* (the pallid falcon of South America). The latter form is now generally believed to be nothing more than a rare color morph of *F. p. cassini* (Cade, 1982; McNutt et al., 1985).

The status of the peregrine is impossible to summarize easily; Hickey (1969) provided a book-length history of the decline of the peregrine, and more recently Cade et al. (1988) have documented their recent recovery, based on a 1985 international symposium on the management of birds of prey sponsored by the Raptor Research Foundation and hosted by the Peregrine Fund, Inc. The races *tundrius* and *anatum* are on the list of endangered species of the U.S. Department of the Interior, and *anatum* is also considered as endangered in Canada as well as by the International Union for Conservation of Nature and Natural Resources (Evans, 1982). Most states and Canadian provinces also include the peregrine on their lists of endangered species. The race *pealei* has been considered stable, and is apparently maintaining its population, at least in Alaska (Evans, 1982; Ambrose et al., 1985). Otherwise, the Alaskan populations of *tundrius* and *anatum* have increased to nearly normal levels along the major rivers after suffering declines of 35 and 45 percent respectively of historical levels through the 1960s and early 1970s (Ambrose et al., 1985).

In the eastern United States the estimated pre-DDT breeding population of 350 pairs was reduced to zero until 1980, when the first successful nesting by released falcons occurred. Over 750 birds had been released in the eastern states between 1975 and 1985, and through 1985 there were 62 known nesting attempts, of which 47 were successful and 128 young hatched, or 2.72 young produced per successful attempt (Barclay, 1985). Temple (1985) has also documented recovery efforts in eastern North America.

In the upper Mississippi Valley and western Great Lakes area, where the pre-DDT peregrine population may have numbered about 30 pairs, releases began in 1976. Through 1988 over 260 birds had been released, 13 nesting sites had subsequently been occupied, and there had been eight known nesting efforts, which produced a total of 13 fledged young. City eyries had a much higher success rate than those outside of cities (Redig and Tordoff, 1985).

In the Rocky Mountains and Colorado Plateau (New Mexico, Idaho, Montana, Wyoming, Utah, and Colorado) known peregrine territories increased from 131 in 1975 to 211 in 1985, and production is now 2.1–2.5 young per successful pair. In the central Rockies occupancy rates are about one-third of normal, but only six pairs are now known to nest in the northern parts of the region, where over 84 historical territories are known (Enderson, Craig, and Berger, 1985). Farther south, in the Chihuahuan desert of Texas (where 9 territories have been found during 12 recent survey years) and in the Sierra Madres of northern Mexico reproduction has either recently declined or been lower than normal (Hunt et al., 1985). On the western coast of mainland Mexico, the Gulf islands, and the Baja peninsula the peregrine population was nearly decimated by 1966, at least on the Pacific side of the peninsula, but now in the "midriff" portion of the Gulf the occupancy rate of known breeding sites approaches 80 percent (Porter et al., 1987).

In California, a very high level of DDE contamination still exists from various sources, including the use of dicofol (containing chloro-DDT) as a miticide in the Central Valley (Risebrough, 1985). By 1970, a 95 percent decline in known active eyries in California (to less than

5) had occurred (Herman, 1971), and similar but less well documented declines occurred in Oregon and Washington (Henny and Nelson, 1981). From 1975 to 1985 there was a substantial recovery in California, so that in 1985 California had 77 known pairs, Oregon 1, and Washington 4. However, many pairs continue to be unable to hatch their eggs, and DDE levels have not lowered (Walton, 1985).

In Canada, where surveys have been made at five-year intervals since 1970 (Fyfe, Temple, and Cade, 1976), early surveys suggested that *anatum* populations had declined or were declining in the prairie regions and in the boreal forest, whereas *tundrius* populations had quite variable levels of pesticide contamination and were just beginning to be affected. However, northerly populations, especially of *tundrius*, have now either stabilized or are recovering. Southern Canadian populations continue to remain at very low levels despite release efforts involving more than 500 birds in eight provinces and two territories. The status of *pealei* in British Columbia has been reviewed by Hodson (1980), while Van Horn, McDonald, and Ravensfeather (1982) reported that in southeastern Alaska 20 nesting sites of this race were located between 1978 and 1981, representing an estimated 70–80 percent of the probable total population of that area.

In Greenland, annual surveys in one area suggest that the population has perhaps been showing an upward trend in the past five years, and at least appears to be healthy and stable (Mattox, 1985).

Cade (1982) estimated that in the early 1980s the number of pairs in North America may have consisted of 2,000–3,000 in the boreal regions of Greenland, Canada, and Alaska, 500 in the Aleutian Islands, 200 in southeastern Alaska and coastal British Columbia, 10 in Washington and Oregon, 35–40 in California, 25 in Baja California, 20–30 in mainland Mexico, 15 in Arizona, 16 in New Mexico, 6 in Texas, 10 in Colorado, 3–4 in Utah, 3 in New Jersey, 1 in Maryland, 1 in Maine, and 1 in southern Quebec. This would total 2,800 to 3,800 breeding pairs in North America. He estimated that during the same period there may also have been 600 pairs in Britain, 1,600–3,600 pairs in continental Europe, a few thousand pairs in the USSR, a few hundred pairs in South America, 200–300 pairs in Africa, 3,000–5,000 pairs in Australia, and a few hundred pairs scattered over the western Pacific islands, or a world population of at least 12,000 to 18,000 breeding pairs.

In addition to a short bibliography provided by Snow (1972), an extensive "working bibliography" of the peregrine containing more than 1,300 references has recently been published (Porter et al., 1987). Further, a book on the biology of the peregrine, with an emphasis on British studies, is also available (Ratcliffe, 1980), and Harris (1979) published a narrative book-length account of peregrine nesting biology in Greenland.

Very rare in winter southwardly

Figure 58. North American breeding range (hatched), usual wintering range (shading), and rare wintering limits (stippling) of the gyrfalcon. The breeding (inked) and wintering (shaded) Old World ranges are shown on the inset map.

Gyrfalcon *Falco rusticolus* Linnaeus 1758

Other Vernacular Names

Black gyrfalcon, gray gyrfalcon, gyr, Icelandic gyrfalcon, jerfalcon, partridge hawk (Canada), white gyrfalcon; faucon gerfaut (French)

Distribution

Breeds in tundra regions of North America from northern Alaska, northern Yukon, and Banks, Prince Patrick, and Ellesmere islands south to the Alaska Peninsula, northwestern British Columbia, southern Yukon, southeastern Mackenzie, southern Keewatin, Southampton Island, northern Quebec, and northern Labrador; also breeds along most of coastal Greenland, in Iceland, and in northern Eurasia.

Winters in North America from the breeding range south irregularly to the Pribilof and Aleutian islands, southern Alaska, southern Canada, and the extreme northern United States, casually to the southern United States; also in Greenland, Iceland, and northern Eurasia. (See Figure 58.)

North American Subspecies

No subspecies recognized by Cramp and Simmons (1980) or Palmer (1988); geographic clines in frequency of color morphs do exist, as well as minor gradations of size, but these are not now regarded as adequate for recognition of subspecies.

Description

Adult (sexes alike). *White morph:* Entire head, body, wings, and tail white, usually with a faint creamy tinge, the crown with narrow black shaft stripes, these broadening into tear-shaped subterminal spots on the nape; upperparts, remiges, and rectrices variably barred with dark sepia to black (sometimes unbarred but with grayish black shaft stripes); the bars on the rump and upper tail coverts narrower; the primaries with broad black tips very narrowly edged with white, and the dark bars disappearing on the inner webs, which are largely immaculate white; rectrices varying from immaculate white to as many as 11 dark bars present on both webs of the median pair, the other pairs with the bars chiefly on the outer webs and becoming reduced in size and number outwardly; chin, throat, breast, and middle of abdomen either immaculate white or with a few small tear-shaped blackish spots on the breast and abdomen; sides and flanks similar to breast and abdomen; thighs either flecked or immaculate white; under tail coverts usually unmarked; under wing coverts usually white with a few tear-shaped spots. Iris dark brown; eyelids pale flesh; cere light yellow, bill pale yellowish gray basally and dusky at the tip; tarsi and toes light yellowish gray, claws pale horn-color.

Dark morph (including gray and sepia variants): Entire upperparts deep neutral gray with or without a dark sepia tinge, the head with darker gray shaft stripes and often with whitish margins to the feathers, producing a white-streaked appearance; scapulars, interscapulars, back, rump, upper wing coverts, and tail coverts with dark shafts, barred and edged narrowly with grayish white to pale buffy white, the light bars widely spaced and variable in their extent; remiges dark sepia with a grayish wash, the primaries indistinctly barred or mottled with grayish white on the outer webs and crossed by 15 or more broad white bars on the inner webs, terminated with a long unbarred sepia area and narrowly tipped with whitish; secondaries incompletely barred with pale grayish to buffy white on both webs; rectrices tipped with white and crossed by 10–12 whitish or grayish bands, the pale bands freckled with slate-gray to grayish sepia; lores, cheeks, and auriculars usually whitish (in some cases a pronounced malar stripe results from the widening of the shaft streaks); sides and underparts white, generally with a wash of buff or cream-color, the chin and upper throat usually immaculate, sometimes with dusky shafts; sides, flanks, and thighs usually with the shaft streaks more pronounced and broadening more extensively, these sometimes forming transverse bars; under tail coverts immaculate or marked like abdomen; under wing coverts white, barred broadly with dark sepia to fuscous. Tarsi, toes, and cere yellowish orange; other softparts as in light morph. (Intermediate or "gray" plumage variants between the dark and white morphs also occur frequently; dark-morph birds that are heavily tinged with sepia are the most dark-plumaged individuals.)

Subadult. Virtually identical to adults after first year, although some small juvenal feathers may

persist, especially on upper wing coverts.
Breeding probably initially normally occurs at
four years (Platt, 1977), although breeding at two
years has also been suggested (Cramp and
Simmons, 1980).

Juvenile (sexes alike). Similar to adults of the
corresponding morph, but more heavily streaked
below. Young of the white and intermediate
morphs are browner dorsally than are adults, and
dark-morph juveniles are darker both above and
below than are adults. Iris dark brown; bill bluish
horn, with a darker tip; cere, tarsi, and toes
bluish to greenish gray; claws brown to black.

Measurements (in millimeters)

Wing, males 340.3–378 (ave. of 42, 364.3), females
368–423 (ave. of 49, 402.6); tail, males 203–244
(ave. of 42, 221.5), females 215–259 (ave. of 49,
239.6) (Friedmann, 1950). Eggs average 59.4 ×
45.3 (Bent, 1938).

Weights (in grams)

Males 960–1304 (ave. of 7, 1170); females 1396–
2000 (ave. of 12, 1752) (Brown and Amadon,
1968). Males 1000–1300, ave. 1135; females 1400–
2100, ave. 1703 (sample sizes unspecified) (Clark
and Wheeler, 1987). Adult males 805–1300 (ave.
of 6, 1115), adult females 1400–2100 (ave. of 19,
1735) (Cramp and Simmons, 1980). Estimated egg
weight 67.3, or 3.8% of female.

Identification

In the hand. The combination of circular nostrils, a
wing length of at least 300 mm, and the tenth
(outermost) primary about the same length as the
eighth (the ninth the longest) serves to identify
this species.

In the field. When perched, the large size (500–
650 mm long) and rather massive bodily
proportions of the gyrfalcon help to distinguish it
from all the smaller falcons, and additionally it
tends to be paler on the head than is the
peregrine, without the dark-crowned and
characteristically "hooded" aspect of that species.
The wings are relatively short, the tips extending
only slightly beyond the midpoint of the tail,
rather than nearly reaching the tip as in
peregrines. The upperparts are highly variable in
darkness, ranging from nearly pure white to

almost sooty gray. The underparts are
comparably variable but tend to have some dark
breast and abdominal streaking (rather than the
dark spotting or barring more common in
peregrines).

In flight, the large size and somewhat less
pointed wings of this species help to separate it
from the peregrine. Although it lacks a two-toned
or "hooded" head pattern like that of the
peregrine, its underwing patterning is often
somewhat two-toned, with the wing linings
averaging somewhat darker than the
undersurfaces of the flight feathers (most
noticeable in darker-plumaged birds). Like the
other falcons, it utters sharp repeated alarm notes
when disturbed, especially near its nest.

Habitats and Ecology

This species occupies a surprising variety of arctic
habitats, from the timberline zone at the northern
edges of the boreal forest to the fell fields of high-
arctic islands, but is absent from areas of flat and
boggy tundra, probably because of the lack of
elevated nesting sites. Such sites are typically
rock ledges 1–2 meters wide and deep, with a
distinct overhang or roof, on cliffs 10–30 meters
high, and the ledge itself 6–25 meters above the
base of the cliff. The eyrie may overlook a variety
of hunting habitats, including tundra, rivers,
lakes, or the ocean (Beebe, 1974).

According to Cade (1982), gyrfalcons use
approximately the same types of habitats for
nesting as do peregrines, namely some sort of
cliff. However, they not only occupy sea cliffs and
river bluffs but also nest well away from larger
rivers, and usually occupy cliffs that are less
accessible to humans. A sample of 21 eyries
averaged about 29 meters above the river, 13
meters below the brink of the cliff, and 15 meters
above the base of the vertical cliff face. Although
these averages were similar to those of peregrines, nearly all of the sites were inaccessible to
humans without the aid of climbing ropes.
Fidelity to the same nesting site is not especially
typical of gyrfalcons, perhaps because of local
variations in prey populations, but possibly nest
changing also helps to lower the incidence of nest
parasites.

Nest sites are often relocated from year to
year, perhaps on the basis of relative nesting
success the prior year. Nesting sites that are
snow-free during the winter seem to be preferred

Gyrfalcon and willow ptarmigan

for new sites, possibly because these sites allow the birds to perform the early courtship displays that occur on the nest ledges, and such displays are needed to "encourage" females to accept a new nesting site (Platt, 1977). In some areas old raven nests or golden eagle nests are favored sites, although at times simple ledges with no stick substrates are used. In one area the average nest height was 12.9 meters (total cliff height 24.5 meters), with an average overhang of 80 percent and with most nests completely covered from above. In that area at least no directional orientation was evident (Poole and Bromley, 1988).

Gyrfalcon densities are extremely low everywhere, even in good habitats such as around the Bering Sea where peregrines are rare. Cade (1982) estimated that perhaps 300–500 pairs nest in all of Alaska, which represents less than a half of the peregrine population of 600–1,200 pairs he had estimated earlier (1960) for Alaska. He noted that in Denali National Park, with an area of about 5,200 square kilometers, five gyrfalcon eyries were present, of which a maximum of three were

known to be active in any single year, or about 1,700 square kilometers per active eyrie. In the Colville River drainage of northern Alaska there may be about 100 pairs, or a breeding density of about a pair per 1,300 square kilometers, and on the Seward Peninsula, where from 13 to 36 pairs have nested recently, the average density in good years was about a pair per 1,000 square kilometers. Observations by Platt (1977) on the North Slope of Yukon Territory indicate an average density there of a nesting pair per 975 square kilometers. In a study area comprising 21 territories in the Northwest Territories, the average internest distance varied annually from 9.5 to 11.4 kilometers, representing an extremely high nesting density of a territory per 125–150 square kilometers (Poole and Bromley, 1988). An even greater density of a pair per 113 square kilometers was estimated in Iceland by Nielsen (1986), but the nesting success in that area (a successful pair per 236 square kilometers) was lower than reported by Poole and Bromley for their study area (a successful pair per 210 square kilometers).

Foods and Foraging

Compared with the peregrine, this species has a distinctly restricted range of dietary intake. Although gyrfalcons have been known to feed on a variety of small mammals and birds ranging in size up to geese and cranes, the single most important source of prey is ptarmigans (*Lagopus* spp.), which alone often comprise about 90 percent of their food biomass. In some areas such as the Alaska Range ground squirrels may make up as much as half of the diet, and in some other areas such as the high-arctic Canadian islands arctic hares (*Lepus timidus*) make up at least 80 percent of the total food biomass (Muir and Bird, 1984). In the Northwest Territories three species, the rock ptarmigan (*L. mutus*), arctic ground squirrel (*Spermophilus parryii*), and arctic hare comprised over 95 percent of the estimated total food biomass during late spring (Poole and Boag, 1988). Finally, in Greenland lemmings (*Dicrostonyx*) are an important food source during years of lemming peaks (Cade, 1982), and the same is true in Norway (Dementiev and Gortschakowkaja, 1945; Dementiev, 1960).

Snyder and Wylie (1976) estimated that almost 60 percent of the gyrfalcon's foods numerically come from avian prey, with mammals comprising the remainder. Cade (1960), summarizing data on seven Alaskan eyries from three different areas, estimated that ptarmigans comprised from 19 to 89 percent of the total food biomass, and ground squirrels from 1 to 79 percent. Other birds, lemmings, and voles made up the rest. Prey from ten eyries on the Colville River had 56 percent of the estimated biomass from ptarmigans and ground squirrels representing 6 percent, with various birds making up most of the remainder. Poole (1987) judged that the importance of ground squirrels increased late in the breeding season as juveniles emerged and became available, while the usage and apparent vulnerability of ptarmigans declined. One recent study (Poole and Boag, 1988) indicated that females take prey weighing significantly more than males (averaging 330 and 250 grams respectively), and average prey weight increased as time away from the nest increased, as predicted by optimal foraging theory.

In contrast to the typical stooping behavior of peregrines, the gyrfalcon relies on its blinding speed to outfly nearly all other birds, including the most swiftly flying of ducks. Such a pursuit may cover three or four kilometers, but a surprise attack using a low and screened approach is the favored method of taking prey. The bird or mammal is normally initially disabled by a blow struck with the feet while flying full speed, after which the falcon makes a circle and returns to complete the kill (Beebe, 1974).

When hunting ptarmigans, gyrfalcons use three methods, including searching while flying high, searching while flying low, and low flight plus observation from temporary perches. When the falcon misses on the first try it may either give up the chase or pursue its prey in level flight. The two species are of such nearly equal abilities in flight that under such circumstances the ptarmigan has a good chance of surviving (White and Weeden, 1966).

Social Behavior

Platt (1976b, 1977) established that gyrfalcons remained on their Yukon breeding grounds during winter months, perhaps because in so doing they derive an advantage over other cliff-nesting raptors by being able to occupy preferred nesting areas before the other species have returned in spring. They are also thereby able to initiate their breeding cycles earlier. Apparently more males than females are involved in territorial defense, and thus have a greater attachment to nest sites. As a result, females may not attend these sites until the onset of courtship, thus avoiding food competition with the males during this difficult period, when ptarmigans (*Lagopus*) provide the only reliable food supply. Additionally, egg laying may be initiated only after the spring arrival of migrating ptarmigans, at least in some high-arctic areas (Poole and Boag, 1988).

Courtship in gyrfalcons seems to consist of several phases, including advertisement by lone males, courtship flights by paired birds, nest ledge displays, food transfer behavior, and copulation. Lone males perform at least five advertisement activities, two of which (eyrie flyby and wail-pluck) were not seen in mated birds, and three (male ledge display, wail, and undulating roll) were similar to those used by paired birds. The eyrie flyby consists of repeated flights parallel to the nest cliff and close to it, the flight path forming a figure-eight pattern, with the central crossing point in front of the eyrie. A constant wailing accompanies the display, and a prey item is often carried. Wail-plucking consists of wailing while slowly plucking a prey item, often pausing to look about. The prey is finally eaten.

Wailing is uttered more often by unmated than paired males, and is a two-syllable call given in a sequence of up to 10 notes. It is usually performed at the bird's primary perching site. Male ledge display is done only at a prospective nest site, and consists of standing with a horizontal stance, the beak pointing down, and uttering a series of sharp "chup" notes. He may also crouch and turn, but apparently does not perform nest scraping. When the ledge display is performed by mated birds of either sex it is often followed by scraping behavior, in which the bird thrusts its feet back, pushing away surface material and eventually forming a depression or nest scrape. During mutual ledge display both adults face one another and utter a rapid series of "chup" calls, after which the male typically leaves the eyrie but the female usually remains and begins scraping (Platt, 1977). Ledge display in male gyrfalcons seems to be more frequent and more vocal than in male peregrines, but female ledge display seems to be less well defined (Wrege and Cade, 1977).

Undulating-roll flights consist of the male starting to glide in level flight, and then rotating the body laterally about 20 degrees in one direction, followed by a 180 degree roll in the opposite direction. When this second roll is half completed the bird begins a steep dive of 30–50 meters. The wings are then returned to their original orientation by rotating the body in the opposite direction, bringing the bird to his original orientation as he terminates the dive. He then banks upward at a steep angle, reaching his original altitude, from which the maneuver is repeated. Two or three such undulations are typically performed in a display series. A roll may also be performed during a long dive, orienting the bird's dorsal surface temporarily toward the ground, then reversing the body position and continuing the dive in normal position. Rarely a roll may be performed during normal level flight. These displays seem to function in drawing a female's attention to the male and the eyrie, and may also stimulate the female in order to assure synchrony in the pair's sexual cycles (Platt, 1977).

During the male's flight display the female sometimes soars above the nest cliff, and occasionally both may perform a mutual floating display, in which the male positions himself above the flying female and both begin a slow

descent, holding their wings slightly above the back, the tails spread and the feet extended. A guttural call is uttered by the male as both birds descend in tandem for about 10–13 seconds (Platt, 1977). At times the male may also fly ahead of a female and begin weaving back and forth in front of her, in a display similar to that observed in peregrine falcons and called "passing and leading" (Nelson, 1970).

Food transfer behavior was observed by Platt to occur from at least ten days before egg laying through the feeding of the young. Transfers occurred more often on perches than in the air. When done in the air, the female flies toward the male, but before reaching him climbs slightly, then dives in front of him and pitches up underneath him, momentarily turning upside down to receive the food from the male's feet. When food transfers are made on a perch the male utters his wailing call on approach, and as he lands he stops wailing and begins his "chup" notes. The female may fly out toward the returning male in a distinctive "sandpiper flight" posture, with shallow wingbeats and the tail fanned, while uttering begging calls. She lands next to the male and quickly accepts the food. The male may then fly off to preen or, if eggs are present, begin to incubate (Platt, 1977).

Either sex can initiate copulation, the male doing so by standing erect in a distinctive curved-neck posture, arching the neck and pointing his beak downward so that the back of the neck is the highest point of the bird. This is a silent display. The female often then assumes the copulation solicitation posture (see Figure 18B), leaning forward with the beak similarly pointed down, the tail slightly raised, and uttering a soft whine. Her wings are slightly opened, and as the male lands on her back the tail is raised and slightly tilted to one side. As the male positions himself on her back his wings are flapped constantly and his toes are balled as he rests on his tarsi, which are wedged between the female's thorax and humeri (see Figure 17D). The female begins a strident copulatory wail, while the male utters irregular chutter calls. Copulations apparently occur wherever the female happens to be at the time, but Platt never observed them performed on the nest site and doubted that they ever occur there. Woodin (1980) also described copulatory behavior in gyrfalcons, and noted that self-preening and mutual billing were sometimes a part of the preliminary activities.

The male's curved-neck posture is also used as a generalized post-landing display when the male lands within sight of the female, as he holds himself vertically on extended legs. In peregrines this same or a similar display is called the "hitched-wing display" (Weaver and Cade, 1973) or the "slow-landing display" (Nelson and Campbell, 1973). However, in peregrines the display is not so clearly limited to the precopulatory situation, and so is less likely to elicit a sexual response from the female (Wrege and Cade, 1977).

Breeding Biology

Although younger birds (two- and three-year-olds) may participate increasingly in sexual display, there is good evidence that gyrfalcons are not fully sexually mature and able to breed successfully until they are four years old (Platt, 1977).

Nest sites used by the birds in wooded areas are often in trees, and in treeless areas are invariably on cliffs. Kuyt (1980) found that, like peregrines, gyrfalcons prefer to use sites with southern or western exposures, and in wooded

areas he found nearly as many nests in trees as on cliffs. Platt (1977) examined 14 nest sites over a two-year period, but never found any sign of nest refurbishment by the falcons. About five days before egg laying begins the female becomes quite lethargic, often visiting the nest and spending 30–50 minutes scraping or lying in the scrape. Eggs are laid about 60 hours apart; thus a clutch of four is completed in about eight days, according to Platt. He believed incubation to begin with the laying of the third egg, and in one case found the incubation period for the final egg of the clutch to be 35 days, a duration he also obtained with artificial incubation, and which Woodin (1980) later confirmed. The female apparently incubates through the night, while the male incubates during morning and late afternoons when the female is eating food brought her by the male. Daytime incubation durations of the female average about twice as long as those of the male.

Clutch sizes of the gyrfalcon are most often of 4 eggs, with a range of 2–7, according to Cade (1960, 1982). There is no good evidence of regional variation in clutch size, but possibly prey abundance influences the number of eggs laid. The average of 60 Finnish clutches (some possibly incomplete) was 3.53 eggs, with a range of 2–5 (Cramp and Simmons, 1980). The average of 20 clutches in the Canadian arctic was 3.8 eggs, with 85 percent of the nests having 4-egg clutches. One renest of 3 eggs was also found in this study (Poole and Bromley, 1988).

Following hatching, the male typically visits the nest site only to feed the young. In one case noted by Platt (1977) the female did not leave her young at all until they were nine days old, and Bente (1981) first observed the female hunting when the young were 17 days old. Of 105 prey items brought to nests studied by Platt during two summers, 61 percent consisted of willow ptarmigans (*L. lagopus*), 15 percent of arctic ground squirrels, and the rest almost entirely of passerine birds, microtine rodents, and sandpipers. At least 90 percent of the total prey biomass brought to the nest consisted of ptarmigans and ground squirrels in Platt's study; similar observations have been made by Cade (1960) and Roseneau (1972), while Poole (1987) estimated that nearly all of the prey taken in his study area during the breeding season consisted of these two species plus arctic hare. Hunting sessions by the male lasted several times longer than those by the female, and prey transfer to the female sometimes occurred more than a kilometer from the nest (Bente, 1981).

The fledging period ranges from 45 to 50 days, with males averaging 46 days and females 48.3 days (Poole and Bromley, 1988). The young become independent about a month later (Cramp and Simmons, 1980). Brooding by the female gradually tapers off as the young develop, but after normal brooding ceases the female will continue to brood nestlings during periods of heat, cold, or precipitation (Jenkins, 1978).

Platt (1977) summarized a good deal of information on productivity for North American gyrfalcon populations, based on mostly unpublished information available to him through 1976. In six different study areas of Alaska, Canada, and Greenland that had been observed over periods of 1–9 years, the average number of young produced per successful nest ranged from 2.5 to 3.1 young. Individual yearly estimates for particular sites ranged from a low of 1.3 to a high of 4.0 young per successful nest. In Platt's Yukon study area the yearly average productivity ranged from 2.8 to 3.2 young, with the number of nests declining substantially (from 26 to 12 on a 19,500-square-kilometer study area) during a four-year period but with the average productivity per successful nest increasing slightly, albeit not significantly.

Evolutionary Relationships and Status

This species is apparently a close relative of the saker (*F. cherrug* including *altaicus*), a semidesert-adapted falcon of Eurasia (Cade, 1982) that is somewhat smaller than the gyrfalcon.

Cade estimated that as of about 1980 there might have been in the range of 7,000–17,000 pairs of gyrfalcons in the world, with a probable increase in Canada and Alaska occurring during the preceeding 30 years. A recent estimate of wintering birds in southern Canada and the 48 contiguous states was 500 (Anonymous, 1986).

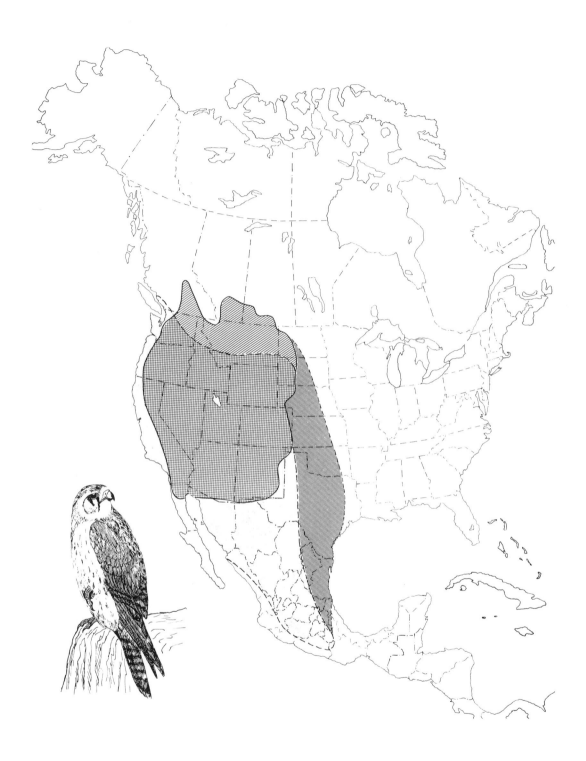

Figure 59. Breeding (wide hatching), residential (cross-hatching), and wintering (shading) ranges of the prairie falcon, plus additional areas often occupied during migration (narrow hatching).

Other Vernacular Names

Faucon des prairies (French); halcon café (Spanish)

Distribution

Breeds from southeastern British Columbia, southern Alberta, and southern Saskatchewan south in open-country habitats (mostly hilly to montane grasslands or semideserts) to Baja California, western Texas, eastern Colorado, western Nebraska, and the western Dakotas; formerly to northwestern Missouri.

Winters in grassland to semidesert habitats from the breeding range in southern Canada south to central Mexico, with some birds wintering both to the east and west of the breeding range. (See Figure 59.)

North American Subspecies

None recognized.

Description

Adult (sexes nearly alike). Forehead white to cream-buff, most upperparts umber to sepia, the dorsal feathers often with distinct blackish shafts, the feathers of the back, rump, and the upper wing coverts edged with tawny-olive and incompletely barred with lighter brown; the feathers of the occiput edged pale buffy white to cartridge-buff, with blackish shafts and tawny-olive tips, forming a fairly distinct occipital band of whitish connecting laterally with pale buffy white superciliary stripes; some of the feathers of the crown with whitish instead of tawny-olive margins; upper wing coverts, scapulars, and back feathers with a grayish cast; remiges mostly dull sepia, the inner webs whitish externally except for a broad sepia tip, this pale area crossed by three to five (on the secondaries) or eight to ten (on the primaries) bands of dull sepia that merge to form a continuous lengthwise sepia area next to the shaft; the innermost secondaries almost devoid of any whitish on either web; upper tail coverts brown, the more anterior ones and rump feathers edged with cinnamon; rectrices plain pale drab to buff tipped with dirty buffy white, the inner webs of all but the middle pair with

eight or nine pale incomplete bars, the bars merging to form a continuous pale edge to the inner web; lores and superciliaries like the forehead but less yellowish; a malar stripe of dark sepia extends back from the anterior angle of the eye to the sides of the throat; cheeks and sides of neck buff, the auriculars becoming wholly sepia posteriorly, sides and underparts whitish washed with buff on the breast and pale pinkish buff on the middle of the abdomen; the chin and throat immaculate; breast and upper abdomen with numerous ovate to elongated, tear-shaped brown spots, which become larger, longer, and much broader on the sides and flanks (in males tending to become bars rather than streaks), often coalescing into a large, somewhat interrupted patch on the flanks; lower abdomen and under tail coverts immaculate or with only a few spots; axillaries plain brown with a few obsolete rusty spots toward the tip, contrasting with the rest of the mostly white under wing coverts, the feathers of which have brown shaft streaks, sometimes broadened to include all the edges of the feathers. Iris dull brown; cere, orbital ring, tarsi, and toes yellow; bill pale greenish at base becoming bluish and dark slate at tip; claws blackish slate.

Subadult. Postjuvenal molt not well known, but young apparently gradually become more adultlike, with the legs and toes becoming yellow, while the iris and bill change little if at all in color (Bent, 1938). Immatures tend to be buffy rather than white below, are streaked on the flanks, and have pale bluish feet. Breeding by first-year birds is only rarely attempted, but one yearling male occupied a nest site although no eggs were laid (Enderson, 1964).

Juvenile (sexes alike). Like the adult, but generally darker, with the scapulars, interscapulars, back, rump, and upper tail coverts mostly ashy to dull grayish cinnamon barred with fuscous, and with more rufous feather edges; the median pair as well as the rest of the rectrices indistinctly barred with brown on both webs; the nuchal collar washed with pale cinnamon-buff; the thighs and underparts with more and heavier dark brown spots and the superciliary line with brownish streaks. Iris brown; bill pearl-gray to bluish white, becoming blacker toward the tip; cere, edge of mouth, and orbital skin pearl-gray to greenish white; tarsi and toes greenish yellow to bluish white or light gray.

Measurements (in millimeters)

Wing, males 289–313 (ave. of 17, 299.4), females 331–357 (ave. of 18, 342.8); tail, males 159–179 (ave. of 17, 169.6), females 185–201 (ave. of 18, 189.7) (Friedmann, 1950). Eggs average 52.3 × 40.5 (Bent, 1938).

Weights (in grams)

Males, ave. of 10, 496; females, ave. of 34, 801 (Snyder and Wiley, 1968). Adult males 500–635, ave. of 15, 554; adult females 760–975, ave. of 31, 863 (Enderson, 1964). Unaged males 420–635, ave. 524; females 675–975, ave. 848 (sample sizes unspecified) (Clark and Wheeler, 1987). Estimated egg weight 47.4, or 5.9% of female.

Identification

In the hand. The combination of circular nostrils, a wing length of 289–357 mm, and axillary feathers that are contrastingly patterned with blackish serves to separate this species from other North American raptors.

In the field. When perched, this species appears rather uniformly medium brown above and mostly white below, with varying amounts of spotting (more on young birds, but generally less than occurs on young of comparably sized falcons). The large dark eyes have a buffy superciliary stripe above and a distinct mustache streak extending down and back below, but the head lacks the "hooded" pattern of the peregrine and the distinctly blackish crown typical of both that species and the aplomado falcon.

In flight, prairie falcons closely resemble both peregrines and aplomado falcons in size and shape, but unlike either of these they have very dark axillary areas, which color often extends outward to include most of the larger under wing coverts, producing a two-toned wing. The tail is not as long as that of the aplomado falcon, and not so distinctly banded with dark and light. Unlike peregrines, nearly vertical stooping from great heights on prey is not common, and instead the birds typically capture their prey on or very near the ground after an approach from almost ground level. Their calls include loud repeated *kik* notes when alarmed.

Habitats and Ecology

The typical habitat of prairie falcons consists of rather open and treeless terrain, such as sagebrush desert, desert grassland, and other generally arid habitats, with nearby cliffs and escarpments suitable for nesting sites. Breeding occurs at various elevations up to at least 3,700 meters in alpine tundra habitats (Marti and Braun, 1975), but is more common at lower elevations. Thus, the birds nest commonly in the Snake River Birds of Prey Natural Area of Idaho. About 200 pairs occur there, or perhaps 5–10 percent of the total North American breeding population (estimated at 2,000–6,000 pairs by various authorities) (J. H. Enderson, quoted by Snow, 1974a). Birds nesting in the Rocky Mountain region drift to lower areas in the winter, and then often hunt over winter wheat fields that support flocks of horned larks (*Eremophila alpestris*). With the dispersal of these flocks in spring the falcons move back to their nesting cliffs (Enderson, 1964).

In Arizona, the species is most common in low- and moderate-elevation communities during winter, and in higher communities such as desert grassland and chaparral during summer. The densest nesting occurs in isolated cliff areas surrounded by vast expanses of unbroken terrain, with nests in lower elevations usually on north-facing cliffs, apparently to avoid overheating (Millsap, 1981). Farther north, as in Wyoming, south-facing slopes seem to be favored.

During the winter periods the birds range fairly widely; Enderson determined the average home range of four males during two winter periods as encompassing 6.1 kilometers (this apparently being a diameter measurement), and that of six females as 11.5 kilometers. Studies of radio-tagged birds in the Snake River canyon indicated that some falcons may fly as far as 26 kilometers to feed, and individuals may have hunting ranges of from 26 to 141 square kilometers, which often overlap broadly (Cade, 1982).

Craighead and Craighead (1956) estimated that the home range of a single pair in Wyoming was 26 square kilometers, and two breeding territories in Utah were 5.7 and 6.5 square kilometers (Smith and Murphy, 1973). There are rather few estimates of breeding density. Webster (1944) reported a density of 23 pairs along 26 linear kilometers of cliffs, and Ogden and Hornocker (1977) found from 56 to 101 pairs along 72 linear kilometers of Snake River canyon in southern Idaho during three successive seasons. Olendorff (1973) estimated a breeding density of a pair per 185 square kilometers in the Pawnee National Grasslands of Colorado, and Boyce, Garrett, and Walton (1986) made an estimate of 520 nesting territories in nearly 400,000 square kilometers of California, averaging a territory per 764 square kilometers. Millsap (1981) reported that nest sites in an Arizona study area averaged 10.5 kilometers apart, and were never closer than 150 meters apart.

Foods and Foraging

Considerable information is available on prairie falcon foods, with the summary by Sherrod (1978) listing 13 studies or compilations, including three unpublished studies. The relative numbers of birds and mammals in these studies are highly variable, with mammals numerically varying from 8 to 66 percent and birds from 33 to 92 percent. With few exceptions, these were spring and summer studies; winter analysis may well show a consistently higher incidence of birds in the diet. Other vertebrate groups as well as invertebrates are relatively inconsequential in all these studies. Similarly, Boyce (1985a) summarized 10 studies (including several of those also listed by Sherrod), and found the mammalian component to vary numerically from 8 to 55 percent (averaging 33.5 percent), and the avian component to vary numerically from 33 to 92 percent (averaging 61 percent). The reptilian component averaged 4 percent and insects 2 percent. In his own area the average prey size was 107 grams, and the mammals tended to average about twice the weight of the avian prey, suggesting that the birds may get greater energetic returns from capturing mammals than birds.

Studies done on biomass analysis provide a similar general perspective on food intake in this species. Smith and Murphy (1973) found that, although black-tailed jackrabbits (*Lepus californicus*) comprised only 8 percent numerically of the food items they identified, the relative prey biomass contribution of jackrabbits was 65 percent. Horned larks were taken more frequently than any other bird species in their study; horned larks were similarly the most frequently found bird species in the items identified by Boyce (1985a), and in unpublished studies by J. B. Platt

and McKinley (cited by Sherrod, 1978). Mourning doves (*Zenaida macroura*) and western meadowlarks (*Sturnella neglecta*) also ranked high in frequency of occurrence in several studies. Millsap (1981) estimated a 52 percent mammalian biomass for prairie falcon food in his Arizona study area; ground squirrels (*Spermophilus treticaudus*) and antelope squirrels (*Ammospermophilus harrisii*) exceeded lagomorphs in prey numbers as well as prey biomass, and about half of the avian biomass component was represented by Gambel's quail (*Callipepla gambelii*), with mourning doves of secondary importance. In the Snake River canyon area the Townsend's ground squirrel (*Spermophilus townsendi*) makes up about 40–80 percent of the prairie falcon's food biomass during the nesting season, and the bird's nesting schedule is closely tied to the life cycle of this rodent (U.S. Dept. of Interior, 1979; Cade, 1982). Richardson's ground squirrels (*S. richardsonii*) and horned larks were the two most important prey animals on Enderson's Wyoming study area during the breeding season.

Of 44 observed capture attempts (Dekker, 1982), all involved prey on the ground or in shallow water, mostly small passerine birds or shorebirds. In 26 cases the prey was apparently seen from afar as the falcon was in soaring or flapping flight, and in 11 it was detected just ahead as the falcon flew low above the ground. The last phase of these 37 hunts was a high-speed, low-level surprise attack, of which only one was successful. Soaring to heights of more than 1,000 meters also occurred, followed by stooping dives at angles of 30 to 60 degrees, with the bird leveling off in the terminal part of the dive and traveling low over ground or water for up to 300 meters without any wing beating. Oblique descents from flapping flights at altitudes up to 150 meters were also seen, with the last phase of such attacks done with the wings partly or completely flexed. Phipps (1979) and Haak (1982) have also described the hunting methods and foraging ecology of prairie falcons.

Social Behavior

Little is known of pair bond length in this species, but regular reuse of the same eyries suggests that pair bonds are long-term. Earle (1984) reported that at least three of 22 known nesting territories in Los Padres National Forest of California were active during five successive years, two more were active in four of the five years, and three in three successive years. Of 25 sites studied during three years by Enderson (1964), 14 were active during all three years, 9 were used in two years, and 2 were used only once. However, use of an eyrie by a single bird has been reported for as long as nine years, and in some cases a female may have different mates in successive years at the same eyrie. Pairs do not appear to remain associated during winter months, but return to their eyries in March in the Rocky Mountain region, with either sex arriving first. When both have arrived the birds go through a courtship period of about a month, during which time they perch for long periods near the nest site and the males occasionally hunt. Copulation is frequent during this period (Enderson, 1964). There apparently are no detailed descriptions available of aerial or ledge displays of this species, but they probably differ but little from those described for the peregrine and gyrfalcon. Hybridization between the prairie falcon and peregrine has been reported (Oliphant, 1985a).

Territories are strongly defended in this species, as indicated by the observations of Holthuijzen, Oostenhuis, and Fuller (1987). They observed nearly 3,000 aggressive interactions involving 37 nesting pairs, with an average of about 6 interactions per observation day. The nesting stage and specific nesting pair influenced the number of aggressive encounters, but the intruder's sex did not. The rate of interactions was high before the eggs were laid, declined during incubation, peaked about the time the eggs hatched, declined again during early brood rearing, and finally increased again late in the brood-rearing stage.

Breeding Biology

Nest site selection is the first stage of breeding in this species, and typically an experienced pair returns to its nest site of the prior year. Typically the nest site is on sheer cliff, has an overhanging ledge, and has a broad vista overlooking a hunting area. Enderson (1964) reported that 36 nesting cliffs ranged from 8 to 39 meters in height, averaging 16 meters, with the nest site an average of 11 meters above the base of the cliff. All but eight were judged inaccessible to mammalian predators, and all but one had a rocky overhang above the nesting ledge. About two-thirds faced south, and most of the ledges consisted of open shelves or "potholes," although a few were in larger caves. Similarly 60 percent of 126 nest sites in Idaho were in rock cavities that offered some overhead protection, and about 20 percent were in abandoned raptor nests that often also had overhead protection.

Nest sites in Millsap's (1981) Arizona study area were entirely on cliffs, the nest sites averaging 14 meters high and the cliffs about 20 meters high. About two-thirds of the 44 sites were in "potholes" less than a meter deep, while 20 percent were in stick nests built by other raptors and the rest were in deeper holes, cracks, or caves. Nest sites averaged 10.5 kilometers apart, with no nests located closer than 150 meters apart.

Runde and Anderson (1986) evaluated nest site data from these and other studies, totaling over 418 sites. They determined a consistent pattern of eyrie placement, with the average eyrie height 63 percent of the mean cliff height, the two collectively averaging 18.5 and 29.3 meters. Nearly half of all the nests were pothole-like sites, including caves, holes, and cavities, and about a quarter were on horizontal ledges, while old stick nests of other raptors and crevice nests each comprised about 15 percent. There are also a few records of prairie falcons nesting in trees (MacLaren, Runde, and Anderson, 1984).

Probably a rather small proportion of paired birds fail to lay each year; Enderson (1964), Ogden and Hornocker (1977), and Millsap (1981) all reported rather low levels of nonbreeding, averaging about 15 percent. Renesting following the loss of the first clutch is apparently infrequent (Enderson, 1964); in the Birds of Prey Area of southern Idaho such renesting perhaps accounted for less than one percent of the total nesting effort (Allen et al., 1986).

Clutch sizes fairly consistently average about 4.5 eggs in most populations, with 5 eggs probably the usual modal number (Cade, 1982). Thus, Ogden and Hornocker (1977) estimated an average clutch of 4.4 eggs for 68 nests in Idaho, Olendorff (1973) found an identical average clutch size for 24 nests in Colorado, and Leedy (1972) reported an average clutch of 4.3 eggs in 20 Montana nests. Millsap (1981) found a lower (3.85) average clutch size for 26 nests in Arizona, suggesting to him that perhaps clutch sizes tend to diminish southwardly.

The egg-laying interval is still uncertain but is probably about two days, meaning that 9 or 10 days are needed to complete a full normal clutch. Incubation apparently lasts 29–33 days, with the female doing most of it and the male normally taking over only while the female is eating prey that has been brought to her. The young hatch over a period of only two or three days, indicating that incubation must not begin until the clutch is nearly completed. The fledging period lasts about 40 days, the usual range being 36–41 days (Enderson, 1964; Snow, 1974a).

In contrast to the high degree of similarity in clutch sizes indicated by most studies, there is a very great amount of variation in successfully hatched nestlings per nest site, and in fledglings per nesting effort (Ogden and Hornocker, 1977). Thus, brood size has ranged from as little as 1.9 young per nest (Enderson, 1964) to as high as 3.9 per nest (Olendorff, 1973), and fledglings produced per nesting attempt from as little as 1.2 (Enderson) to as high as 3.4 (Olendorff). In at least some of these areas reproductive success was influenced by losses of nestlings to falconers, but at least in one study (Ogden and Hornocker, 1977) there was a high loss of eggs to mammalian predation, unexplained disappearance, or infertility. To some degree this variation in productivity may be the result of reduced reproductive success associated with pesticide-induced eggshell thinning; Leedy (1972) thus found that sample eggs from one Montana population with low reproductive success had unusually thin eggshell characteristics.

Mortality rates of young prairie falcons are quite high, perhaps as much as 74 percent during their first year, with a significant amount of first-year mortality apparently associated with shooting (Enderson, 1969; Shor, 1975). Adults have a considerably lower annual mortality rate, but one that may nevertheless be substantially higher than that of the peregrine falcon, so that the additional life expectancy of an adult prairie falcon may only be about 2–2.8 years (Shor, 1975). Denton (1975) estimated a first-year mortality of about 65 percent, followed by an annual adult mortality rate of about 35 percent. Countering this high immature and adult mortality is an apparently high incidence of breeding by young females (Platt, 1981).

Evolutionary Relationships and Status

This is part of a group of "desert falcons" of nearly worldwide distribution. Among these are the lagger (F. jugger) of the Indian subcontinent, the lanner (F. biarmicus) of Africa, and the saker (F. cherrug) of Eurasia. The last-named species in turn is certainly a close relative of the gyrfalcon; thus the lagger and lanner are more clearly counterparts of the prairie falcon and at times these three have been suggested as perhaps comprising parts of a single, widely ranging species (Cade, 1982).

The prairie falcon is among the species that the Audubon Society has placed on its Blue List of apparently declining species, but Cade (1982) believes that its status is generally secure on a continental basis. He judged the total population to be in the range of 5,000–6,000 pairs. Many local problems do exist, such as the effects of agricultural chemicals on reproduction, and conversions of grassland to cropland in central California. State estimates of breeding populations in the early 1970s included about 300 pairs in Colorado, less than 300 pairs in New Mexico, 200 pairs in Montana, 200 pairs in Wyoming, and 60–70 pairs in California (Snow, 1974a). Utah and Nevada probably support good populations, but their numbers are unknown; the Arizona population is at least locally substantial (Millsap, 1981). Local populations also exist in Washington and Oregon (Denton, 1975), as well as in the Dakotas, western Nebraska, and the Oklahoma panhandle. North Dakota currently supports about 100 pairs (Allen, 1987). Probably the best single state population is in Idaho, where about 200 pairs are resident in a stretch of about 130 kilometers of the Snake River canyon (Cade, 1982), plus an unknown number elsewhere in the state.

The situation in Canada is still rather poorly documented, but small breeding populations exist in British Columbia, Alberta, and Saskatchewan. Fyfe et al. (1969) reported a 34 percent reduction in known territories in Alberta and Saskatchewan during a ten-year period, apparently in association with agricultural pesticide use. However, Oliphant et al. (1976) noted that 19 pairs under study in Saskatchewan had better than average reproductive success in 1975, and this represented the largest number of known

active territories ever reported for one year in that province.

Given all these uncertainties, it is impossible to calculate a total North American population, but Cade's estimate of 5,000–6,000 pairs would seem reasonable. There is a recent winter estimate of 13,000 birds in the early 1980s (Anonymous, 1986). My own estimate, based on the Audubon Society's 1986 Christmas Counts, is of 7,800 birds, with the highest numbers in California, Arizona, and Colorado, which had estimated winter populations of 1,200, 1,200, and 900 birds respectively.

Key to the Species of North American Falconiform Birds

A. Nostrils circular, with a central tubercle, or slitlike, with the anterior end
the lower one .... Falconidae, 7 spp.
  B. Nostrils slitlike, tarsus scutellated ....
    .... Polyborini; crested caracara the only North American species
  BB. Nostrils circular, tarsus reticulated .... Falconini (*Falco*), 6 spp.
    C. Outermost primary (10th) as long as or longer than 8th
      D. Axillaries not blackish, upper mandible not strongly
        "toothed" .... peregrine falcon
      DD. Axillaries blackish, upper mandible edge with distinct "tooth"
        .... prairie falcon
    CC. Outermost primary shorter than 8th
      D. Most of upper tarsus feathered, wing at least 340 mm
        .... gyrfalcon
      DD. Tarsus no more than half feathered, wing to 302 mm
        E. Wing at least 248 mm, flanks dark brown to black
          .... aplomado falcon
        EE. Wing no more than 215 mm, flanks not dark
          F. Dorsal base of toes reticulated like tarsus, back not
            barred with rufous .... merlin
          FF. Toes fully scutellated dorsally, back barred with rufous
            .... American kestrel
AA. Nostrils usually oval and lacking central tubercles, or if slitlike then the
posterior end the lower one .... Accipitridae, 24 spp.

B. Tarsus covered with small spicules; nostrils slitlike, the claws deeply rounded on undersides .... Pandioninae, osprey the only species
BB. Tarsus smoothly scaled; nostrils usually oval, claws usually grooved below .... Accipitrinae, 23 spp.
   C. No shelf above the eye; tarsus partially feathered above and scales reticulated (networklike) or weakly scutellated below .... pernine kites, 2 spp.
     D. Tail black, deeply forked; bill not compressed nor enlarged .... American swallow-tailed kite
    DD. Tail banded, square-tipped; bill strongly compressed and very large .... hook-billed kite
  CC. The eye overhung above with a distinct shelf; tarsal scales usually aligned (scutellated) in front, or tarsus sometimes feathered to base of toes
     D. Facial feathers forming a distinct ruff around the large ear orifice; tarsus over twice the length of the middle toe ....
     ... harriers, northern harrier the only North American species
    DD. Facial feathers not forming a ruff, and the tarsus not twice the length of the middle toe
       E. Tail slightly notched (emarginate) and usually less than half as long as the wing; tarsus no more than 15 percent as long as the wing .... elanine and milvine kites, 3 spp.
         F. Talons rounded below; tarsus reticulated all around .... black-shouldered kite
        FF. Talons grooved below; tarsus scutellated on anterior surface .... milvine kites
          G. Upper mandible edge strongly notched; tail dark basally and wings pointed .... Mississippi kite
         GG. Upper mandible edge smooth; tail white basally and wings rounded .... snail kite
      EE. Tail square-tipped to rounded in outline and at least half as long as the wing, tarsus more than 15 percent as long as the wing .... typical hawks and eagles, 17 spp.
        F. Tail length at least 70 percent the wing length
          G. Lores nearly naked; wing coverts rufous above and below .... Harris' hawk
         GG. Lores well feathered; wing coverts never rufous .... Accipiter, 3 spp.
           H. Wing to 210 mm; tarsus very compressed; tail square-tipped .... sharp-shinned hawk
          HH. Wing over 210 mm; tarsus more robust; tail slightly rounded (middle rectrices up to 26 mm longer than lateral ones)
            I. Wing to 278 mm, less than half of the tarsus feathered .... Cooper's hawk
            II. Wing at least 280 mm, at least half of the tarsus feathered .... northern goshawk
        FF. Tail no more than 65 percent the wing length
          G. Wing over 500 mm .... eagles, 2 spp.
           H. Tarsus feathered to base of toes ..... golden eagle
           HH. Lower tarsus unfeathered .... bald eagle
          GG. Wing under 450 mm ...... buteonine hawks, 11 spp.

H. Lores nearly naked; secondaries 40–70 mm shorter
than primaries .... common black-hawk
HH. Lores well feathered; secondaries much shorter
than primaries .... *Buteo*, 10 spp.
I. Tarsus feathered to base of toes; spread
primaries with a "window" area above, formed
by white bases of inner webs
J. Bill to 38 mm wide at base, its profile slightly
concave as viewed from above ....
.... rough-legged hawk
JJ. Bill at least 42 mm wide at base, and bill
slightly convex as viewed from above
.... ferruginous hawk
II. Lower tarsus unfeathered; primaries usually
lacking white "window" basally
J. Four outer primaries strongly notched on
inner webs
K. Primaries distinctly barred on outer web;
wing coverts reddish ....
.... red-shouldered hawk
KK. Primaries plain or only faintly barred on
outer web; wing coverts not reddish
L. Wing under 290 mm; adults gray-barred
below .... gray hawk
LL. Wing over 330 mm; never barred on
underparts with gray
M. Body and wings mostly black; tail
with up to seven whitish bands
.... zone-tailed hawk
MM. Body and wings rarely all black; tail
lacking white banding but sometimes
white at tip .... red-tailed hawk
JJ. Only 3 outer primaries so notched
K. Tail up to half the wing length and mostly
white in adults; lesser wing coverts rusty
.... white-tailed hawk
KK. Tail over half the wing length and never
mostly white; no rust-color on lesser wing
coverts
L. Wing over 360 mm, the flight feathers
dark below; tail only weakly barred
.... Swainson's hawk
LL. Wing under 340 mm, the flight feathers
pale basally; tail with distinct barring
M. Forehead white; under wing coverts
not tinted rufous ....
.... short-tailed hawk
MM. Forehead not white; under wing
coverts tinted rufous ....
.... broad-winged hawk

Origins of Vernacular and Scientific Names of North American Falconiform Birds

Accipiter: Latin, a kind of hawk.

cooperii: after William Cooper (c. 1798–1864), American naturalist who helped found the New York Lyceum of Natural History.

gentilis: Latin, "noble." The vernacular name "goshawk" derives from the Old English *goshavoc,* "goose hawk."

apache: after the Apache Indian tribe.

atricapillus: Latin, "black-capped."

laingi: after Hamilton Laing, who helped in the discovery of this form.

striatulus: Latin, "streaked."

striatus: Latin, "streaked." The vernacular term "sharp-shinned" refers to the compressed tarsus.

chionogaster: Latin, "white-bellied."

fringilloides: Latin, "fringe-like."

madrensis: Latin, "of the Sierra Madre range" of southern Mexico.

perobscurus: Latin, "very dusky."

suttoni: after George Sutton (1898–1982), American ornithologist and bird artist.

velox: Latin, "swift."

venator: Latin, "a hunter."

Aquila: Latin, "an eagle."

chrysaetos: Greek, "a golden eagle."

canadensis: "of Canada."

Buteo: Latin, a kind of buzzard.

    albicaudatus: Latin, "white-tailed." "Sennett's white-tailed hawk" refers to George B. Sennett (1840–1900), whose specimens were used for describing *sennetti,* a form now regarded as invalid.

        hypospodius: Greek, "lighter gray."

    albonotatus: Latin, "white-marked."

    brachyurus: Greek, "short-tailed."

        fuliginosus: Latin, "sooty."

    jamaicensis: "of Jamaica," the type locality.

        alascensis: "of Alaska."

        borealis: Latin, "northern."

        calurus: from the Greek *calos,* "beautiful," and *oura,* "tailed."

        costaricensis: "of Costa Rica."

        fuertesi: after L. A. Fuertes (1874–1927), American bird artist.

        fumosus: Latin, "smoked or smoky."

        hadropus: from the Greek *hadros,* "thick," and *podos,* "a foot."

        harlani: after Richard Harlan (1796–1843), a friend of J. J. Audubon's who was a Philadelphia naturalist and physician.

        kemsiesi: after Emerson Kemsies, museum curator at the University of Cincinnati.

        kriderii: after John Krider, who collected the type specimens in Iowa.

        socorroensis: "of Socorro Island."

        solitudinus: Latin, "alone."

        umbrinus: Latin, "dark."

    lagopus: Greek, "hare-footed." The vernacular name "rough-legged hawk" also refers to the feathered tarsi.

        sancti-johannis: "of St. John's" (Newfoundland).

    lineatus: Latin, "striped."

        alleni: after J. A. Allen (1838–1921), first president of the American Ornithologists' Union.

        elegans: Latin, "elegant."

        extimus: Latin, "of the most remote distribution."

        texanus: "of Texas."

    nitidus: Latin, "bright."

        maximus: Latin, "great."

        micrus: Latin, "small."

        plagiatus: Greek, "oblique."

    platypterus: Greek, "broad-winged."

        brunnescens: Latin, "dark brown."

        cubanensis: "of Cuba."

    regalis: Latin, "regal." The vernacular name "ferruginous hawk" refers to the rusty plumage coloration.

    swainsoni: after William Swainson (1789–1855), English-born naturalist and lithographer, for whom a thrush and a warbler were also named.

Buteogallus: Latin, "a cock-hawk or chicken-hawk."

    anthracinus: Greek, "coal-colored."

        gundlachii: after Dr. Jean Gundlach (1810–1896), authority on Cuban ornithology.

    subtilis: Latin, "slender."

        bangsi: after Outram Bangs (1863–1932), curator of Harvard's Museum of Comparative Zoology.

        rhizophorae: Latin, "bearing roots" (referring to plumage patterning that resembles roots).

Chondrohierax: from the Greek *chondros*, "cartilage," and *hierakos*, "a hawk or falcon."

    uncinatus: Latin, "hooked or barbed" (referring to the bill).

        aquilonis: Latin, "the north wind."

        mirus: Latin, "wonderful."

        wilsonii: after Alexander Wilson (1766–1813), pioneer American naturalist for whom the Wilson Ornithological Society was named.

Circus: from the Greek *kirkos*, "a circle." The vernacular name "harrier" is from the Old English *hergian*, and refers to its chasing or harrying of prey.

    cyaneus: Greek, "bluish."

        hudsonius: "of Hudson Bay."

Elanoides: Greek, "resembling a kite." The vernacular term "kite" refers to the birds' buoyant flying behavior ("kiting"), and derives from the the Old English *cyta*.

    forficatus: Latin, "forked."

        yetapa: apparently derived from a native Paraguayan word.

Elanus: Latin, "a kite."

    caeruleus: Latin, "dark-colored or dark blue."

        leucurus: Greek, "white-tailed."

        majusculus: Latin, "somewhat greater."

Falco: Latin, in reference to its sickle-shaped (falcate) bill and talons.

    columbarius: Latin, "pertaining to doves or pigeons," which is also the basis for the older vernacular name "pigeon hawk." The preferred name "merlin" is apparently from the Old French vernacular name *esmerillon.*

        bendirei: after Major C. E. Bendire (1836–1897), Honorary Curator of the U. S. National Museum.

        richardsonii: after Sir John Richardson (1787–1865), English explorer of arctic Canada.

        suckleyi: after Dr. George Suckley (1830–1869), American physician and naturalist.

    femoralis: Latin, "pertaining to the thigh" (color). The vernacular name "aplomado" falcon is from Spanish, meaning "lead-colored."

        septentrionalis: Latin, "northerly."

    mexicanus: "of Mexico."

    peregrinus: Latin, "wandering." The vernacular name "peregrine" has the same origin.

        anatum: Latin, "pertaining to a duck" (*Anas*).

        pealei: after Titian Peale, Philadelphia artist and taxidermist on Major S. Long's expedition to the Rocky Mountains.

        tundrius: Latin, "of the tundra."

    rusticolus: Latin, "a country-dweller." The vernacular name "gyrfalcon" may be a corruption of the Greek *gyros* ("circle") and "falcon," or more probably may come from the Old High German *giri*, "greedy," and *Valke*, "a falcon."

    sparverius: Latin, "in reference to sparrows." The vernacular name "kestrel" is from a French word *crecerelle* that refers to the "rattling" call of the related Eurasian species.

        dominicensis: "of Santa Domingo."

        guadalupensis: "of Guadalupe Island."

        nicaraguensis: "of Nicaragua."

        paulus: Latin, "little."

        peninsularis: Latin, "peninsular."

        phalaena: Greek, "a devouring monster."

sparverioides: Latin, "sparrow-like."

tropicalis: from the Greek *tropicos,* "of the tropics."

Haliaeetus: from the Greek *haliaetos,* "a sea-eagle."

leucocephalus: Greek, "white-headed."

alascanus: of Alaska.

Ictinia: Greek, "a kite."

mississippiensis: "of Mississippi."

Pandion: Greek, after a legendary king of Athens who along with his daughters Procne and Philomela became transformed into birds. The vernacular name "osprey" derives indirectly from the Latin *ossifraga,* "a bone-breaker," originally applied to another species of raptor.

haliaetus: Greek, "a sea-eagle."

carolinensis: "of Carolina."

ridgwayi: after Robert Ridgway (1850–1929), famous American ornithologist and organizer of the American Ornithologists' Union.

Parabuteo: Latin, "similar to or close to *Buteo.*"

unicinctus: Latin, "single-banded" (referring to the tail pattern).

harrisi: after Edward Harris (1799–1863), a friend and patron of J. J. Audubon.

superior: Latin, "higher or larger."

Polyborus: from the Greek *poly,* "many," and *boros,* "foods," thus "omnivorous or gluttonous." The vernacular name "caracara" is of Tupian or Guarani Indian (of Brazil) origin, and is onomatopoeic.

plancus: Latin, "flat-footed."

audubonii: after J. J. Audubon (1785–1851).

cheriway: Apparently based on an Indian name, possibly onomatopoeic.

lutosus: Latin, "muddy" (in reference to its darker plumage tones).

pallidus: Latin, "pale."

Rostrhamus: Latin, "a hooked bill."

sociabilis: Latin, "sociable."

major: Latin, "greater."

plumbeus: Latin, "lead-colored."

APPENDIX 3

Glossary

ABDOMEN. That part of the underparts between the breast and under tail coverts; sometimes called the "belly."

ACCIDENTAL. An individual occurring well beyond its species's normal geographic range, sometimes called a "vagrant."

ACCIPITERS. Normally used as a collective vernacular term to designate members of the genus *Accipiter,* sometimes also called sparrowhawks.

ACCIPITRES. Alternative taxonomic name for the Falconiformes.

ACCIPITRIDAE. The family of hawks, kites, eagles, and Old World vultures; "accipitrids" are members of this family.

ADAPTIVE RADIATION. The divergent patterns of evolution shown in a single phyletic line that result from varying speciation patterns and local evolutionary adaptations to differing environments.

ADULT. A collective age category (composed of an indefinite number of age classes) of sexually mature individuals in their definitive plumage. See also DEFINITIVE PLUMAGE.

AERIE. See EYRIE.

AGE CLASS (*or* YEAR CLASS). A population subgroup including all those individuals hatched or born during the same year (thus belonging to the same population cohort). See also COHORT.

AGONISTIC BEHAVIOR. Social behavior associated with attack and escape situations, including intermediate stages of social dominance and submission.

ALBINISM. Absence (locally or overall) of melanins or other pigments in individuals that normally are more highly pigmented. See also LEUCISM.

ALLOPATRY (*adjectival form* ALLOPATRIC). Populations occupying completely separated geographic areas, at least during breeding. See also SYMPATRY.

ALLOSPECIES. Two or more populations (comprising a superspecies) that appear to have the necessary criteria to be considered separate species, but that are allopatric and thus cannot be tested for the presence of possible reproductive isolating mechanisms. Taxonomically, allospecies may be signified in trinomials by placing the name of the nominate form of the superspecies in parentheses between the names of the genus and the form(s) under consideration.

ALTRICIAL. Refers to species whose young are hatched blind, relatively helpless, and often naked. See also NIDICOLOUS and PRECOCIOUS.

ALULA. The group of miniature flight feathers (or "bastard wing") associated with the wrist area (actually inserting on the first of the discernible digits, usually called the "thumb") (see Figure 64).

ANISODACTYL. An avian foot arrangement in which one toe (the first) is oriented posteriorly, and the other three anteriorly. Typical of all species included in this book except ospreys. See also ZYGODACTYL.

ANNUAL MORTALITY RATE. A statistic obtained by dividing the number of deaths during a year by the number of individuals in that group that had been alive at the start of the year. The group composition is often specified by age and/or sex. See also MORTALITY.

ANNUAL SURVIVAL RATE. A statistic obtained by dividing the number of individuals alive at the end of the year by the number in that group that had been alive at the start of the year. The group composition is often specified by age and/or sex. See also SURVIVAL.

ANTERIOR. Toward the front, as opposed to posterior.

AOU. Abbreviation for the American Ornithologists' Union.

ARBOREAL. Frequenting trees.

ARBORESCENT. Treelike.

ARVICOLINE RODENTS. Rodents such as voles and lemmings of the subfamily Arvicolinae; synonymous with "microtine."

ASCENDENT. A molting pattern that (at least in hawks) begins at the fourth primary and ascends toward the secondaries, simultaneously descending toward the wingtip. See also DESCENDENT.

ASYNCHRONOUS. Nonsimultaneous, such as the staggered hatching of eggs of a clutch over a period of several days.

ATTENUATED. Becoming slender toward the tip.

AURICULARS. Feathers of the ear region (see Figure 64).

AVERAGE. The arithmetic mean of a sample.

AXILLARIES. Feathers of the "armpit" area between the underside of the wing and body (see Figure 64).

BASIC. A term for the plumage that (at least in raptors) follows the juvenal plumage and thereafter is renewed without intervening molts and plumages.

BAT-HAWK. The vernacular name of an Old World falconlike hawk (*Machaeramphus*) adapted to catching bats in flight.

BINOMIAL. A two-parted name, such as one consisting of genus and species names. See also TRINOMIAL.

BIOMASS. The total weight of organisms (of a specified species or collectively) occupying an area of land, or (if used in food analysis) consumed per individual or sample.

BOOTED. Refers to (1) a relatively or completely unbroken tarsal surface pattern (see also RETICULATE and SCUTELLATE), or (2) a fully feathered tarsus (as in "booted" eagles), which is alternatively referred to as "rough-legged."

BRANCHER. A young raptor that is still unfledged but able to clamber about on branches ("branching").

BROODING. Parental sitting on or over the young, as opposed to incubation (the sitting on eggs).

CALL. An acoustically simple avian vocalization, often not sex- or age-limited.

CERE. A relatively soft or horny and often distinctly colored covering of the base of the upper mandible, typical of all hawklike birds, within which the nostrils are located (see Figure 64).

CHORD. An unflattened wing measurement, from the tip of the longest primary to the most distant point on the wrist, with the feathers in normally curved position. Unless otherwise indicated, this is the measurement method used in this book.

CINNAMOMEOUS. Cinnamon-colored.

CLADAGRAM. A line diagram derived by cladistic techniques (see below).

CLADE. A strictly monophyletic group of organisms.

CLADISTICS. A method of reconstructing phylogenies from attempts to determine the sequence of evolutionary branching points of related lines (clades), using information based on shared derived characters (synapomorphies), but not on shared primitive characters (symplesiomorphies), and with no reference to external similarities. Sometimes called phylogenetic systematics, as opposed to evolutionary systematics. See also EVOLUTIONARY SYSTEMATICS and TAXONOMY.

CLINE. A graded geographic trend in one or more characters among members or demes of a species; such trends may be continuous (as in unbroken populations) or discontinuous (as in variously isolated populations).

CLUTCH. The normal number of eggs laid and simultaneously incubated by a female during nesting.

COHORT. A component of a population consisting of individuals of the same age class.

COLEOPTERAN. A member of the beetle order Coleoptera.

COMMUNICATION. Behavior patterns of an individual that alter the probability of subsequent behavior by another individual in an adaptive manner.

COMMUNITY. The collective array of organisms occupying a particular habitat.

CONFLUENT. Running together and merging, as in the colors of a feather.

CONGENERIC. Members of the same genus.

CONSPECIFIC. Members of the same species.

CONVERGENT EVOLUTION. The evolution of structural or behavioral similarities in organisms that are not closely related, often because of analogous ecological adaptations.

COSMOPOLITAN. Referring to a taxon represented on nearly all of the major continental regions.

COURTSHIP. Communication between individuals of opposite sexes of a species that facilitates pair bonding, pair maintenance, or fertilization. See also DISPLAY.

COVERTS. Small feathers covering the wings (wing coverts) or tail (tail coverts); also sometimes used for feathers of some other body areas, such as ear coverts (auriculars) (see Figure 64).

CREPUSCULAR. Active during the dawn and dusk hours, as opposed to strictly nocturnal or diurnal (q.v.).

CROSS-FOSTERING. The use of foster parents of another avian species to reestablish young birds into the wild. See also HACKING.

CROWN. The top of the head; also used to refer to a group of lengthened or distinctive feathers on the crown region (see Figure 64).

CULMEN. The dorsal ridge of the upper mandible of the bill; normally measured as a straight line from the base of the bill (or, if so defined, from the forehead feathering or the edge of the cere) to its tip (see Figure 64).

DDT. Abbreviation for a widely used type of chlorinated hydrocarbon pesticide (a pesticide group that also includes aldrin, dieldrin, and heptachlor), noted particularly, along with its metabolic breakdown products DDE and DDD, for having eggshell-thinning effects in affected birds. See also PESTICIDE and FOOD CHAIN.

DEFINITIVE PLUMAGE. Plumage attained and renewed without further change by adult birds. In hawks there is apparently only a single annual ("prebasic") molt, and thus no distinction between breeding and non-breeding plumages, although wear and fading may cause minor seasonal appearance changes. See also ADULT and BASIC.

DEME. A local interbreeding (Mendelian) population.

DESCENDENT. A molting pattern that begins at the innermost primaries (or secondaries) and proceeds ("descends") toward the outermost.

DIHEDRAL. Refers to a variable uptilting of wings while gliding or soaring, increasing aerodynamic stability.

DIMORPHISM. A population containing individuals of two discrete types, as shown by their mensural or structural traits (typical dimorphism), colors ("dichromatism"), and/or behaviors ("diethism"), these differences often but not always being associated with sex. See also POLYMORPHISM.

DISPERSAL. Movements (usually multidirectional and unpredictable) of organisms away from some point of origin or concentration; in predatory birds often occurring shortly after fledging (juvenile dispersal) or by adults after breeding (postbreeding dispersal). See also MIGRATION.

DISPLAY. Behavior patterns ("signals") that have been evolved ("ritualized") to provide communication functions for an organism, often through stereotypical performance (thereby reducing signal confusion) and variable exaggeration (increasing signal conspicuousness). Such signals may be visual, tactile, acoustic, or combinations of these.

DISTAL. Toward the periphery of the body or its appendages, as opposed to proximal.

DIURNAL. Active during the day, as opposed to nocturnal.

DNA. Abbreviation for the genetic material of a species's chromosomes.

DNA HYBRIDIZATION. A biochemical technique for estimating phyletic and taxonomic relationships indicated by the chemical similarities of the genetic material (DNA) from two species, through determining the average temperature required to melt half of the hydrogen bonds formed in "hybrid DNA" that has been obtained by chemically combining single-stranded DNAs of the two component species.

DORSAL. In the direction of or pertaining to the upperparts, as opposed to ventral.

EAGLE. A nontaxonomic vernacular name applied to various large raptorial species of the Accipitridae, and specifically in North America to species of the genera *Aquila* and *Haliaeetus;* the name includes the true "booted" eagles as well as fish eagles and sea eagles.

ECOLOGICAL (NICHE) SEGREGATION. The process by which competition between potentially competing individuals or populations is reduced by development of niche differences, such as behavioral or morphological differences associated with food getting. Ecological release is the opposite process, by which selection for such segregation is removed in areas where such competition does not occur, allowing for broader niche utilization than in areas of competitive interaction. See also HABITAT SEGREGATION.

EGGSHELL THINNING. The laying of abnormally thin-shelled eggs as a result of disease, nutritional deficiency, stress, or the ingestion of some pesticides. See also PESTICIDE POISONING.

EMARGINATED. Either (1) slightly notched or forked, when used in reference to tail outline as seen in profile (see Figure 65), or (2) abruptly narrowed or tapered toward the tip, when used in reference to the vane shape of individual remiges, involving the gradual or abrupt narrowing of one or both vanes (see Figure 64). See also NOTCHED and SINUATED.

ENDANGERED. A conservation category defined by the ICBP (q.v.) as including those taxa that are in danger of extinction and whose survival is unlikely if the factors causing their decline continue operating. Defined by the U.S. Endangered Species Act as including taxa in danger of extinction throughout a significant portion of their ranges, based on the best available information. See also THREATENED and RARE.

ENDEMIC. A taxon that is native to and limited to a particular area.

ERYTHRISM. Having a high level of rufous (phaeomelanin) pigments present in the plumage, sometimes virtually replacing the darker eumelanins. Produces "rufous morphs" in some hawks.

EURYOECIOUS. Having a broad range of ecological tolerances, as opposed to stenoecious.

EVOLUTIONARY SYSTEMATICS. A method of reconstructing the phylogeny of a taxon by analyzing the evolution of its major features, along with the distribution of its shared primitive and derived characteristics. See also CLADISTICS.

EXTINCT. A taxon that is no longer alive anywhere. See also EXTIRPATED.

EXTIRPATED. A taxon that has been eliminated from part of a previously occupied range. See also EXTINCT.

EXTRALIMITAL. Occurring beyond the stated limits of an area or range.

EYASS. Used in falconry to describe a nestling (from the French *niais*, or nest) or a hawk obtained as a nestling. See also GENTLE and HAGGARD.

EYRIE (*or* AERIE). A hawk, falcon, or eagle nest, especially a lofty one; sometimes also used to refer to the brood in the nest (probably derived from the Latin *area*, an open space or location).

FALCON. As used here, indicating a typical member ("falconid") of the Falconidae, especially the true falcons of the genus *Falco*. Also used in falconry to refer to a female falcon (the male being called a "tiercel").

FALCONETTE. The vernacular name of some very small and mainly insectivorous falcons (*Microhierax* and *Spiziapteryx*) found in tropical Asia and South America.

FALCONIFORMES. The order of hawklike birds, including falcons; "falconiforms" are members of this order. Also called Accipitres.

FAMILY. A taxonomic category representing a group of one or more related genera; consistently spelled with an "-idae" suffix, as in Accipitridae.

FERRUGINOUS. Rusty brown.

FISH EAGLE (*or* FISHING EAGLE). A vernacular name for certain kinds of eagles (*Haliaeetus* and *Icthyophaga* spp.) that feed largely on fish, some of which are also called sea eagles.

FITNESS. The relative genetic contribution of an individual toward future generations of its species. Sometimes expanded to include inclusive fitness, the effect of the individual on the reproduction of all its genetic relatives.

FLEDGE. To attain the power of flight.

FLEDGING PERIOD. The period of time between hatching and fledging. See also POSTFLEDGING PERIOD.

FLEDGING SUCCESS. An estimate of the percentage of hatched young that fledge successfully. See also HATCHING, NESTING, and REPRODUCTIVE SUCCESS.

FLEDGLING. A recently fledged but still dependent bird.

FLIGHT FEATHERS. The collective primary and secondary feathers of the wing (see Figure 64).

FOOD CHAIN. A sequence of energy transformations through successive "trophic levels" of a community by successive consumption of various of its members by one another. Food chains are actually only component parts of much more complex "food web" interactions, both of which end in "top-level" predators such as ospreys and eagles. Sometimes pesticides or other alien substances are "biologically magnified" during their passage through successive organisms comprising food chains or food webs, and thus accumulate in top-level predators. See also PESTICIDE POISONING.

FORB. A broad-leaved (nongrasslike) herbaceous plant.

FOREST FALCON. The vernacular name of certain woodland falcons (*Micrastur* and *Herpetotheres*) that are endemic to South America.

FORM. A taxonomically neutral term for a species or some subdivision of a species; the term lacks nomenclatural significance and is often used to avoid implying specific taxonomic meaning or rank when referring to a particular individual or population.

FRATRICIDE. The killing and consumption of younger or weaker nestmates by their siblings. Sometimes also called "siblicide."

FUSCOUS. Brownish black.

GAPE. The lateral distance across the mouth at the base of the opened bill; also refers to the skin associated with the open mouth, as in "gape color."

GENTLE. A falconry term for a young hawk caught after leaving the nest but before migrating. See also EYASS and HAGGARD.

GENUS (*plural* GENERA, *adjectival form* GENERIC). A taxonomic category representing one or more species that are believed to be more closely related to one another than to any other species. Consistently italicized and capitalized, as in *Accipiter,* and comprising the first half of a binomial scientific name. See also SPECIES.

GRADUATED. Showing a progressive increase in length, as in the feathers of a somewhat pointed tail.

GULAR. Pertaining to the throat area.

GUTTATE. Having the shape of teardrops.

HABITAT. The physical, chemical, and biotic characteristics of a specific environment.

HABITAT SEGREGATION. A physical subdivision of available habitat resources by two or more of the resource users (such as different species, age classes, or sexes); the resources may be subdivided by space, time of usage, or both. See also ECOLOGICAL SEGREGATION, of which habitat segregation is one example.

HABITAT SELECTION. The ability of an organism to assess environmental variations in such a way as to be able to locate itself within desirable habitats, whether by behavioral responses, physiological tolerances, or both.

HACKING. A process of allowing captive hawks progressively more freedom, usually to reestablish them in the wild. "Tame hacking" refers to cases where the control is regulated by the falconer; "wild hacking" refers to the use of wild foster parents to provide food and protection.

HAGGARD. In falconry, a wild, intractable hawk captured after molting its juvenal plumage. See also EYASS and GENTLE.

HARRIER. The vernacular name for members of the genus *Circus* of the family Accipitridae, a group noted for their usually long wings and tails, as well as their long legs and rather owllike head shapes. Additionally, some unrelated raptors are also called harrier-hawks, including two long-legged African and Madagascan eaglelike forms (*Polyboroides*).

HATCHING SUCCESS. An estimate of the percentage of eggs in a sample that hatch successfully. See also NESTING, FLEDGING, and REPRODUCTIVE SUCCESS.

HAWK. A vernacular term implying a member of the Accipitridae, especially raptors other than forms specifically referred to as eagles, kites, or harriers. Also sometimes used for falconiform birds in general.

HAWKING. Catching insects or other prey while in full flight.

HECTARE. A metric surface measure (10,000 square meters) equal to 2.47 acres.

HERB. Any nonwoody plant, including forbs (q.v.).

HOBBY. The vernacular name for a group of small, long-winged, and mostly insectivorous members of the genus *Falco*, not represented in North America.

HOLARCTIC. The land areas and islands of the northern hemisphere, including North America north of Mexico's central highlands, Africa north of the Sahara, and Eurasia north of the Himalayas. See also NEARCTIC and PALEARCTIC.

HOME RANGE. An area regularly used (but not necessarily defended) by an individual or social group over some defined time period, such as a day (daily home range), the nesting season (breeding home range), or a year (annual home range). The home range may include a more restricted area (territory) that is defended against incursions by other individuals or groups. See also TERRITORY.

HONEY-BUZZARD. A vernacular name for a group of pernine hawks, the most typical species of which (*Pernis apivorus*) feeds largely on the larvae of bees and wasps.

HOVER. A form of flapping flight during which the bird remains nearly stationary.

HYBRID. An individual produced by the crossing of taxonomically different populations, usually between subspecies (intraspecific), less often between species (interspecific), and rarely between genera (intergeneric).

ICBP. International Council for Bird Protection.

IMMATURE. Generally referring to the period in a bird's life from the time it fledges until it is sexually active; conveniently used here to indicate any fledged raptor that has not yet reached its definitive plumage (although some individuals may initially breed while still "immature"). Possible distinction between juveniles (immatures carrying most or all of their juvenal feathers) and older but still sexually immature birds not yet in definitive plumage (here called subadults) depends largely on age-related differences in plumages and associated softpart colors. Bird banders sometimes use "second-year" (SY) and "after-second-year" (ASY) as similar age class categories. See also JUVENILE and SUBADULT.

IMPRINTING. A type of seemingly irreversible learning typical of very young animals (such as newly hatched or newly fledged birds), during which such things as parental recognition or learning the location of a natal nesting area may be achieved.

INFRATAXON. An infrequently used taxonomic category below that of the subtaxon, such as an infraclass below a subclass.

INNATE. Inherited, such as instinctive patterns of behavior.

INSTINCTIVE BEHAVIOR. Innate activities that typically are more complex than simple reflexes or directed movements, and that are dependent upon specific external stimuli ("releasers") as well as on variable and specific internal states ("tendencies") for their expression.

INTERGRADE. An individual or population having transitional characteristics genetically and phenotypically linking different subspecies.

INTERSPECIFIC. Occurring between separate species.

INTRASPECIFIC. Occurring within a single species.

INTROGRESSION. A movement of genes across population boundaries as a result of hybridization between populations.

IRRUPTIVE. Descriptive of nomadic or migratory individuals or species whose magnitudes of movements differ markedly from time to time.

ISABELLINE. Brown, tinged with reddish yellow.

ISOLATING MECHANISM. An innate property of an individual that prevents successful mating and subsequent offspring with genetically unlike individuals, including both premating (occurring before mating, such as courtship differences) and postmating (occurring after mating, such as hybrid sterility) mechanisms.

JUVENAL. The plumage stage typical of juvenile birds, when the first generation of flight feathers erupts and fledging occurs.

JUVENILE. A bird exhibiting part or all of its first (juvenal) plumage of nondowny feathers. Bird banders often use the "hatching-year" (HY) designation for birds showing diagnostic juvenile traits, as opposed to the "after-hatching-year" (AHY) category for all older age classes. See also IMMATURE.

K-SELECTED SPECIES. Those normally long-lived and slowly reproducing species (such as eagles) that have evolved reproductive rates, mating systems, and other adaptations that tend to keep their populations fairly constant and near the carrying capacity ("K") of relatively stable environments. By comparison, r-selected species (such as some kites) have evolved potentially high reproductive ("r") rates, flexible mating systems, and other adaptations such as efficient dispersal mechanisms that tend to allow for rapid changes of population sizes and locations, enabling them to exploit environments that vary in carrying capacity, duration of existence, and distribution.

KESTREL. A vernacular name applied to various small species of *Falco* noted for their tendency to hover while searching for prey on the ground.

KETTLE. A term used to describe an aggregation of flying (often circling in thermals) hawks; also sometimes used as a verb ("kettling").

KITE. A vernacular name of no taxonomic significance, but most often applied to various relatively agile and sometimes insectivorous members of the subfamilies Milvinae, Elaninae, and Perninae.

KITING. Remaining in place aloft while facing the wind without flapping the wings.

LAGOMORPH. Rabbits, hares, and pikas of the order Lagomorpha.

LATERAL. Toward the right or left side of the body, as opposed to medial.

LEARNING. The adaptive modification of behavior in an individual based on its past experience.

LEUCISM. A paler than normal ("diluted") plumage variation, resulting from reduced melanin deposition in the feathers.

LONGEVITY. Records or estimates of lifespan. Includes maximum recorded longevity, mean longevity (average life expectancy after hatching), and mean afterlifetime (expectancy of additional life as of some later point, such as fledging, after the first year of life, etc).

LORE (*adjectival form* LORAL). The area between the eye and the base of the upper mandible in birds, which in hawks is often bare or nearly bare (see Figure 64).

MALAR. Referring to the area extending from the base of the bill backward below and behind the eye, sometimes forming a mustachelike streak (see Figure 64).

MANDIBLE. In plural, the upper and lower halves of the bill; in singular, the lower half only (the upper being the maxilla).

MATE SELECTION. The tendency for individuals of a species selectively to "choose" appropriate mates from the population in an adaptive manner, whether by innate tendencies or learned clues as to individual differences in probable relative fitness among the available choices.

MATING SYSTEM. Patterns of mating within a population, including length and strength of pair bond, the number of mates, degree of inbreeding, and the like.

MAXILLA. The upper half of the bill (see Figure 64).

MEDIAL. Toward that plane that separates right and left halves of the body, as opposed to lateral.

MELANISM. Descriptive of an unusually high level of blackish brown pigments (eumelanins) in the plumage or skin of an individual. See also ALBINISM, ERYTHRISM, and LEUCISM.

MIGRANT. A species or individual that regularly moves ("migrates") on a seasonal basis from one area to another and back, usually between breeding and nonbreeding areas.

MODE. The most frequently occurring value in a sample. See also AVERAGE.

MOLT. The process by which feathers are individually or collectively lost and replaced, eventually producing a change in plumage. See also PLUMAGE.

MONOGAMY. A mating system involving the coordinated reproductive efforts of a single male and female through at least a single breeding cycle (single-brood and seasonal monogamy), and sometimes extended indefinitely (sustained monogamy), even though the pairs may be separated during the nonbreeding season. See also POLYGAMY.

MONOTYPIC. Referring to a taxonomic category having only a single subordinate member, such as a genus having only a single species. See also POLYTYPIC.

MORPH. A term referring to an example of one of the recognized types of dimorphism or polymorphism characteristic of an organism (e.g., melanistic morph, albinistic morph, etc.). As used here, essentially synonymous with "phase" (q.v.).

MORTALITY RATE ("M"). The estimated rate (expressed as a percentage or decimal fraction) of death among individuals of a population, usually calculated on an annual (12-month) basis and sometimes defined by age or sex. See also SURVIVAL RATE.

NAPE. The area immediately behind and below the occipital portion of the skull (see Figure 64). See also NUCHAL.

NARES. The nasal cavities, opening via the nostrils.

NEARCTIC. The New World portion of the Holarctic (q.v.).

NEOTROPICAL. The land masses and associated islands of the New World south of the Nearctic, including South America plus Central America north to Mexico's central highlands, and the West Indies.

NEST SITE. The actual nest location; sometimes also called an eyrie (q.v.) in eagles and falcons. An occupied nest site is one that is actively used by a pair during a breeding season. Site tenacity (or fidelity) refers to a year-to-year reoccupation of the same nest site.

NESTING SUCCESS. An estimate of the proportion of initiated nests that successfully hatch one or more young.

NESTLING. Descriptive of unfledged birds still in the nest ("nidicoles"). The nestling period is the length of time from hatching to fledging, and is comparable to the fledging period ("age at first flight") in nidicolous forms such as hawks and eagles.

NICHE. Structural, physiological, and behavioral adaptations of a species to its environment.

NICHE SEGREGATION. See HABITAT SEGREGATION.

NIDICOLOUS. Referring to species whose altricial young remain in the nest until attaining the ability to fly (fledging) or nearly so. See also ALTRICIAL and NIDIFUGOUS.

NIDIFUGOUS. Referring to species whose precocial young leave the nest very soon after hatching, usually within a day or so. See also NIDICOLOUS and PRECOCIAL.

NOMAD: A species or individual whose movements are relatively unpredictable as to their timing, direction, or duration.

NOMENCLATURE. The taxonomic principles and procedures by which scientific names are applied to organisms. See also TAXONOMY.

NOMINATE. Referring to a taxon that is the nomenclatural basis for the name of the larger taxonomic group to which it belongs; e.g., the genus *Accipiter* of the family Accipitridae.

NONBREEDER. A term used to describe seemingly paired birds that nevertheless fail to breed, even if they may be defending a territory.

NOTCHED. Having a distinct indented outline, such as a slightly forked tail profile (the central rectrices somewhat shorter than the lateral ones), or the abruptly emarginated inner vanes of the flight feathers in many raptors (see Figure 64). See also EMARGINATED and SINUATED.

NUCHAL. Pertaining to the nape area or occiput.

OBSOLETE. Nearly invisible or lacking, in reference to plumage pattern or structure.

OCCUPIED NEST. A nest site used by a potentially breeding pair, whether or not eggs are laid. An "active nest" is one containing eggs or young.

OCHRACEOUS. Ochre-colored, or yellowish brown.

OPTIMAL FORAGING. Foraging behavior that maximizes energy gain per unit of hunting time.

ORBITAL RING. An area of bare, often colorful skin circumscribing the eye.

ORDER. A taxonomic category immediately subordinate to that of the class (or subclass), consisting of one or more related families, and normally identified (at least in the taxonomic procedures used by the AOU) by the suffix "-iformes," as in Falconiformes.

ORTHOPTERAN. A member of the insect order Orthoptera, such as grasshoppers.

PAIR BOND. A prolonged and individualized social association between members of a mated pair in association with breeding. See also MONOGAMY.

PALATE. The upper roof of the mouth, including both the bony portion (bony palate) and its covering (soft or horny palate).

PALEARCTIC. The Old World component of the Holarctic (q.v.).

PARALLEL EVOLUTION. Traits or taxa that evolve in parallel fashion, showing neither convergent nor divergent trends. See also CONVERGENT EVOLUTION.

PCB. Abbreviation for the polychlorinated biphenol group of chemicals.

PESTICIDE POISONING. Poisoning of an individual by the presence of metabolic poisons applied to control "pests," and effected either directly or indirectly, the latter by eating prey that have themselves concentrated ("biologically magnified") the poisons by eating contaminated foods from organisms lower in the food chain. Mortality from pesticides may also be indirect, such as through eggshell thinning (q.v.) and hatching failures. See also DDT and FOOD CHAIN.

PHASE. A term traditionally used in ornithology to describe nontransient plumage variations, such as a (usually) genetically determined pigmentary or other plumage deviation from the norm of the species, that are typically independent of sex and age (unless sex-linked or age-related, respectively). See also MORPH, which is a more recently applied (and more appropriate) substitute term encompassing these phenomena.

PHENOTYPE. Pertaining to an organism's appearance, irrespective of the genetic basis of this appearance.

PHYLETIC. Referring to a pattern of evolutionary lineage, or phylogeny. A phylogeny is typically a hypothetical representation or "phylogram" of evolutionary descent within a single phyletic line. It may be illustrated in traditional treelike representations (dendrograms), in the form of branching evolutionary clades (cladograms), or as diagrams showing degrees of phenotypic differences based on numerical analyses (phenograms). See also CLADE and PHENOTYPE.

PLUMAGE. A single generation of feathers. See also BASIC, DEFINITIVE PLUMAGE, and JUVENAL.

PLUMBEOUS. Lead-colored, grayish blue.

POLYGAMY. A collective (and imprecise) mating system category involving more than one individual of one sex mating with one individual of the other sex in conjunction with a single reproductive cycle; polygamy includes polygyny (the mating of a male with more than one female simultaneously or successively) and polyandry (the mating of a female with more than one male simultaneously or successively). Polygamy also usually includes bigamy and, if no pair bonding occurs between members of copulating pairs, promiscuity. See also MONOGAMY.

POLYMORPHISM. A population occurring in two or more forms or colors ("morphs") that are typically genetically controlled and often independent of sex (but are sometimes sex-limited).

POLYTYPIC. A taxon having more than one member in the category immediately subordinate to it, such as a genus with two or more species. See also MONOTYPIC.

POSTFLEDGING PERIOD. The period between fledging and complete independence of the juveniles. Also called "postfledging dependency period."

POSTOCULAR. Behind the eye.

PRECOCIAL. Referring to species whose young are hatched in a condition allowing them to leave the nest and often feed for themselves very shortly after hatching. See also NIDIFUGOUS and ALTRICIAL.

PREDATION. The killing of one species by another for food. Predatory birds are usually called "birds of prey" or "raptors," although some may also scavenge for foods.

PRIMARIES. Those large contour feathers attached to the bones of the hand and digits that, together with the secondaries, comprise the flight feathers (remiges).

PROXIMAL. Toward the body's axis, as opposed to distal.

PYGMY FALCON. A vernacular name applied to one of two African species of *Polihierax*. See also FALCONETTE.

R-SELECTED SPECIES. See under K-SELECTED SPECIES.

RACE. An alternative name for the subspecies category.

RANGE. The highest and lowest numerical values in a sample.

RAPTOR. A bird of prey or predator (from the Latin *raptor,* a robber or plunderer), typically one with sharp and strong talons and a pointed, curved bill. Also used in the adjectival form "raptorial," to describe these and associated predatory traits. See also PREDATION.

RARE. A conservation category defined by the ICBP (q.v.) as including those taxa having world populations that are small but not currently considered to be either endangered (q.v.) or vulnerable.

RECTRICES (*singular* RECTRIX). The tail feathers.

REMIGES (*singular* REMIX). The primary and secondary wing feathers, also called "flight feathers."

REPRODUCTIVE ISOLATION. A genetic barrier (anatomical, ecological, or behavioral) that helps to prevent matings between species (premating isolation) or, if such matings do occur, tends to prevent the hatching, survival, and reproduction of the resulting hybrid genotypes (postmating isolation). Such "intrinsic" (rather than extrinsic or environmental) barriers to gene exchange are called "reproductive isolating mechanisms."

REPRODUCTIVE SUCCESS. As used here, an estimate of the percentage of eggs laid in a population that produce successfully fledged young. Sometimes also defined as the average number of fledged young raised per reproductive pair, or as otherwise stipulated. See also FLEDGING, HATCHING, and NESTING SUCCESS.

RESIDENT (*adjectival form* RESIDENTIAL). A sedentary (nonmigratory or non-nomadic) population.

RESOURCE. A particular feature of the environment, the control of which contributes to an organism's fitness (q.v.).

RETICULATE. Referring to a weblike or networklike pattern of rounded or hexagonal scales on the surface of the tarsus (see Figure 64).

REVERSED SEXUAL DIMORPHISM. The phenomenon typical of many raptorial birds in which the female is the larger and more powerful sex, the reverse of the usual situation in birds.

RICTUS. The facial area that is at the base of the gape, which in hawks is often nearly featherless or may be somewhat bristly ("rictal bristles").

RIPARIAN. Associated with rivers or shorelines.

RITUALIZATION. The evolutionary development of signaling ("display") behavior and associated signaling devices in an animal species, thereby providing an effective innate communication system.

RUFESCENT. Tinged with rufous brown.

SCAPULARS. Those feathers located in the shoulder region, just medial to the upper wing coverts (see Figure 64).

SCIENTIFIC NAME. The (usually binomial) combination of general (generic) and specific (species-level) names that collectively uniquely identify an organism, such as *Accipiter gentilis*. To be complete, the name of the describer of the species and the year of its initial valid description should be added, such as *A. gentilis* (Wilson) 1812 (the parentheses around Wilson's name indicating that he originally described it as a member of some genus other than the one to which it is currently allocated). See also BINOMIAL and TRINOMIAL.

SCUTELLATE. Referring to a vertically aligned pattern of overlapping scales on the surface (often only the front and sometimes rear edges) of the tarsus (see Figure 64).

SEARCH IMAGE. A learned foraging method that facilitates visual location of easily overlooked foods.

SECONDARIES. Those flight feathers attached to the forearm (ulna) of the wing.

SECRETARY-BIRD. A terrestrial and long-legged African raptor (*Sagittarius serpentarius*).

SEDENTARY. Descriptive of nonmigratory and nonnomadic populations.

SEMISPECIES. A term sometimes used conveniently to designate allopatric populations that may be either subspecies or full species, there being no way of determining the exact level of speciation.

SEPIA. Dark reddish brown, like the color of ink from *Sepia* cuttlefish.

SEXUAL DIMORPHISM. The situation common in sexually reproducing animals for at least mature individuals of the sexes to differ in appearance ("dichromatism"), behavior ("diethism"), and/or size (dimorphism, *sensu stricto*). See also REVERSED SEXUAL DIMORPHISM.

SINUATED. A type of feather emargination in which the web is gradually rather than abruptly narrowed. See also NOTCHED.

SINUOUS. Gracefully curved, similar to the form of extended sine waves.

SPECIATION. The process of species proliferation through the gradual development of reproductive isolation between separated populations.

SPECIES (*abbreviated* SP.; *same spelling in plural but abbreviated* SPP.; *adjectival form* SPECIFIC). Taxonomically, the category below that of the genus and above that of the subspecies. It is written as the subsidiary (following the generic) component of a two-parted (binomial) name, in italics but not capitalized, as in *Accipiter gentilis*. Biologically, one or more populations of actually or potentially interbreeding organisms that are reproductively isolated from all other such populations.

SPECIES GROUP. A group of two or more closely related species whose components usually have partially overlapping (sympatric) ranges. See also SUPERSPECIES.

SPICULES. Here used to refer to the sharply pointed scutes on the toes of an osprey.

STENOECIOUS. Having a narrow range of ecological tolerances, as opposed to euryoecious.

STOOPING. A falconry term used to describe a steep dive toward prey from a substantial height ("steep" and "stoop" having common English semantic ties).

STRAGGLER. Here used to refer to individuals remaining in a region well beyond their normal seasonal period of occurrence. See also VAGRANT.

STRATEGY. The evolved niche adaptations of a population that are associated with its fitness (q.v.) in a particular environment. Includes mating strategies, foraging strategies, etc.

STRIGIFORM. A member of the owl order Strigiformes.

SUBADULT. As used here, a preadult individual in a plumage condition transitional between its juvenal and definitive plumages, especially in taxa requiring more than a year to attain their definitive plumage. The term is also sometimes used more loosely to refer to all preadult individuals (including those in juvenal plumage), here described as "immatures" (q.v.).

SUBFAMILY. A taxonomic category representing an initial subdivision of a family (above that of an infrafamily or tribe); identified by the suffix "-inae," as in Accipitrinae.

SUBGENUS. An optional taxonomic category below that of the genus, used to associate the most closely related members of some more inclusive genus.

SUBORDER. A taxonomic category sometimes interpolated between the order and family (or superfamily) levels of classification, having no universally consistent suffix ending, but typically using "-i," as in Accipitri.

SUBSPECIES (*abbreviated* SPP. *singular;* SSPP. *plural*). A taxonomic category that is defined as a recognizable geographic subdivision of a species; also called race. Taxonomically identified as the final part of a three-parted (trinomial) scientific name, e.g., *Accipiter gentilis atricapillus.*

SUPERCILIARY. Above the eye, such as a superciliary stripe.

SUPERFAMILY. A taxonomic category immediately higher than that of the family but below that of order or suborder; identified by its "-oidea" suffix, as in Accipitroidea.

SUPERSPECIES. A set of two or more species with largely or entirely non-overlapping ranges ("allospecies") that are clearly derived from a common ancestor but are too distinct to be considered a single species. If significant sympatry is present among the included species they are usually instead called a "species group." See also ALLOSPECIES.

SURVIVAL RATE ("S"). The probability (given as a percentage or decimal fraction, the latter equal to $1 - M$) of an individual surviving for a given period; usually defined for a 12-month period (annual survival), and often differentiated as to sex or age. See also MORTALITY RATE.

SYMPATRY (*adjectival form* SYMPATRIC). Two or more populations coexisting in the same area, especially during the breeding season. See also ALLOPATRY.

SYRINX (*plural* SYRINGES). The sound-producing organ of birds, located at the junction of the trachea and bronchi.

SYSTEMATICS. The practices of taxonomy concerned with the erection of taxonomic classifications and determining phylogenies. See also TAXONOMY and NOMENCLATURE.

TAIL. As used ornithologically, the collective rectrices of a bird. Measured (unless otherwise indicated) from the point of insertion of the central rectrices to their tips.

TALONS. The sharply pointed and curved claws of a raptorial bird.

TARSUS. A term collectively applied to the tarso-metatarsus of birds; sometimes also called the "leg" or "foot," but actually consisting of the fused ankle and foot bones.

TAXON (*plural* TAXA). As used here, any taxonomic unit (category), or a particular example of that category.

TAXONOMY. The science of biological classification, which is the basis for providing appropriate biological names (nomenclature) and the establishment of systematic hierarchies (systematics) that best reflect probable lines of phyletic descent. Various contemporary taxonomic techniques include cladistics (the study and analysis of definable monophyletic units, or clades) and numerical taxonomy (the use of "operational taxonomic units" for estimating degrees of phenotypic differences in related groups).

TERRESTRIAL. Frequenting or associated with the ground.

TERRITORY. A definable area having resources that are consistently controlled or defended by an individual against others of its species (intraspecific territories) or, less often, against individuals of other species (interspecific territories), at least for some part of the year. Included are inclusive resource territories as well as breeding territories, foraging territories, etc. Often territories comprise part of more inclusive home ranges (q.v.). Occupied territories are those with a resident pair present during the breeding season, even if eggs are not laid. Breeding territories are those that contain at least one active nest site. Productive territories are those within which at least one chick has been raised at least to an advanced state of development and presumably fledged.

TERRITORIALITY. The advertisement and agonistic behavior associated with territorial establishment and defense.

THREATENED. A conservation category of the federal U.S. and Canadian wildlife agencies (and some individual states and provinces) for designating those taxa that are not yet believed to be endangered, but whose known numbers place them at risk of falling into that category. Similar to the "Vulnerable" category (q.v.) of the ICBP. See also ENDANGERED and RARE.

TIERCEL (or TERCEL). A falconry term used to refer to a male hawk, especially a male falcon (the female then being referred to as the "falcon"). From Latin, and implying a size that is about one-third smaller (than the female).

TOOTHED. Used to describe a condition in falcons and some kites in which the cutting edges of the mandibles form toothlike projections (see Figure 64).

TRIBE. A taxonomic category representing a suprageneric subdivision of a subfamily and containing one or more genera; identified by the suffix "-ini," as in Accipitrini.

TRINOMIAL. A three-parted name, typically consisting of genus, species, and subspecies names. See also BINOMIAL.

TRUNCATE. Square-tipped, as in a feather shape or tail profile.

TYMPANIFORM (or TYMPANIC) MEMBRANES. Paired membranes of the syrinx that, when set into motion by the passage of air, are the basis for bird vocalizations.

VAGRANT. An individual occurring well outside its population's normal migratory or nomadic limits. See also ACCIDENTAL.

VENTRAL. In the direction of or pertaining to the underside, as opposed to dorsal.

VERMICULATIONS. Small wavy lines of pigment in feathers, reminiscent of marks made by the movements of worms.

VERNACULAR NAME. The "common" name of a taxon (usually a species) or morph in some native language.

VULNERABLE. Defined by the ICBP (q.v.) as those taxa believed likely to move into the endangered category in the near future if the causal factors responsible for their declines continue operating.

WING. The arm and associated feathers of birds, measured from the bend (wrist) of the folded wing to the tip of the longest primary, either with the feathers flattened, or as the chord (unflattened) distance (see Figure 64). See also CHORD.

XERIC. Desert-adapted.

ZYGODACTYL. An avian foot arrangement in which two toes (first and fourth) are facultatively or permanently oriented posteriorly and the other two (second and third) anteriorly; typical only of ospreys among the species in this book. See also ANISODACTYL.

*Field Identification Views
and Anatomical Drawings
of North American
Falconiform Birds*

Eagles

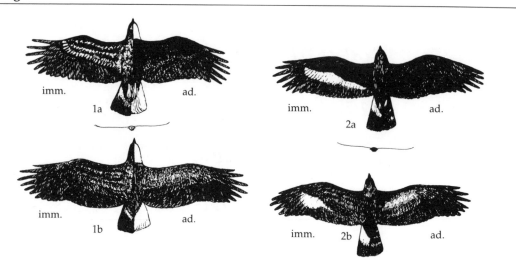

Eaglelike Raptors

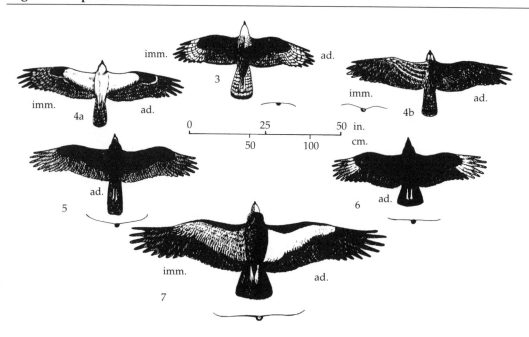

Figure 60. Field identification sketches of bald eagle (1a from below, 1b from above); golden eagle (2a from below, 2b from above); crested caracara, from below (3); osprey (4a from below, 4b from above); and views from below of turkey vulture (5), black vulture (6), and California condor (7).

Accipiters

Woodlands

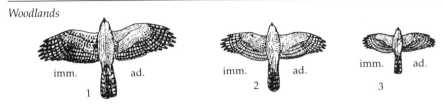

Falcons

Mostly open country

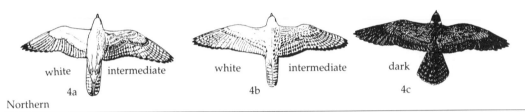

Northern

Southern or Widespread

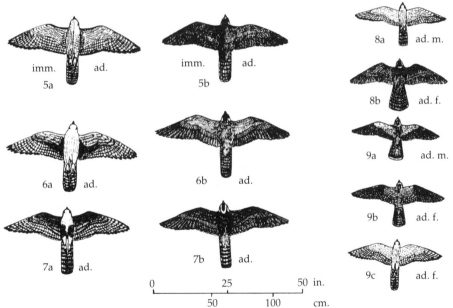

Figure 61. Field identification sketches of northern goshawk (1), Cooper's hawk (2), sharp-shinned hawk (3), gyrfalcon (4a from below, 4b and 4c from above), peregrine falcon (5a from below, 5b from above), prairie falcon (6a from below, 6b from above), aplomado falcon (7a from below, 7b from above), merlin (8a from below, 8b from above), and American kestrel (9a and 9b from above, 9c from below).

Kites

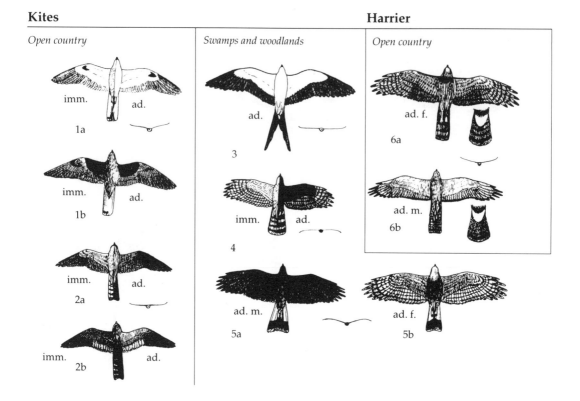

Harrier

Southern Buteos

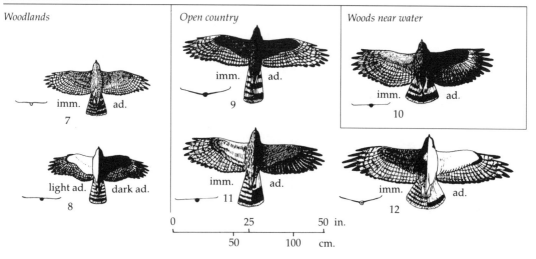

Figure 62. Field identification sketches of black-shouldered kite (1a from below, 1b from above), Mississippi kite (2a from below, 2b from above), American swallow-tailed kite (3), hook-billed kite (4), snail kite (5), northern harrier (6), gray hawk (7), short-tailed hawk (8), zone-tailed hawk (9), common black-hawk (10), Harris' hawk (11), and white-tailed hawk (12).

Widespread Buteos

Open country

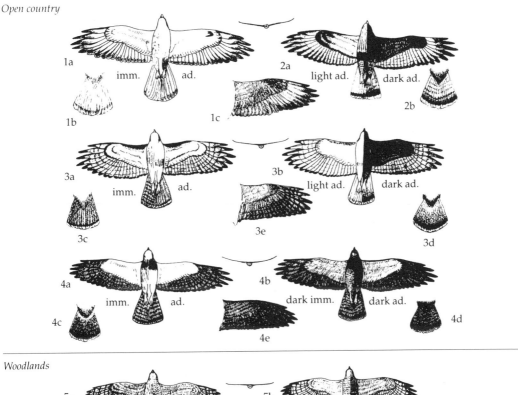

Woodlands

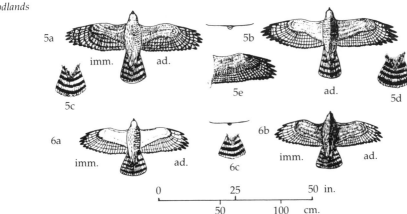

Figure 63. Field identification sketches of ferruginous hawk (1a; adult tail and wing from above, 1b and 1c), rough-legged hawk (2a; adult male tail from above 2b), red-tailed hawk (typical phase 3a, Krider's and Harlan's races 3b; adult tail from above, typical phase 3c, Harlan's race 3d; wing from above 3e), Swainson's hawk (typical phase 4a, dark phase 4b; typical and dark-phase adult
tails from above, 4c and 4d; wing from above 4e), red-shouldered hawk (typical eastern 5a, adults of pale-bodied and dark-bodied variants 5b; typical and dark-phase adult tails from above, 5c and 5d; adult wing from above 5e), and broad-winged hawk (typical phase 6a, dark phase 6b; adult tail from above 6c).

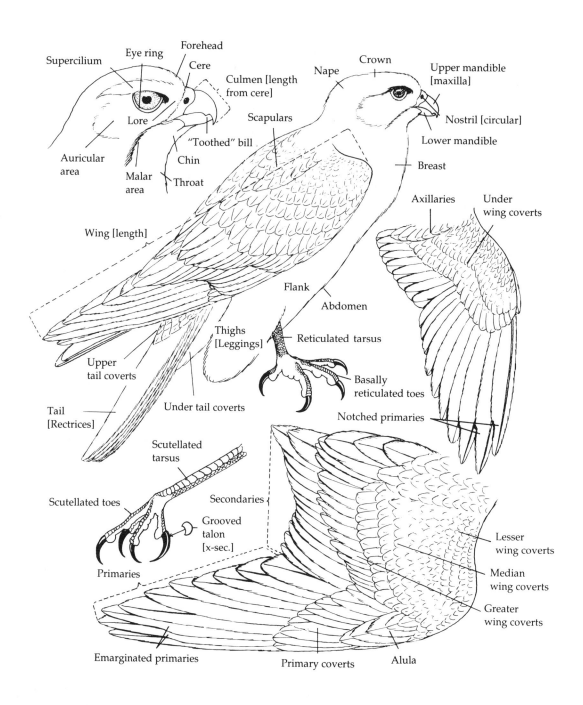

Figure 64. External anatomy of hawks, showing anatomical structures or measurements mentioned in the text and used in the identification key.

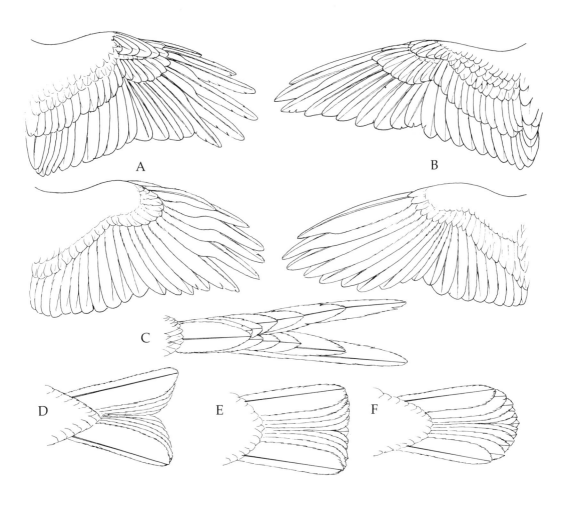

Figure 65. Terminology of wing and tail shapes of falconiform birds, including (A) rounded wing, (B) pointed wing, (C) forked tail, (D) emarginate tail, (E) square-tipped tail, and (F) rounded tail. In part after Friedmann (1950).

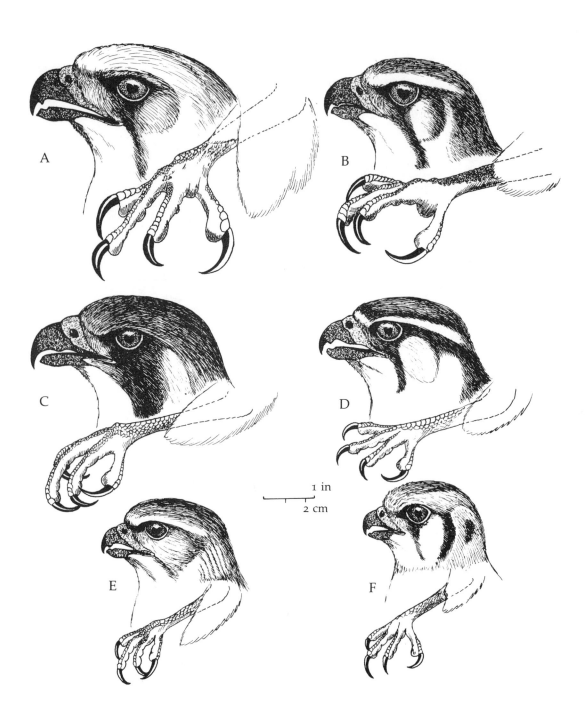

Figure 66. Head and foot detail, gyrfalcon (A), prairie falcon (B), peregrine falcon (C), aplomado falcon (D), merlin (E), and American kestrel (F). Adapted from Friedmann (1950).

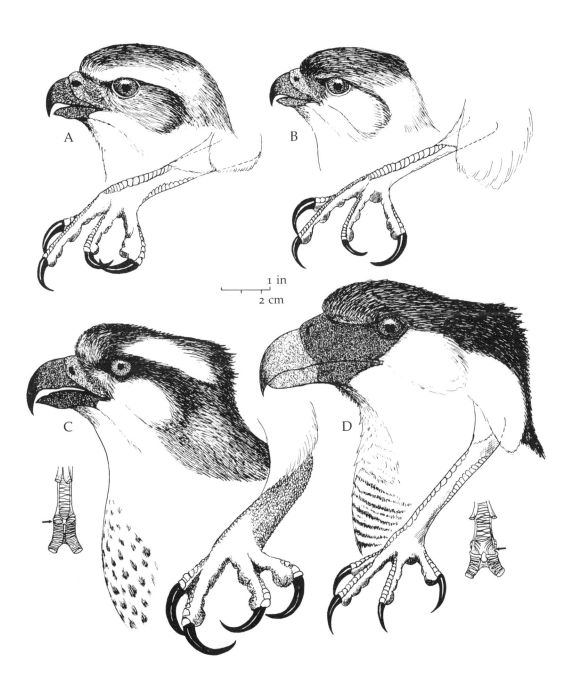

Figure 67. Head and foot detail, northern goshawk (A), Cooper's hawk (B), osprey (C), and crested caracara (D). Adapted from Friedmann (1950). Also shown are syringes (ventral view) of osprey and crested caracara, with locations of lateral tympanic membranes indicated by arrows (after Jollie, 1977).

Figure 68. Head and foot detail, American swallow-tailed kite (A), black-shouldered kite (B), hook-billed kite (C), snail kite (D), and northern harrier (E). Adapted from Friedmann (1950). The location of the ear opening of the northern harrier is shown by stippling. Also shown is syrinx (ventral view) of black-shouldered kite, with location of lateral tympanic membranes indicated by arrow (after Jollie, 1977).

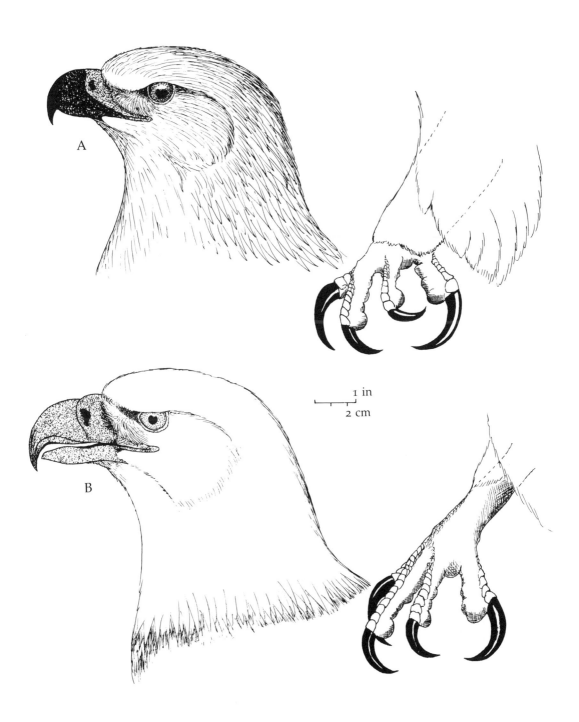

1 in

2 cm

*Figure 69. Head and foot detail, golden eagle (A) and
bald eagle (B). Adapted from Friedmann (1950).*

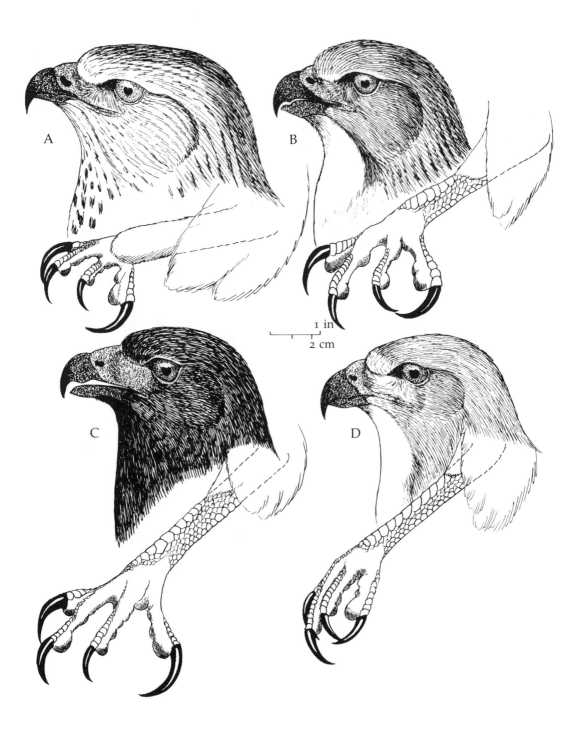

Figure 70. Head and foot detail, red-tailed hawk (A),
Swainson's hawk (B), Harris' hawk (C), and white-tailed
hawk (D). Adapted from Friedmann (1950).

Figure 71. Head and foot detail, ferruginous hawk (A), gray hawk (B), common black-hawk (C), and short-tailed hawk (D). Adapted from Friedmann (1950).

References

Note: Most citations to journal article titles have been abbreviated to conserve space. Such shortened titles are indicated by parentheses.

Abramson, I. J. 1976. (Black-hawk in Florida.) *Amer. Birds* 30:661–62.

Albuquerque, J. 1986. The peregrine falcon (*Falco peregrinus*) in southern Brazil: Aspects of winter ecology in an urban environment. M.S. thesis, Brigham Young Univ., Provo.

Alerstam, T. 1987. (Stooping speed of peregrine and goshawk.) *Ibis* 129:267–73.

Allen, B. A. 1978. Nesting ecology of the goshawk in the Adirondacks. M.S. thesis, State Univ. N.Y., Syracuse.

Allen, G. T. 1987. (Prairie falcons and golden eagles in North Dakota.) *J. Wildl. Manage.* 51:739–44.

Allen, H. L., R. Friesz, and R. E. Fitzner. 1985. (Status of ferruginous hawk in Washington.) Proc. Raptor Res. Found. Symp., Management of Birds of Prey, Nov. 3–6, Sacramento (abstract).

Allen, H. L., R. K. Murphy, K. Steenhof, and S. W. Platt. 1986. (Late nesting by prairie falcons.) *Wilson Bull.* 98:463–65.

Allen, R. P., and R. T. Peterson. 1936. (Hawk migration at Cape May.) *Auk* 53:393–404.

Amadon, D. 1975. (Reversed sex dimorphism in raptors.) *Raptor Res.* 9:3–11.

———. 1982. (Revision of sub-buteo hawks.) *Amer. Mus. Novitates* 2741:1–20.

Ambrose, R. E., R. J. Ritchie, P. F. Schempf, T. Swem, and R. Dittrick. 1985. (Status of peregrines in Alaska.) Proc. Raptor Res. Found. Symp., Management of Birds of Prey, Nov. 3–6, Sacramento (abstract).

American Ornithologists' Union (AOU). 1957. *Check-list of North American birds.* 5th ed. Baltimore: American Ornithol. Union. (6th ed., 1983.)

Angell, T. 1969. (Nesting behavior of ferruginous hawk.) *Living Birds* 8:225–41.

Anonymous. 1985. (Endangered status of aplomado falcon.) *Eyas* 8:4–5.

————. 1986. (Raptor population estimates.) Coop. Res. Newsletter, Lab. of Ornithology, Ithaca. 22(36):3.

Anthony, R. G. 1983. (Synopsis of workshop.) Pp. 43–47, in Anthony, Isaacs, and Frenzel, 1983.

Anthony, R. G., F. B. Isaacs, and R. W. Frenzel (eds.). 1983. *Proceedings of a workshop on habitat management for nesting and roosting bald eagles in the western United States.* Corvallis: Coop. Wildl. Res. Unit, Oregon State Univ.

Apanus, V. 1977. (Red-shouldered hawks breeding in immature plumage.) *Auk* 94:585.

Apfelbaum, S. I., and P. Seelbach. 1983. (Nest tree, habitat selection and productivity in seven North American raptors.) *Raptor Res.* 17:99–104.

Arlott, N. 1984. (Aerial display of black-shouldered kite.) *Brit. Birds* 77:22–23.

Asay, C. E. 1980. Habitat, biology and productivity of Cooper's hawks nesting in the oak woodlands of California. M.S. thesis, Univ. Calif., Davis.

————. 1987. (Habitat and productivity in Cooper's hawks.) *Calif. Fish & Game* 73:80–87.

Bailey, A. M., and R. J. Niedrach. 1965. *Birds of Colorado.* 2 vols. Denver: Denver Mus. of Nat. Hist.

Baird, G. (comp.). 1970. Everglade kite (*Rostrhamus sociabilis plumbeus*). USDI Office Library Serv., Bibliogr. Ser. no. 18.

Balfour, E., and C. J. Cadbury. 1979. (Polygyny and spacing in northern harriers.) *Ornis. Scand.* 10:133–41.

Balgooyen, T. G. 1976. (Behavior and ecology of American kestrel.) *Univ. Calif. Publ. Zool.* 103:1–83.

Bangs, E. E., N. Bailey, and V. D. Burns. 1982. (Nesting bald eagles on Kenai Peninsula.) Pp. 47–54, in Ladd and Schempf, 1982.

Banks, R. C., R. M. McDiarmid, and A. L. Gardner. 1987. Checklist of vertebrates of the United States, the U.S. Territories, and Canada. USDI, Fish & Wildl. Serv., Resource Publ. 166.

Barclay, J. H. 1985. (Restoration of peregrine in eastern USA.) Proc. Raptor Res. Found. Symp., Management of Birds of Prey, Nov. 3–6, Sacramento (abstract).

Barnard, P. E. 1982. Foraging behaviour and energetics of breeding northern harriers *Circus cyaneus* (L.). Honors thesis, Acadia Univ., Nova Scotia.

Barnard, P. E., B. MacWhirter, R. Simmons, G. A. Hansen, and P. C. Smith. 1987. (Seasonal importance of passerine prey to northern harriers.) *Can. J. Zool.* 65:1942–46.

Barnard, P. E., and R. Simmons. 1986. (Leg-lowering in raptors.) *Ostrich* 57:107–09.

Barnard, M. J. 1980. Factors affecting nest productivity of Swainson's hawk (*Buteo swainsoni*) in southeastern Washington. Ph.D. diss., Wash. State Univ., Pullman.

————. 1981. (Nesting of ferruginous hawk in Manitoba.) *Can. Field-Nat.* 95:467–69.

Bechard, M. J., and K. D. Hague. 1985. (Nesting of ferruginous hawk in

Idaho.) Proc. Raptor Res. Found. Symp., Management of Birds of Prey, Nov. 3–6, Sacramento (abstract).

Becker, D. M. 1984. Reproductive ecology and habitat utilization of Richardson's merlins in southeastern Montana. M.S. thesis, Univ. Montana, Bozeman.

Becker, D. M., and C. H. Sieg. 1985. (Reproduction of merlins.) *Raptor Res.* 19:52–55.

———. 1987. (Home ranges and habitats of merlins in Montana.) *Can. Field-Nat.* 101:398–403.

Bednarz, J. C. 1987. (Autumn breeding in Harris' hawk.) *Auk* 104:85–96.

———. 1988a. (Cooperative hunting in Harris' hawk.) *Nature* 239:1525–27.

———. 1988b. (Comparative ecology of Harris' and Swainson's hawks.) *Condor* 90:311–23.

Bednarz, J. C., and J. J. Dinsmore. 1981. (Red-shouldered hawks in Iowa.) *J. Wildl. Manage.* 45:236–41.

———. 1982. (Red shouldered and red-tailed hawks in Iowa.) *Wilson Bull.* 94:31–45.

Bednarz, J. C., and T. J. Hayden. 1987. (Cooperative hunting in Harris' hawk.) Proc. Annual Meeting Raptor Res. Found., Oct. 28–31, Boise (abstract).

Bednarz, J. C., and J. D. Ligon. 1988. (Cooperative breeding in Harris' hawk.) *Ecology* 69:1176–87.

Beebe, F. J. 1960. (Peregrines of Pacific Northwest.) *Condor* 62:145–89.

———. 1974. (Falconiformes of British Columbia.) *Brit. Col. Proc. Mus. Occ. Papers* 17:1–163.

Beecham, J. J., and M. N. Kochert. 1975. (Golden eagle breeding biology, Idaho.) *Wilson Bull.* 87:506–13.

Beissinger, S. R. 1983. (Hunting behavior of snail kite.) *Auk* 100:84–92.

———. 1984. Mate desertion and reproductive efforts in the snail kite. Ph.D. diss., Univ. Mich., Ann Arbor.

———. 1986. (Demography and mating behavior of snail kite.) *Ecology* 67:1445–59.

———. 1988. (Mating behavior of snail kite.) *Natural Hist.* 91(1):42–50.

Beissinger, S. R., and S. F. R. Snyder. 1987. (Mate desertion in snail kite.) *Anim. Behav.* 35:477–87.

Beissinger, S. R., A. Sprunt IV, and R. Chandler. 1983. (Snail kite in Cuba.) *Amer. Birds* 37:262–65.

Beissinger, S. R., and J. E. Takekawa. 1983. (Snail kite during Florida drought.) *Fla. Field Nat.* 11:89–106.

Belyea, G. Y. 1976. A nesting study of the red-tailed hawk in southern Michigan. Ph.D. diss., Mich. State Univ., East Lansing.

Bennetts, R. E., M. W. Collopy, and S. R. Beissinger. 1988. (Nesting ecology of snail kite.) Gainesville: Fla. Coop. Fish & Wildl. Res. Unit, Tech. Rept. 31.

Bent, A. C. 1937, 1938. (Life histories of North American birds of prey.) *U.S. Natl. Mus. Bull.* 167:1–398; 170:1–482.

Bente, P. J. 1981. Nesting behavior and hunting activity of the gyrfalcon in south-central Alaska. M.S. thesis, Univ. Alaska, College.

Berger, D. D., and C. H. Mueller. 1970. (Prey preference in sharp-shinned hawk.) *Auk* 87:452–57.

Bergo, G. 1987. (Territoriality in golden eagles.) *Brit. Birds* 80:361–76.

Beske, A. E. 1978. Harrier radio-tagging techniques and local and migratory movements of radio-tagged juvenile harriers. M.S. thesis, Univ. Wisc., Stevens Point.

Bider, J. R., and D. M. Bird. 1983. (Osprey populations in Quebec.) Pp. 223–30, in Bird, 1983.

Bildstein, K. L. 1987. (Behavioral ecology of red-tailed hawk, rough-legged hawk, northern harrier and American kestrel in south-central Ohio.) *Ohio Biol. Surv., Biol. Notes* 18:1–53.

Binford, L. 1979. (Fall raptor migration, California.) *Western Birds* 10:1–16.

Bird, D. M. (ed.). 1983. *Biology and management of bald eagles and ospreys.* Macdonald Raptor Res. Cent., McGill Univ., and Raptor Res. Found. Ste. Anne de Bellevue, Quebec: Harpell Press.

Bird, D. M., and P. C. Lague. 1976. (Captive breeding of rough-legged hawks.) *Raptor Res.* 10:1–5.

Bird, D. M., and J. D. Weaver. 1985. (Peregrine populations in Ungava Bay.) Proc. Raptor Res. Found. Symp., Management of Birds of Prey, Nov. 3–6, Sacramento (abstract).

Black, E. A. 1979. (Possible double-brooding by American kestrel.) *Bull. Okla. Ornith. Soc.* 12:29–30.

Blair, C. L. 1978. Breeding biology and prey selection of the ferruginous hawk in South Dakota. M.S. thesis, Univ. South Dakota, Vermillion.

Blair, C. L., and F. Schitoskey, Jr. 1982. (Ferruginous hawk in South Dakota.) *Wilson Bull.* 94:45–54.

Blake, E. R. 1977. *Manual of neotropical birds.* Vol. 1. Chicago: Univ. Chicago Press.

Bloom, P. H., and S. J. Hawks. 1982. (Food habits of golden eagle.) *Raptor Res.* 16:110–15.

Bloom, P. H., S. J. Hawks, S. W. Janes, R. W. Risebrough, and B. Woodbridge. 1985. (Swainson's hawk in California.) Proc. Raptor Res. Found. Symp., Management of Birds of Prey, Nov. 3–6, Sacramento (abstract).

Bock, C. E., and L. W. Lepthien. 1976. (Geographic ecology of *Buteo* and *Parabuteo.*) *Condor* 78:554–57.

Boeker, E. L., and T. D. Ray. 1971. (Golden eagles in Southwest.) *Condor* 73:463–67.

Bond, R. M. 1939. (Raptors of northern California.) *Condor* 41:54–61.

———. 1946. (Peregrines of western North America.) *Condor* 48:101–16.

Boshoff, A. F., and N. G. Palmer. 1983. (Osprey biology in South Africa.) *Ostrich* 54:189–204.

Bourne, G. R. 1983. Snail kite feeding ecology: Some correlates and tests of optimal foraging. Ph.D. diss., Univ. Mich., Ann Arbor.

Bowman, R. 1987. Sex dimorphism in mated pairs of American kestrels. *Wilson Bull.* 99:465–67.

Bowman, R., and D. M. Bird. 1985. (Replacement clutches of American kestrels.) *Can. J. Zool.* 25:90–93.

Boyce, D. A., Jr. 1985a. (Prairie falcon prey in Mojave Desert.) *Raptor Res.* 19:128–34.

———. 1985b. (Merlins and wintering shorebirds.) *Raptor Res.* 19:94–96.

Boyce, D. A., Jr., R. L. Garrett, and B. J. Walton. 1986. (California prairie falcon populations.) *Raptor Res.* 20:71–74.

Brannion, J. D. 1980. The reproductive ecology of a Texas Harris' hawk (*Parabuteo uncinatus harrisi*) population. M.S. thesis, Univ. Texas, Austin.

Braun, C. E., J. H. Enderson, C. J. Henny, H. Meng, and A. G. Nye, Jr. 1977. (Committee report on falconry's effects on raptors.) *Wilson Bull.* 89:360–69.

Braun, C. E., F. Hamerstrom, and C. M. White. 1975. (Committee report on eagle populations.) *Wilson Bull.* 87:140–43.

Breckenridge, W. J. 1935. (Ecology of northern harriers in Minnesota.) *Condor* 37:266–76.

Brodkorb, P. 1964. Catalogue of fossil birds. Pt. 2. (Anseriformes through Galliformes.) *Bull. Fla. State Mus.* 8:195–335.

Broley, C. L. 1947. (Florida bald eagle biology.) *Wilson Bull.* 59:3–20.

Brown, L. H. 1976a. *British birds of prey.* London: Collins.

————. 1976b. *Birds of prey: Their biology and ecology.* London: Hamlyn.

————. 1976c. *Eagles of the world.* Newton Abbot: David & Charles.

Brown, L. H., and D. Amadon. 1968. *Eagles, hawks and falcons of the world.* 2 vols. New York: McGraw-Hill.

Brown, L. H., E. K. Urban, and K. Newman. 1982. *The birds of Africa.* Vol. 1. New York: Academic Press.

Brown, L. H., and A. Watson. 1964. (Golden eagle and its foods). *Ibis* 106:78–100.

Brown, P. 1979. *The Scottish ospreys: From extinction to survival.* London: Heinemann.

Brown, W. 1971. (Winter populations of red-shouldered hawks.) *Amer. Birds* 25:813–17.

Browning, M. R. 1974. (Wintering Swainson's hawks in North America.) *Am. Birds* 28:865–67.

Bruce, A. M., R. J. Anderson, and G. T. Allen. 1982. (Golden eagles in Washington.) *Raptor Res.* 16:132–33.

Bryant, A. A. 1986. (Selective logging and red-shouldered hawks in Ontario.) *Can. Field-Nat.* 100:520–25.

Buchanan, J. B., C. T. Schick, L. A. Brennan, and S. G. Herman. 1988. (Merlin predation on dunlins.) *Wilson Bull.* 100:8–18.

Bull, J. 1974. *Birds of New York state.* Garden City, N.Y.: Doubleday.

Burke, C. J. 1979. Effect of prey and land use on mating systems of harriers. M.S. thesis, Univ. Wisc., Stevens Point.

Burnham, W. A. 1975. Breeding biology and ecology of the peregrine falcon in Greenland. M.S. thesis, Brigham Young Univ., Provo.

Burns, F. L. 1911. (Broad-winged hawk monograph.) *Wilson Bull.* 23:139–320.

Burrows, D. B. 1917. (Nesting of white-tailed hawk.) *Oologist* 34:78–81.

Cade, T. J. 1955. (Winter territoriality of kestrel.) *Wilson Bull.* 67:5–17.

————. 1960. (Ecology of peregrine and gyrfalcon in Alaska.) *Univ. Calif. Publ. Zool.* 63:151–290.

————. 1982. *The falcons of the world.* Ithaca: Cornell Univ. Press.

Cade, T. J., J. H. Enderson, C. G. Thelander, and C. M. White (eds.). 1988. *Peregrine falcon populations: Their management and recovery.* (Proceedings of the 1985 Peregrine Conference, Management of Birds of Prey Symposium, Sacramento.) Boise: The Peregrine Fund.

Cade, T. J., C. M. White, and J. R. Haugh. 1968. (Peregrine and pesticides in Alaska.) *Condor* 70:170–78.

Calef, G. W., and D. C. Heard. 1979. (Peregrines in the Northwest Territories.) *Auk* 96:662–74.

Caminzind, F. J. 1969. (Nesting ecology of the golden eagle.) *Brigham Young Univ. Sci. Bull. Biol.* 10(4):4–15.

Cash, K. H., P. J. Austin-Smith, D. Banks, D. Harris, and P. C. Smith. 1985. (Bald eagle foods in Nova Scotia.) *J. Wildl. Manage.* 49:223–25.

Cely, J. E. 1979. (Status of swallow-tailed kite.) Pp. 144–50, in D. M. Forsythe and W. B. Ezell (eds.), Proc. 1st. S. Carolina Endangered Species Symposium. Columbia: S. Carol. Wildl. & Marine Res. Dept.

———. 1987. (Swallow-tailed kite nesting in S. Carolina.) *Chat* 51:48–50.

Chancellor, R. D. (ed.). 1977. *Proceedings World Conference on Birds of Prey, Vienna, 1975.* London: Intern. Counc. Bird Preserv.

Chandler, R., and J. M. Anderson. 1974. (Snail kite reproduction.) *Amer. Birds* 28:856–58.

Chrest, H. R. 1965. Nesting of the bald eagle in the Karluk Lake drainage of Kodiak Island, Alaska. M.S. thesis, Colo. State Univ., Fort Collins.

Clark, R. J. 1972. (Northern harrier nesting in Manitoba.) *Blue Jay* 30:43–48.

Clark, W. S. 1984. (Whirling by zone-tailed hawks.) *Condor* 86:488.

———. 1985a. (Swainson's hawk population estimate.) Proc. Raptor Res. Found. Symp., Management of Birds of Prey, Nov. 3–6, Sacramento (abstract).

———. 1985b. (Swainson's hawk in Great Plains.) Proc. Raptor Res. Found. Symp., Management of Birds of Prey, Nov. 3–6, Sacramento (abstract).

———. 1985c. (Migration of merlin in New Jersey.) *Raptor Res.* 19:85–93.

Clark, W. S., and B. K. Wheeler. 1987. *A field guide to the hawks of North America.* Boston: Houghton Mifflin.

Clarke, R. G. 1982. (Nest-selection by sharp-shinned hawks in Alaska.) Pp. 155–62, in Ladd and Schempf, 1982.

———. 1984. The sharp-shinned hawk (*Accipiter striatus* Vieillot) in interior Alaska. M.S. thesis, Univ. Alaska, College.

Cochran, W. W., and R. D. Applegate. 1986. (Flying speeds of merlins and peregrines.) *Condor* 88:397–98.

Cohen, S. F. 1970. The distribution of the western red-shouldered hawk (*Buteo lineatus elegans* Cassin). M.A. thesis, Calif. State College, Long Beach.

Collopy, M. W. 1981. Food consumption and growth energetics of nestling golden eagles. Ph.D. diss., Univ. Mich., Ann Arbor.

———. 1983. (Foraging success of golden eagles.) *Auk* 100:747–49.

———. 1984. (Foraging ecology of golden eagles.) *Auk* 101:753–60.

Collopy, M. W., and T. C. Edwards, Jr. 1989. (Territoriality, activity budget and undulating flight in golden eagle.) *J. Field Ornith.* 60:43–51.

Conner, R. N. 1974. (Aerial courtship of red-tailed hawks.) *Bird-Banding* 45:269.

Cottrell, M. J. 1981. Resource partitioning and reproductive success of three species of hawks (*Buteo* spp.) in an Oregon prairie. M.S. thesis, Oregon State Univ., Corvallis.

Coues, E. 1903. *Key to the North American birds.* 5th ed. Boston: Page Co.

Court, G. S. 1986. Some aspects of the reproductive biology of tundra peregrine falcons. M.S. thesis, Univ. Alberta, Edmonton.

Court, G. S., C. C. Gates, and D. A. Boag. 1988. (Peregrine natural history, Northwest Territories.) *Arctic* 41:17–30.

Craig, E. H., T. H. Craig, and L. R. Powers. 1986. (Golden eagle and rough-legged hawk habitat use in Idaho.) *Raptor Res.* 20:69–71.

Craig, T. H., E. H. Craig, and J. S. Marks. 1982. (Talon-grappling by northern harriers.) *Condor* 84:239.

Craighead, J., and F. Craighead. 1940. (Nesting merlin observations.) *Wilson Bull.* 52:241–48.

———. 1956. *Hawks, owls and wildlife.* Harrisburg: Stackpole Co.

Cramp, S., and K. E. L. Simmons (eds.). 1980. *Handbook of the birds of Europe, the Middle East and North Africa.* Vol. 2. Oxford: Oxford Univ. Press.

Crenshaw, J. G., and B. R. McClelland. 1983. (Bald eagle roosts in Glacier National Park.) Proc. Workshop on Habitat Management for Nesting and

Roosting Bald Eagles in the Western United States, Sept. 7–9, Corvallis (abstract).

Czechura, G. V. 1984. (Peregrines in Queensland.) *Raptor Res.* 18:81–91.

Davies, R. G. 1985. (Population of bald eagles in British Columbia.) Pp. 63–66, in Gerrard and Ingram, 1985.

Dawson, J. W. 1987. (Breeding biology of Harris' hawk.) Proc. Annual Meeting, Raptor Res. Found., Oct. 28–31, Boise (abstract).

———. 1988. (Territoriality and sociality of Harris' Hawk.) Proc. Annual Meeting, Raptor Res. Found., Oct. 27–29, Minneapolis (abstract).

Dekker, D. 1982. (Hunting behavior of prairie falcon.) *Can. Field-Nat.* 96:477–78.

———. 1985. (Hunting behavior of golden eagle.) *Can. Field-Nat.* 99:383–85.

Delnicki, D. 1978. (Hook-billed kite nesting in Texas.) *Auk* 95:427.

Dementiev, G. P. 1960. *Der Gerfalke.* Neue Brehm-Bucherei, No. 264. Wittenberg: A. Ziemsen Verlag.

Dementiev, G. P., and N. Gortschakowkaja. 1945. (Biology of gyrfalcon in Norway.) *Ibis* 87:559–65.

Dennis, R. H. 1983. (Osprey populations in Scotland.) Pp. 207–17, in Bird, 1983.

Denton, S. J. 1975. Status of the prairie falcons breeding in Oregon. M.S. thesis, Oregon State Univ., Corvallis.

Diamond, J. 1985. (Ecological segregation in New Guinea avifauna.) In T. Case and J. M. Diamond (eds.), *Community Ecology.* New York: Harper & Row.

Dickey, D. R., and van Rossem, A. J. 1938. The birds of El Salvador. Field Mus. Natural History, Zool. Ser. No. 23.

Dickson, R. C. 1988. Habitat preferences and prey of merlins in winter. *Brit. Birds* 81:269–74.

Ditto, L. R. 1983. (Nesting white-tailed hawks.) *Raptor Res.* 17:91.

Dixon. J. B. 1937. (Golden eagle in San Diego County.) *Condor* 39:49–56.

Dixon, J. B., and R. M. Bond. 1937. (Raptors of Lava Beds National Monument.) *Condor* 39:97–102.

Dixon, J. B., R. E. Dixon, and J. E. Dixon. 1957. (Black-shouldered kite biology in San Diego County.) *Condor* 59:156–65.

Duke, G. E., A. A. Jaegers, G. Loft, and D. A. Evanson. 1975. (Digestion in raptors.) *Comp. Biochem. Physiol.* 50A:649–56.

Dunkle, S. W. 1977. (Swainson's hawk in Wyoming.) *Auk* 94:65–71.

Dunning, J. B., Jr. 1984. Body weights of 686 species of North American birds. Western Bird Banding Assoc. Monograph No. 1. Cave Creek, Ariz.: Eldon Publ. Co.

Dunstan, T. C. 1973. (Biology of ospreys in Minnesota.) *Loon* 45:108–13.

———. 1974. (Foraging by ospreys in Minnesota.) *Wilson Bull.* 86:74–76.

Dunstan, T. C., and J. F. Harper. 1975. (Bald eagle foods in Minnesota.) *J. Wildl. Manage.* 39:140–43.

Dunstan, T. C., and B. E. Harrell. 1973. (Red-tailed hawks and great horned owls in South Dakota.) *Raptor Res.* 7:49–54.

Earle, W. L. 1984. (Nesting prairie falcons in California.) *Raptor Res.* 10:31–33.

Early, T. 1982. (Breeding raptors of the Aleutians.) Pp. 99–111, in Ladd and Schempf, 1982.

Edwards, C. C. 1969. Winter behavior and population dynamics of American eagles in western Utah. Ph.D. diss., Brigham Young Univ., Provo.

Edwards, T. C., Jr. 1987. Temporal and seasonal aspects of the foraging ecology of a piscivore, the osprey (*Pandion haliaetus*.) Ph.D. diss., Univ. Fla., Gainesville.

Edwards, T. C., Jr., and M. N. Kochert. 1986. (Predictors of sex in golden eagles.) *J. Field Ornithol.* 57:317–19.

Eisenmann, E. 1971. (Range expansion of black-shouldered kite.) *Amer. Birds* 25:529–36.

Ellis, D. H. 1976. (First Montana breeding record for merlin.) *Condor* 78:112–14.

————. 1979. (Development of golden eagle behavior.) *Wildl. Monogr.* 70:1–94.

————. 1985. (Peregrine management in Arizona.) Proc. Raptor Res. Found. Symp., Management of Birds of Prey, Nov. 3–6, Sacramento (abstract).

Ellis, D. H., and L. Powers. 1982. (Mating behavior of golden eagle.) *Raptor Res.* 16:134–36.

Enderson, J. H. 1960. (American kestrel population study.) *Wilson Bull.* 72:222–31.

————. 1964. (Prairie falcon population study, central Rockies.) *Auk* 81:332–52.

————. 1969. (Peregrine and prairie falcon life tables.) Pp. 505–08, in Hickey, 1969.

Enderson, J. H., G. R. Craig, and D. D. Berger. 1985. (Peregrine status in Rockies and Colorado Plateau.) Proc. Raptor Res. Found. Symp., Management of Birds of Prey, Nov. 3–6, Sacramento (abstract).

Enderson, J. H., and M. N. Kirven. 1983. (Telemetry tracking of peregrines.) *Raptor Res.* 17:33–37.

Eng, R. L., and G. W. Gullion. 1962. (Goshawk predation on ruffed grouse.) *Wilson Bull.* 74:227–42.

Ensign, J. T. 1983. Nest site selection, productivity and food habitats of ferruginous hawks in southeastern Montana. M.S. thesis, Montana State Univ., Missoula.

Errington, P. L., and W. J. Breckenridge. 1936. (Northern harrier food habits, north-central states.) *Am. Midl. Nat.* 17:831–48.

————. 1938. (Buteo food habits, north-central states.) *Wilson Bull.* 50:113–21.

Evans, D. L. 1982. Status reports on twelve raptors. USDI, Fish and Wildl. Serv., Wildlife 238.

Evans, D. L., and R. N. Rosenfield. 1985. (Minnesota-banded sharp-shinned hawk recoveries.) Pp. 311–27, in Newton and Chancellor, 1985.

Evans, S. A. 1981. Ecology and behavior of the Mississippi kite in southern Illinois. M.S. thesis, Southern Ill. Univ., Carbondale.

Faaborg, J. 1986. (Cooperative polyandry in Galapagos hawk.) *Ibis* 128:337–47.

————. 1988. *Ornithology: An ecological approach.* Englewood Cliffs, N.J.: Prentice Hall.

Falk, K., and S. Moller. 1985. (Peregrine reproduction in Greenland.) Proc. Raptor Res. Found. Symp., Management of Birds of Prey, Nov. 3–6, Sacramento (abstract).

Farquahar, C. 1986. Ecology and breeding behavior of the white-tailed hawk on the northern coastal prairies of Texas. Ph.D. diss., Texas A&M Univ., College Station.

Fast, A. H., and L. H. Barnes. 1950. (American kestrel behavior.) *Wilson Bull.* 62:38.

Feduccia, A. 1980. *The age of birds.* Cambridge: Harvard Univ. Press.

Feldsine, J. W., and L. W. Oliphant. 1985. (Courtship behavior of merlin.) *Raptor Res.* 19:60–67.

Fielder, P. C. 1982. (Bald eagle foods in Washington.) *Murrelet* 63:46–50.

Fischer, D. L. 1982. The seasonal abundance, habitat use and foraging behavior of wintering bald eagles *Haliaeetus leucocephalus* in west-central Illinois. M.S. thesis, Western Ill. Univ., Macomb.

———. 1984. (Breeding of sharp-shinned hawks in immature plumage.) *Raptor Res.* 18:155–56.

Fischer, D. L., K. L. Ellis, and R. J. Meese. 1984. (Raptor habitat selection in Utah.) *Raptor Res.* 19:98–102.

Fisher, A. K. 1893. Hawks and owls of the United States in their relation to agriculture. *Bull. U.S. Dept. Agr., Div. Ornithol. & Mammal.* 3:1–210.

Fitch, H. S. 1958. (Home ranges, territories and seasonal movements of vertebrates in Kansas.) *Univ. Kansas Publ., Mus. Nat. Hist.* 11:63–226.

———. 1963. (Mississippi kite in Kansas.) *Univ. Kansas Publ., Mus. Nat. Hist.* 12:505–19.

———. 1974. (Foods of nesting broad-winged hawks, Kansas.) *Condor* 76:331–33.

Fitch, H. S., B. Glading, and V. House. 1946. (Cooper's hawk nesting and predation.) *Calif. Fish & Game* 32:144–54.

Fitch, H. S., F. Swenson, and D. F. Tillotson. 1946. (Food habits of red-tailed hawk.) *Condor* 48:205–37.

Fitzner, R. E. 1978. The ecology and behavior of the Swainson's hawk (*Buteo swainsoni*) in southeastern Washington. Ph.D. diss., Washington State Univ., Pullman.

Fleetwood, R. J. 1967. (Nesting of the hook-billed kite in Texas.) *Auk* 87:598–601.

Fox, G. A. 1964. (Notes on western merlins.) *Blue Jay* 22:140–47.

———. 1971. (Reproductive success of merlins.) *J. Wildl. Manage.* 35:122–28.

Fox, N. C. 1977. The biology of the New Zealand falcon (*Falco novaeseelandiae* Gmelin 1788). Ph.D. diss., Univ. Canterbury, New Zealand.

Fox, R., S. W. Lehmkuhle, and D. H. Westendorf. 1976. Falcon visual acuity. *Science* 192:263–65.

Fraga, R. M., and S. A. Salvador. 1986. (Reproductive biology of the chimango.) *El Hornero* 12:223–29.

Frenzel, R. 1983. (Nest-site spacing of bald eagles.) P. 18, in Anthony, Isaacs, and Frenzel, 1983 (abstract).

Friedmann, H. F. 1950. The birds of North and Middle America. Pt. XI. *U.S. Natl. Mus. Bull.* 50:1–793.

Fyfe, R. W. 1985. (Canadian peregrine populations.) Proc. Raptor Res. Found. Symp., Management of Birds of Prey, Nov. 3–6, Sacramento (abstract).

Fyfe, R. W., J. Campbell, B. Hayson, and K. Hodson. 1969. (Peregrine declines and pesticides.) *Can. Field.-Nat.* 83:191–200.

Fyfe, R. W., S. A. Temple, and T. J. Cade. 1976. (North American peregrine survey.) *Can. Field-Nat.* 90:228–73.

Garber, D. P. 1972. Osprey nesting ecology in Lassen and Plumas counties, California. M.S. thesis, Humboldt State Univ., Arcata.

Gates, J. M. 1972. (Red-tailed hawk ecology, Wisconsin.) *Wilson Bull.* 84:421–33.

Gatz, T. A., M. D. Jakle, R. L. Glinski, and G. Monson. 1985. (Black-shouldered kites in Arizona.) *Western Birds* 16:57–61.

Gerrard, J. M. 1983. (Bald eagle status in North America. Pp. 5–21, in Bird, 1983.

———. 1985. (Status of bald eagles in Saskatchewan.) Pp. 58–62, in Gerrard and Ingram, 1985.

Gerrard, J. M., and T. N. Ingram (eds.). 1985. *The bald eagle in Canada: Proceedings of bald eagle days, 1983.* Apple River, Ill.: The Eagle Foundation.

Gerrard, P. N., S. N. Wiemeyer, and J. M. Gerrard. 1979. (Bald eagle nesting behavior.) *Raptor Res.* 13:57–64.

Gifford, J. L. 1985. (Common black-hawk in Utah.) *Utah Birds* 1:43–50.

Gilmer, D. S., P. M. Konrad, and R. E. Stewart. 1983. (Red-tailed hawks and great horned owls in North Dakota.) *Prairie Nat.* 15:133–43.

Gilmer, D. S., and R. E. Stewart. 1983. (Ferruginous hawk ecology, North Dakota.) *J. Wildl. Manage.* 47:146–57.

―――. 1984. (Swainson's hawk nesting biology, North Dakota.) *Condor* 86:12–18.

Glazener, W. C. 1964. (Caracara food habits.) *Condor* 66:162.

Glinski, R. L. 1985a. (Ferruginous hawk status, Arizona.) Proc. Raptor Res. Found. Symp., Management of Birds of Prey, Nov. 3–6, Sacramento (abstract).

―――. 1985b. (Swainson's hawk status, Arizona.) Proc. Raptor Res. Found. Symp., Management of Birds of Prey, Nov. 3–6, Sacramento (abstract).

―――. 1985c. Southwest. *Eyas* 8:6.

―――. 1988. Southwest. *Eyas* 11:5–7.

Glinski, R. L., and R. D. Ohmart. 1983. (Mississippi kite ecology, Arizona.) *Condor* 85:200–07.

Glutz von Blotzheim, U., K. M. Bauer, and E. Bezzel. 1971. *Handbuch der Vögel Mitteleuropas.* Bd. 4. Frankfurt: Akademische Verlagsgesellschaft.

Godfrey, R. D., Jr., and A. M. Fedynich. 1987. (Harrier predation on waterfowl.) *J. Raptor Res.* 21:72–73.

Godfrey, W. E. 1970. *The birds of Canada.* Ottawa: National Museums of Canada. (Rev. ed., 1986.)

Goslow, G. E. 1971. (Attack behavior of raptors.) *Auk* 88:15–27.

Grassé, P.-P. (ed.). 1950. *Traité de Zoologie.* XV (Oiseaux). Paris: Masson et Cie.

Green, N. F. 1983a. (Bald eagle status, United States.) P. 9, in Anthony, Isaacs, and Frenzel, 1983 (abstract).

―――. 1983b. (Bald eagle status, United States.) Pp. 89–97, in T. N. Ingram (ed.), *Bald eagle restoration.* Apple River, Ill.: Eagle Valley Environmentalists.

―――. 1985. (Bald eagle.) Pp. 509–31, in R. L. DiSilvestro (ed.), *Audubon Wildlife Report, 1985.* New York: National Audubon Society.

Green, R. 1976. (Osprey breeding behavior, Scotland.) *Ibis* 118:475–90.

Griffin, C. R. 1976. (Texas and Arizona Harris' hawk populations.) *Raptor Res.* 10:50–54.

―――. 1987. (Galapagos and Hawaiian hawk breeding strategies.) Proc. Annual Meeting, Raptor Res. Found., Oct. 28–31, Boise (abstract).

Grossman, M. L., and H. Hamlet. 1964. *Birds of prey of the world.* New York: Bonanza Books.

Grubb, T. G., and R. J. Hensel. 1978. (Bald eagle food habits, Kodiak Island.) *Murrelet* 59:70–72.

Grubb, T. G., R. L. Knight, D. M. Rubink, and C. H. Nash. 1983. (Productivity of bald eagles, Washington.) Pp. 35–45, in Bird, 1983.

Haak, B. A. 1982. Foraging ecology of prairie falcons in northern California. M.S. thesis, Oregon State Univ., Corvallis.

Haines, S. 1986. The feeding, roosting and perching behavior of the bald eagle (*Haliaeetus leucocephalus*) of Mason Neck, Virginia with special reference to the development of Mason Neck Park. M.S. thesis, George Mason Univ., Fairfax, Va.

Hall, P. 1984. Characterization of nesting habitat of goshawks (*Accipiter gentilis*) in northwestern California. M.S. thesis, Humboldt State Univ., Arcata.

Hall, R. S. 1985. (Ferruginous hawk breeding ecology, Arizona.) Proc. Raptor Res. Found. Symp., Management of Birds of Prey, Nov. 3–6, Sacramento (abstract).

Hamerstrom, F. 1969. A harrier population study. Pp. 367–85, in Hickey, 1969.

———. 1986. *Harrier, hawk of the marshes*. Washington: Smithsonian Inst. Press.

Hamerstrom, F., and F. Hamerstrom. 1978. (Harris' hawk sex characteristics.) *Raptor Res.* 12:1–14.

Hamerstrom, F., F. N. Hamerstrom, and C. J. Burke. 1985. (Harrier mating systems and voles.) *Wilson Bull.* 97:332–46.

Hamerstrom, F., and M. Kopeny. 1981. Harrier nest-site vegetation. *Raptor Res.* 15:86–88.

Hamerstrom, F., and D. D. L. Wilde. 1973. (Harrier roosts and cruising ranges.) *Inl. Bird Banding News* 45:123–27.

Hammond, M. C., and C. J. Henry. 1949. (Harrier nesting success, North Dakota.) *Auk* 66:271–74.

Hancock, D. 1964. (Wintering bald eagles, British Columbia.) *Wilson Bull.* 76:111–20.

Hansen, A. J. 1987. (Bald eagle reproduction, Alaska.) *Ecol.* 68:1387–92.

Hansen, A. J., and J. I. Hodges, Jr. 1985. (Non-breeding bald eagles, Alaska.) *J. Wildl. Manage.* 49:454–58.

Hardaswick, V. J., D. G. Smith, and T. J. Cade. 1984. (Peregrine × prairie falcon hybrid development.) *N. Am. Falc. Assoc. J.* 23:13–21.

Hardin, M. E., J. W. Hardin, and W. D. Klimstra. 1977. (Mississippi kite nesting, Illinois.) *Trans. Ill. State Acad. Sci.* 70:341–48.

Hardin, M. E., and W. D. Klimstra. 1976. An annotated bibliography of the Mississippi kite, *Ictinia mississippiensis*. Carbondale: Coop. Wildl. Res. Unit, Southern Ill. Univ.

Harmata, A. R. 1982. (Function of undulating flight in golden eagles.) *Raptor Res.* 16:103–09.

Harris, J. T. 1979. *The peregrine falcon in Greenland: Observing an endangered species*. Columbia: Univ. Missouri Press.

Hartman, F. A. 1961. Locomotor mechanisms in birds. *Smith. Inst. Misc. Coll.* 143:1–91.

Hatch, D. R. M. 1968. Golden eagle hunting tactics. *Blue Jay* 26:78–80.

Haugh, J. R. 1972. (Hawk migration, eastern N. America.) *Search* 2(16):1–60.

Haugh, J. R., and T. J. Cade. 1966. (Spring hawk migration, Lake Ontario.) *Wilson. Bull.* 78:88–110.

Haukioja, E., and M. Haukioja. 1970. (Mortality rates of Finnish and Swedish goshawks.) *Finnish Game Res.* 31:13–20.

Haverschmidt, F. 1962. (Foods and foraging, Surinam hawks.) *Condor* 64:154–58.

———. 1964. (Hook-billed kite observations, Surinam.) *J. f. Ornithologie* 105:64–66. (In German.)

———. 1965. (Eggs of hook-billed kite.) *J. f. Ornithologie* 106:223.

———. 1968. *Birds of Surinam*. Edinburgh: Oliver & Boyd.

———. 1970. (Snail kite notes, Surinam.) *Auk* 87:580–84.

Hawbecker, A. C. 1942. (Life history, black-shouldered kite.) *Condor* 44: 267–76.

Haywood, D. R., and R. D. Ohmart. 1983. (Bald eagle breeding habitat, Arizona.) Pp. 87–94, in Bird, 1983.

Hecht, W. R. 1951. (Northern harrier nesting, Manitoba.) *Wilson Bull.* 63:167–76.

Hector, D. P. 1980. (Aplomado falcon notes.) *Birding* 12:93–102.

———. 1981. The habitat, diet and foraging behavior of the aplomado falcon, *Falco femoralis* (Temminck). M.S. thesis, Okla. State Univ., Stillwater.

———. 1985. (Diet of aplomado falcon, Mexico.) *Condor* 87:336–42.

———. 1987. (Decline of aplomado falcon, U.S.A.) *Amer. Birds* 41:381–89.

Heintzelman, D. S. 1964. (American kestrel food habits.) *Wilson Bull.* 76:323–30.

———. 1986. *The migrations of hawks.* Bloomington: Indiana Univ. Press.

Heintzelman, D. S., and A. C. Nagy. 1968. (American kestrel reproduction, Pennsylvania.) *Wilson Bull.* 80:306–11.

Heinzmann, G. 1970. (Caracara survey, Florida.) *Fla. Nat.* 43:149.

Henderson, C. 1988. Midwest. *Eyas* 11:8–10.

Hennessy, S. P. 1978. Ecological relationships of accipiters in northern Utah, with special emphasis on the effects of human disturbance. M.S. thesis, Utah State Univ., Logan.

Henny, C. J. 1972. An analysis of the population dynamics of selected avian species. Bur. Sport Fisheries & Wildlife, Res. Rept. 1. Washington: Govt. Printing Office.

———. 1977. (Status of osprey in North America.) Pp. 199–222, in Chancellor, 1977.

———. 1983. (Osprey distribution, United States.) Pp. 175–86, in Bird, 1983.

———. 1986. Osprey (*Pandion haliaetus*). Washington: Environmental Impact Res. Rept., U.S. Army Corps of Engineers Tech. Rpt. EL-86-5.

Henny, C. J., and D. W. Anderson. 1979. (Osprey distribution, Baja California.) *Bull. S. Calif. Acad. Sci.* 78:89–106.

Henny, C. J., and J. T. Annear. 1978. (Black-shouldered kite breeding, Oregon.) *Western Birds* 9:131–33.

Henny, C. J., and J. E. Cornely. 1985. (Red-shouldered hawk distribution, Oregon.) *Murrelet* 66:29–31.

Henny, C. J., and M. W. Nelson. 1981. (Peregrine falcon status, Oregon.) *Murrelet* 62:43–53.

Henny, C. J., F. C. Schmid, E. L. Martin, and L. L. Hood. 1973. (Red-shouldered hawk population ecology, Maryland.) *Ecology* 54:545–54.

Henny, C. J., and H. M. Wight. 1969. (Osprey mortality and productivity.) *Auk* 86:188–98.

———. 1972. (Population ecology of red-tailed and Cooper's hawks.) Pp. 229–50, in Population ecology of migratory birds. USDI, Fish & Wild. Serv., Symposium Vol., Patuxent Wildl. Res. Center, Laurel, Md.

Henzel, R. J., and W. A. Troyer. 1964. (Bald eagle nesting, Alaska.) *Condor* 66:282–86.

Herbert, R. A., and K. G. S. Herbert. 1965. (Peregrine behavior, New York vicinity.) *Auk* 82:92–94.

Heredia, B., and W. S. Clark. 1984. (White-tailed hawk kleptoparasitism on black-shouldered kites.) *Raptor Res.* 18:30–31.

Herman, S. G. 1971. (Peregrine status in California.) *Amer. Birds* 25:818–20.

Herrick, F. H. 1934. *The American eagle.* New York: Appleton-Century Co.

Herron, G. B. 1985. (Ferruginous hawk status, Nevada.) Proc. Raptor Res. Found. Symp., Management of Birds of Prey, Nov. 3–6, Sacramento (abstract).

Hickey, J. J. 1942. (Eastern peregrine populations.) *Auk* 59:176–204.

——— (ed.). 1969. *Peregrine falcon populations, their biology and decline.* Madison: Univ. Wisconsin Press.

———. 1985. (Peregrine population crash, eastern N. America.) Proc. Raptor Res. Found. Symp., Management of Birds of Prey, Nov. 3–6, Sacramento (abstract).

Hickey, J. J., and D. W. Anderson. 1969. (Peregrine life history and literature.) Pp. 3–42, in Hickey, 1969.

Hobbie, J. E., and T. J. Cade. 1962. (Golden eagle breeding, Alaska.) *Condor* 64:235–37.

Hodges, J. I., J. G. King, and R. Davies. 1983. (Bald eagle survey, British Columbia.) P. 321, in Bird, 1983 (abstract).

———. 1984. (Bald eagle survey, British Columbia.) *J. Wildl. Manage.* 48:993–98.

Hodges, J. I., and F. C. Robards. 1982. (Observations on 3850 bald eagle nests, Alaska.) Pp. 37–46, in Ladd and Schempf, 1982.

Hodson, K. 1975. Some aspects of the nesting ecology of Richardson's merlin (*Falco columbarius richardsonii*) on the Canadian prairies. M.S. thesis, Univ. Brit. Columbia, Vancouver.

———. 1978. (Merlin prey, Alberta.) *Can. Field-Nat.* 92:76–77.

———. 1980. Peregrine falcons in British Columbia. Pp. 85–87, in Threatened and endangered species and habitats in British Columbia and the Yukon, ed. R. Stace-Smith, L. Johns, and P. Joslin. Victoria: B. C. Ministry of Environment.

Hoffman, M. L., and M. W. Collopy. 1988. (American kestrel status, Florida.) *Wilson Bull.* 100:91–107.

Holthuijzen, A. M. A., A. R. Ansell, M. N. Kochert, L. S. Young, and R. D. Williams. 1987. (Territoriality in prairie falcons.) Proc. Ann. Meeting, Raptor Res. Found., Oct. 28–31, Boise (abstract).

Holthuijzen, A. M. A., L. Oostenhuis, and M. R. Fuller. 1985. (Habitat use by migrating sharp-shinned hawks, New Jersey.) Pp. 317–27, in Newton and Chancellor, 1985.

Houston, C. S. 1974. (South American recoveries of banded Swainson's hawks.) *Blue Jay* 32:156–57.

———. 1975. (Red-tailed hawk and great horned owl nest proximity.) *Auk* 92:612–14.

———. 1987. (Ferruginous hawk nesting success, Saskatchewan.) Proc. Ann. Meeting, Raptor Res. Found., Oct. 28–31, Boise (abstract).

Houston, C. S., and M. J. Bechard. 1983. (Red-tailed hawk distribution, Saskatchewan.) *Blue Jay* 41:99–109.

———. 1985. (Ferruginous hawk decline, Saskatchewan.) Proc. Raptor Res. Found. Symp., Management of Birds of Prey, Nov. 3–6, Sacramento (abstract).

Howard, R. P. 1975. Breeding ecology of the ferruginous hawk in northern Utah and southern Idaho. M.S. thesis, Utah State Univ., Logan.

———. 1980. (Artificial nest structures for grassland raptors.) Pp. 117–23, in R. P. Howard and J. F. Gore (eds.), Workshop on Raptors and Energy Developments. Boise: Bonneville Power Admin., Idaho Power Service, U.S. Fish & Wildlife Service, and Idaho Chapter, Wildlife Society.

Howell, J., B. Smith, J. B. Holt, and D. R. Osborne. 1978. (Red-tailed hawk habitat and productivity.) *Bird-Banding* 49:162–71.

Howell, T. R. 1972. (Birds of pine savannas, Nicaragua.) *Condor* 74:316–40.

Hubbard, J. P. 1965. (Black-hawk status.) *Amer. Field Notes* 19:474.

———. 1974a. (Gray hawk status, New Mexico.) *Auk* 91:163–66.

———. 1974b. (Flight display of zone-tailed hawk.) *Condor* 76:214–15.

Hubbard, J. P., J. W. Shipman, and S. O. Williams III. 1988. (Roadside counts of raptors in New Mexico.) Pp. 204–09, in R. L. Glinski et al. (eds.), Proceedings Southwest Raptor Management Symposium and Workshop. Washington, D.C.: National Wildlife Federation Sci. & Tech. Series No. 11.

Huey, L. M. 1962. (Weight-lifting by golden eagle.) *Auk* 79:485.

Hughes, J. H. 1982. (Bald eagles on Stikine River, Alaska.) P. 82, in Ladd and Schempf, 1982 (abstract).

Humphrey, P. S., D. Bridge, P. W. Reynolds, and R. T. Peterson. 1970. *Birds of Isla Grande (Tierra del Fuego.)* Washington: Smithsonian Inst.

Hundertmark, C. A. 1974. (Gray hawk in New Mexico.) *Wilson Bull.* 86:298–300.

Hunt, W. G., J. H. Enderson, B. S. Johnson, D. Lanning, and D. Ukrain. (Nesting peregrines, Chihuahuan desert.) Proc. Raptor Res. Found. Symp., Management of Birds of Prey, Nov. 3–6, Sacramento (abstract).

Hurdle, M. T. 1974. (Apple snail and snail kite ecology.) *Proc. 27th Ann. Conf. S.E. Assoc. Game & Fish Comm.*, pp. 215–24.

Husain, K. Z. 1959. (Taxonomy and zoogeography of *Elanus*.) *Condor* 61:153–54.

Imler, R. H., and E. R. Kalmbach. 1955. The bald eagle and its economic status. Washington: USDI, Fish & Wildl. Serv. Circular 30.

Ingram, T. N. (ed.). 1979. *Wintering eagles.* Proceedings of bald eagle days. Apple River, Ill.: Eagle Valley Environmentalists.

———. 1983. *Bald eagle restoration.* Proceedings of bald eagle days, Rochester, N.Y., 1982. Apple River, Ill.: Eagle Valley Environmentalists.

Jacobs, J. P., E. A. Jacobs, and T. C. Erdman. 1988. (Nesting ecology of red-shouldered hawks in Wisconsin.) Proc. Annual Meeting, Raptor Res. Found., Oct. 27–29, Minneapolis (abstract).

Jaksic, F. M. 1985. (Raptor community ecology.) *Raptor Res.* 19:107–12.

James, P. C. 1985. (Breeding biology of merlins.) Proc. Raptor Res. Found. Symp., Management of Birds of Prey, Nov. 3–6, Sacramento (abstract).

James, P. C., and A. R. Smith. 1987. (Food habits of urban merlins in Alberta.) *Canad. Field-Nat.* 101:592–94.

Janes, S. W. 1984a. (Red-tailed hawk breeding ecology.) *Ecology* 65:862–70.

———. 1984b. (Red-tailed hawk breeding territory fidelity.) *Condor* 86:200–03.

———. 1987. (Status of Swainson's hawk in Oregon.) *Oregon Birds* 13:165–79.

Janik, C. A., and J. A. Mosher. 1982. (Breeding biology of raptors in Appalachians.) *Raptor Res.* 16:18–24.

Jehl, J. R., Jr., and B. G. Murray, Jr. 1983. The evolution of normal and reversed sexual size dimorphism in shorebirds and other birds. Pp. 1–86, in *Current Ornithology* 3, R. F. Johnson, ed. New York: Plenum Press.

Jenkins, M. A. 1978. (Gyrfalcon nesting behavior.) *Auk* 95:122–27.

Jenkins, M. A., and R. A. Joseph. 1984. ("Triplets" in golden eagles.) *Raptor Res.* 18:111–13.

Johnsgard, P. A. 1983. *The hummingbirds of North America.* Washington: Smithsonian Inst. Press.

———. 1988. *North American owls: Biology and natural history.* Washington: Smithsonian Inst. Press.

Johnson, A. W. 1965. *The birds of Chile and adjacent regions of Argentina, Bolivia and Peru.* Vol. 1. Buenos Aires: Platt Est. Graf. S.A.

Johnson, C. G., L. A. Nickerson, and M. J. Bechard. 1987. (Grasshoppers and non-breeding Swainson's hawks.) *Condor* 89:676–78.

Johnson, N. K., and H. J. Peeters. 1963. (Systematics of *Buteo* hawks.) *Auk* 80:417–46.

Johnson, S. J. 1986. (Development of hunting behavior in red-tailed hawks.) *Raptor Res.* 20:29–34.

Johnson, W. J., and J. A. Coble. 1967. (Food habits of merlins.) *Jack-Pine Warbler* 45:97–98.

Jollie, M. 1947. Plumage changes in the golden eagle. *Auk* 64:549–76.

———. 1953. Are the Falconiformes a monophyletic group? *Ibis* 95:369–71.

———. 1976, 1977. (Contribution to the morphology and phylogeny of the Falconiformes.) *Evol. Theory* 1:285–98; 2:115–300; 3:201–342.

Kale, H. W., II (ed.). 1978. *Rare and endangered biota of Florida. II. Birds.* Gainesville: Univ. Presses of Florida.

Kalla, P. I., and F. J. Alsop III. 1983. (Mississippi kite in Tennessee.) *Amer. Birds* 37:146–49.

Kalmbach, E. R., R. H. Imler, and L. W. Arnold. 1964. The American eagles and their economic status. Washington: USDI, Fish & Wildl. Serv.

Keir, J. R., and D. D. L. Wilde. 1976. (Swainson's hawks nesting in Illinois.) *Wilson Bull.* 88:658–59.

Keister, G. P., Jr., and R. G. Anthony. 1983. (Bald eagle roost site characteristics.) P. 12, in Anthony, Isaacs, and Frenzel, 1983 (abstract).

Kennedy, P. I., and D. R. Johnson. 1986. (Prey size in nesting Cooper's hawks.) *Wilson Bull.* 98:110–15.

Kenward, R. E. 1982. (Goshawk hunting behavior and home range.) *J. Anim. Ecol.* 51:69–80.

Keran, D. 1978. (Broad-winged hawk nest-site selection, Minnesota.) *Raptor Res.* 12:15–20.

Kerlinger, P., and S. A. Gauthreaux, Jr. 1985a. (Spring raptor migration behavior.) *J. Field Ornith.* 56:394–407.

———. 1985b. (Broad-winged hawk migratory behavior.) *Auk* 102:735–43.

Kiff, L. F. 1981. (Eggs of hook-billed kite.) *Bull. Brit. Orn. Club* 101:318–23.

Kilham, L. 1979. (Caracara courtship behavior.) *Raptor Res.* 13:17–19.

———. 1980. (Swallow-tailed kite pre-nesting behavior.) *Raptor Res.* 14:29–31.

———. 1981. (Whirling by red-shouldered hawks.) *Raptor Res.* 15:123–24.

Kiltie, R. A. 1987. (Red-tailed and red-shouldered hawk abundance, Florida.) *Fla. Field Nat.* 15:45–51.

King, B. 1982. (Talon-clasping by bald eagles.) *Fla. Field Nat.* 10:19.

King, R. L. 1987. (Nesting of black-shouldered kites in Florida.) *Fla. Field Nat.* 15:106–07.

King, W. B. (comp.). 1981. *Endangered birds of the world. The ICBP bird red data book.* Washington: Smithsonian Inst. Press.

Kirkley, J. S., and M. A. Springer. 1980. (Nesting red-tailed hawks and great horned owls in Ohio.) *Raptor Res.* 14:22–28.

Knapton, R. W., and C. A. Sanderson. 1985. (Foods of merlins at Churchill.) *Can. Field-Nat.* 99:375–77.

Knight, R. L., G. E. Allen, M. V. Stalmaster, and C. W. Servheen (eds.). 1980. Proceedings of the Washington Bald Eagle Symposium. Seattle: Nature Conservancy.

Knight, R. L., V. Marr, and S. K. Knight. 1983. (Roosting of bald eagles in Washington.) P. 11, in Anthony, Isaacs, and Frenzel, 1983 (abstract).

Konrad, P. M., and D. S. Gilmer. 1986. (Ferruginous hawk post-fledging behavior.) *Raptor Res.* 20:35–39.

Koonz, W. 1985. (Status of bald eagle in Manitoba.) Pp. 55–57, in *Bald eagle days, 1983.* Apple River, Ill.: Eagle Valley Environmentalists.

Kopeny, M. 1988. Effect of thornbrush on distribution and nest site selection of white-tailed hawks (*Buteo albicaudatus*). Ph.D. diss., North Dak. State Univ., Fargo.

Koplin, J. R. 1973. (Sexual differences in habitat use, American kestrel.) *Raptor Res.* 7:39–42.

Kushlan, J. A., and O. L. Bass, Jr. 1983. (Florida osprey populations.) Pp. 187–200, in Bird, 1983.

Kuyt, E. 1967. (Banding returns of golden eagle and peregrine.) *Bird-banding* 38:78–79.

———. 1980. (Breeding biology of raptors, Northwest Territories.) *Can. Field-Nat.* 94:121–30.

Ladd, W. N., and P. F. Schempf (eds.). 1982. Proceedings of a symposium and workshop: Raptor management and biology in Alaska and western Canada, February 17–20, 1981, Anchorage. USDI, Fish & Wildl. Serv., FWS/AK/PROC-82.

Laing, K. 1985. (Biology of merlins in Alaska.) *Raptor Res.* 19:42–51.

Larson, D. 1980. (Black-shouldered kites in California and Texas.) *Amer. Birds* 34:689–90.

Lawrence, L. de K. 1949. (Merlin nesting in Ontario.) *Wilson Bull.* 61:15–25.

Layne, J. N. 1982. (Caracara in Florida.) ENFO Report (Environmental Information Center, Florida Conservation Foundation, Winter Park), pp. 10–12.

Layne, J. N., F. E. Lohrer, and C. E. Winegarner. 1977. (Predators of cattle egrets in Florida.) *Florida Field Nat.* 5:1–4.

Leedy, R. R. 1972. The status of the prairie falcon in western Montana: special emphasis on possible effects of chlorinated hydrocarbon insecticides, Montana. M.S. thesis, Univ. Montana, Bozeman.

LeFranc, M. N., Jr., and W. S. Clark. 1983. *Working bibliography of the golden eagle and the genus* Aquila. Washington: National Wildlife Federation.

LeFranc, M. N., Jr., and K. W. Cline. 1983. (Raptors at active bald eagle nests.) Pp. 79–86, in Bird, 1983.

LeFranc, M. N., Jr., and B. A. Millsap. 1984. (State and federal raptor programs.) *Wildl. Soc. Bull.* 12:274–82.

Leighton, F. A., J. M. Gerrard, P. Gerrard, D. W. A. Whitfield, and W. J. Maher. 1979. (Census of bald eagles in Saskatchewan.) *J. Wildl. Manage.* 43:61–69.

Lincer, J. L., W. C. Clark, and M. N. LeFranc, Jr. 1979. *Working bibliography of the bald eagle.* Washington: National Wildlife Federation.

Linden, H., and M. Wikman. 1983. (Goshawk predation on tetraonids.) *Jr. Anim. Ecol.* 52:953–68.

Linner, S. 1980. Resource partitioning in breeding populations of marsh hawks and short-eared owls. M.S. thesis, Utah State Univ., Logan.

Lish, J. W., and W. G. Voelker. 1986. (Field identification of red-tailed hawk subspecies.) *Amer. Birds* 40:197–202.

Lockhart, J. M., and G. R. Craig. 1985. (Ferruginous hawk status in Colorado.) Proc. Raptor Res. Found. Symp., Management of Birds of Prey, Nov. 3–6, Sacramento (abstract).

Lokemoen, J. T., and H. P. Duebbert. 1976. (Ferruginous hawk ecology, North Dakota.) *Condor* 78:464–70.

Luttich, S. N., L. B. Keith, and J. D. Stephenson. 1971. (Population dynamics of red-tailed hawk, Alberta.) *Auk* 88:75–87.

Luttich, S., D. H. Rusch, C. Meslow, and L. B. Keith. 1970. (Ecology of red-tailed hawks, Alberta.) *Ecology* 51:190–203.

Lyons, D. M., and J. A. Mosher. 1987. (Broad-winged hawk development.) *J. Field Ornith.* 58:334–44.

Lyons, D. M., K. Titus, and J. A. Mosher. 1986. (Broad-winged hawk sprig delivery.) *Wilson Bull.* 98:469–71.

McClelland, B. R., L. S. Young, D. S. Shea, P. T. McClelland, H. L. Allen, and E. B. Spettique. 1983. (Bald eagles at Glacier National Park.) Pp. 69–77, in Bird, 1983.

McCollough, M. 1986. Post-fledging ecology of bald eagles in Maine. Ph.D. diss., Univ. Maine, Orono.

———. 1989. Molting sequence and aging of bald eagles. *Wilson Bull.* 101:1–10.

McCrary, M. D. 1981. Space and habitat utilization by red-shouldered hawks (*Buteo lineatus elegans*) in southern California. M.S. thesis, Calif. State Univ., Long Beach.

McEwan, L. C., and D. H. Hirth. 1980. (Bald eagle foods, Florida.) *Condor* 82:229–31.

McGahan, J. 1968. Ecology of the golden eagle. *Auk* 85:1–12.

McGovern, M., and J. M. McNurney. 1986. (Red-tailed hawk nest densities, Colorado.) *Raptor Res.* 20:43–45.

McGowan, J. D. 1975. Distribution, density and productivity of goshawks in interior Alaska. Juneau: Alaska Dept. of Fish & Game Report.

McInvaille, W. B., Jr., and L. B. Keith. 1974. (Ecology of red-tailed hawk and great horned owl, Alberta.) *Can. Field-Nat.* 88:1–20.

MacLaren, P. A., D. E. Runde, and S. H. Anderson. 1984. (Tree-nesting by prairie falcons, Wyoming.) *Condor* 86:487–88.

McNutt, J. W., C. P. Garat, D. H. Ellis, and C. M. White. 1985. (Peregrine status in South America.) Proc. Raptor Research Found. Symp., Management of Birds of Prey, Nov. 3–6, Sacramento (abstract).

Macrander, A. M. 1983. The ecology of wintering American kestrels (*Falco sparverius*) in west central Alabama. Ph.D. diss., Univ. Alabama, University.

MacWhirter, R. B. 1985. Breeding ecology, prey selection and provisioning of northern harriers *Circus cyaneus* (L.). Honors thesis, Mt. Allison Univ. and Acadia Univ., Nova Scotia.

Mader, W. J. 1975a. (Biology of Harris' hawk, Arizona.) *Living Bird* 14:59–85.

———. 1975b. Extra adults at Harris' hawk nests. *Condor* 77:482–85.

———. 1978. (Red-tailed hawk and Harris' hawk ecology, Arizona.) *Auk* 95:327–37.

———. 1979. (Polyandrous breeding, Harris' hawk.) *Auk* 96:776–88.

———. 1982. (Ecology of the savanna hawk.) *Condor* 84:261–71.

Marti, C. D., and C. E. Braun. 1975. (Prairie falcons in tundra habitats, Colorado.) *Condor* 77:213–14.

Martin, G. R. 1985. Eye. Pp. 311–73, in A. S. King and J. McLelland (eds.), *Form and function in birds.* Vol. 3. London: Academic Press.

Martin, J. W. 1986. Behavior and habitat use of breeding northern harriers in southwestern Idaho. M.S. thesis, Brigham Young Univ., Provo.

Matchett, M. R., and B. W. O'Gara. 1987. (Controlling golden eagle sheep depredation.) *J. Raptor Res.* 21:85–94.

Mathews, W. 1902. The Night Chant: A Navaho ceremony. *Amer. Mus. Nat. Hist. Memoirs* 6.

Mathisen, J. E. 1968. (Bald eagle and osprey nest site characteristics.) *Loon* 40:113–14.

———. 1977. (Status of ospreys in Chippewa National Forest.) Pp. 175–79, in Ogden, 1977.

———. 1983. (Nest-site selection by bald eagles.) Pp. 95–100, in Bird, 1983.

Matray, P. F. 1974. (Nesting ecology of broad-winged hawks.) *Auk* 91:307–24.

Matteson, S. W., and J. O. Riley. 1981. (Zone-tailed hawk distribution, Texas.) *Wilson Bull.* 93:282–84.

Mattox, W. G. 1985. (Greenland peregrine surveys.) Proc. Raptor Res. Found. Symp., Management of Birds of Prey, Nov. 3–6, Sacramento (abstract).

Mearns, R. 1985. (Hunting ranges of peregrines.) *Raptor Res.* 19:20–26.

Melquist, W. E., and D. R. Johnson. 1973. (Osprey status, Idaho and eastern Washington.) Raptor Research Foundation: Raptor Res. Rep. 3:121–23.

Mendelsohn, J. 1981. A study of the black-shouldered kite *Elanus ceruleus*. Ph.D. diss., Univ. of Natal, Pietermaritzburg, South Africa.

————. 1983. (Social behavior of black-shouldered kite.) *Ostrich* 54:1–18.

Meng, H. 1951. The Cooper's hawk. Ph.D. diss., Cornell Univ., Ithaca.

————. 1959. (Foods of Cooper's hawks and goshawks.) *Wilson Bull.* 71:169–74.

Menkens, G. E., Jr., and S. H. Anderson. 1987. (Nest sites of tree-nesting golden eagles.) *J. Field Ornith.* 58:22–25.

Mersmann, T. J., D. A. Buehler, and J. D. Fraser. 1987. (Bald eagle food habits and techniques.) Proc. Annual Meeting, Raptor Res. Found., Oct. 28–31, Boise (abstract).

Meserve, P. 1977. (Black-shouldered kite foods in Chile.) *Condor* 79:263–66.

Meyer, K. 1987. Sexual size dimorphism and the behavioral ecology of breeding and wintering sharp-shinned hawks (*Accipiter striatus*). Ph.D. diss., Univ. North Carolina, Chapel Hill.

Meyer, K., and M. W. Collopy. 1988. (Study of swallow-tailed kite in Florida.) Gainesville: Annual Progress Rept., Fla. Nongame Wildlife Program (mimeo).

Meyer, K., M. Collopy, and S. McGhee. 1988. (Breeding biology of swallow-tailed kite, Florida.) Proc. Annual Meeting, Raptor Res. Found., Oct. 27–29, Minneapolis (abstract).

Mills, G. S. 1976. American kestrel sex ratios and habitat separation. *Auk* 93:740–48.

Millsap, B. A. 1981. (Distribution of Arizona Falconiformes.) USDI, Bur. Land Manage., Tech. Note 355:1–102.

————. 1984. (Aplomado falcon status, Southwest.) *Eyas* 6(2):2.

————. 1986a. Biosystematics of the gray hawk. M.S. thesis, Geo. Mason Univ., Fairfax, Va.

————. 1986b. (Wintering bald eagles, United States.) *Wildl. Soc. Bull.* 14:433–40.

————. 1987. (Swallow-tailed kites at Lake Okeechobee.) *Fla. Field Nat.* 15:85–92.

Millsap, B., and S. L. Vana. 1984. (Golden eagle winter distribution, eastern U.S.) *Wilson Bull.* 96:692–701.

Mindell, D. P. 1983. (*Buteo jamaicensis harlani* as valid subspecies.) *Auk* 100:161–69.

————. 1985. (Plumage variation and range of *B. j. harlani*.) *Am. Birds* 39:1272–33.

Mockford, E. L. 1951. (Courtship of male Cooper's hawk.) *Indiana Aud. Quarterly* 29:58–59.

Molinero, F. C., and J. J. F. Cantisan. 1985. (Ecology of black-shouldered kite, Spain.) Pp. 137–41, in Newton and Chancellor, 1985.

Moller, A. P. 1987. (Copulation in goshawk.) *Anim. Behav.* 35:755–63.

Mollhagen, T. R., R. W. Wiley, and R. L. Packard. 1972. (Prey remains in golden eagle nests.) *J. Wildl. Manage.* 36:784–92.

Moore, J. 1987. (Breeding falcons of west Greenland.) *J. Raptor Res.* 21:111–15.

Moore, J. C., L. A. Stimson, and W. B. Robertson. 1953. (Short-tailed hawk in Florida.) *Auk* 70:470–78.

Moore, K. R., and C. J. Henny. 1983. (Accipiter nest sites in Oregon.) *Raptor Res.* 17:65–76.

———. 1984. (Productivity and nest-sites of Cooper's hawks.) *Northwest Sci.* 58:290–99.

Morris, M. M. J. 1980. Nest site selection by the red-shouldered hawk (*Buteo lineatus*) in southwestern Quebec. M.S. thesis, McGill Univ., Montreal.

Morris, M. M. J., B. L. Penak, R. E. Lemon, and D. M. Bird. 1982. (Red-shouldered hawk nest sites in Quebec.) *Can. Field-Nat.* 96:139–42.

Morrison, M. L. 1980. (White-tailed hawk populations, Texas.) *Bull. Texas Orn. Soc.* 11:35–40.

Morrison, M. L., and B. J. Walton. 1980. (Replacement clutches in raptors.) *Raptor Res.* 14:79–85.

Mosher, J. A., and P. F. Matray. 1974. (Size dimorphism in broad-winged hawks.) *Auk* 91:325–41.

Mosher, J. A., and C. M. White. 1976. (Orientation of golden eagle nests.) *Can. Field-Nat.* 90:356–59.

Mueller, H. C. 1970. (Courtship in broad-winged hawks.) *Auk* 87:580.

———. 1971. (Displays of the American kestrel.) *Wilson Bull.* 83:249–54.

———. 1972. (Zone-tailed hawk and turkey vulture mimicry.) *Condor* 74:221–22.

Mueller, H. C., and D. D. Berger. 1970. (Prey preferences of sharp-shinned hawk.) *Auk* 87:452–57.

Mueller, H. C., and K. Meyer. 1985. (Reversed sexual dimorphism in Palearctic raptors.) *Current Ornithol.* 2:65–101.

Mueller, H. C., N. S. Mueller, and P. G. Parker. 1981. (Sexual dimorphism in sharp-shinned hawks.) *Wilson Bull.* 93:83–92.

Muir, D., and D. M. Bird. 1984. (Gyrfalcon foods, Ellesmere Is.) *Wilson Bull.* 96:464–67.

Murie, O. J. 1959. Fauna of the Aleutian Islands and the Alaska Peninsula. USDI, Fish & Wildl. Serv., N. Am. Fauna No. 61.

Murphy, J. R. 1977. (Status of eagles in western U.S.) Pp. 57–63, in Chancellor, 1977.

Murphy, J. R., F. J. Caminzind, D. G. Smith, and J. B. Weston. 1969. (Nesting ecology of raptors in Utah.) *Brigham Young Univ. Sci. Bull., Biol. Ser.* 10(4):1–36.

Murphy, R. K., M. W. Gratson, and R. N. Rosenfield. 1987. (Habitat use by Cooper's hawk.) Proc. Annual Meeting, Raptor Res. Found, Oct. 28–31, Boise (abstract).

Nagy, A. C. 1963. (Population density of kestrels, Pennsylvania.) *Wilson Bull.* 75:93.

Neilsen, O. K. 1986. Population ecology of the gyrfalcon in Iceland, with comparative notes on the merlin and raven. Ph.D. diss., Cornell Univ., Ithaca.

Nelson, M. W. 1962. Hunting characteristics of eagles. In J. Yoakim (ed.), Trans. 1962 Interstate Antelope Conference. Reno: Interstate Antelope Conf.

Nelson, R. W. 1970. Some aspects of the breeding behavior of peregrine falcons on Langara Island, B. C. M.S. thesis, Calgary Univ.

———. 1972. (Incubation period of Peale's peregrine.) *Raptor Res.* 6:11–15.

Nelson, R. W., and J. A. Campbell. 1973. (Breeding behavior of captive peregrines.) *Hawk Chalk* 12:39–54.

Nethersole-Thompson, D., and C. Nethersole-Thompson. 1944. Nest-site selection by birds. *Brit. Birds* 37:70–74, 88–94, 108–113.

Newton, I. 1979. *Population ecology of raptors.* Berkhamsted: T. & A. D. Poyser.

———. 1985. (Population dynamics of peregrines.) Proc. Raptor Res. Found.
Symp., Management of Birds of Prey, Nov. 3–6, Sacramento (abstract).

———. 1986. *The sparrowhawk.* Calton: T. & A. D. Poyser.

Newton, I., and R. D. Chancellor (eds.). 1985. *Conservation studies on raptors.*
London: International Council for Bird Preservation.

Nichols, J. D., G. L. Hensler, and P. W. Sykes, Jr. 1980. (Demography of snail
kite.) *Ecol. Modelling* 9:215–32.

Nieboer, E. 1973. Geographical and ecological differentiation in the genus
Circus. Ph.D. diss., Free Univ. Amsterdam, Netherlands.

Oakleaf, R. J. 1985a. (Wyoming raptor research.) *Eyas* 8:18–21.

———. 1985b. (Ferruginous hawk status, Wyoming.) Proc. Raptor Research
Found. Symp., Management of Birds of Prey, Nov. 3–6, Sacramento
(abstract).

Oberholser, H. C. 1974. *The bird life of Texas,* ed. E. B. Kincaid. 2 vols. Austin:
Univ. Texas Press.

Ogden, J. C. 1974. (Short-tailed hawk in Florida.) *Auk* 91:95–110.

———. 1975. (Bald eagle territoriality and ospreys.) *Wilson Bull.* 87:496–505.

——— (ed.). 1977. *Transactions of the North American osprey research conference.*
Washington: U. S. Natl. Park Serv. Trans. Proc. Ser. no. 2.

Ogden, V. T. 1973. Nesting density and reproductive success of the prairie
falcon in southwestern Idaho. M.S. thesis, Univ. Idaho, Moscow.

Ogden, V. T., and M. G. Hornocker. 1977. (Prairie falcon density, Idaho.)
J. Wildl. Manage. 41:1–11.

Olendorff, R. R. 1971. Falconiform reproduction: a review. Part I. The pre-
nestling period. *Raptor Research Found. Res. Rept.* no. 1:1–111.

———. 1972. The large birds of prey of the Pawnee National Grassland:
nesting habits and productivity, 1969–1971. Fort Collins: Tech. Report 151,
USIBP, Grassland Biome.

———. 1973. The ecology of the nesting birds of prey of northeastern
Colorado. Fort Collins: Tech. Report 211, USIBP, Grassland Biome.

———. 1974. (Courtship flight of Swainson's hawk.) *Condor* 76:215.

———. 1976. (Food habits of golden eagles.) *Am. Midl. Nat.* 95:231–36.

Olendorff, R. R., and S. E. Olendorff. 1968, 1969, 1970. *An extensive bibliogra-
phy of falconry, eagles, hawks, falcons and other diurnal birds of prey.* In 3 parts. I.
Falconry and eagles. II. Hawks and miscellaneous. III. Indexes. Fort Collins:
Published by the authors.

Oliphant, L. W. 1974. (Merlins in Saskatoon.) *Blue Jay* 33:140–47.

———. 1985a. Canada, west. *Eyas* 8:16.

———. 1985b. North American merlin breeding survey. *Raptor Res.* 19:37–
41.

Oliphant, L. W., and E. Haug. 1985. (Population changes of Saskatoon
merlins.) *Raptor Res.* 19:56–59.

Oliphant, L. W., and W. J. P. Thompson. 1976. (Food-caching in merlins.)
Can. Field-Nat. 90:364–65.

Oliphant, L. W., W. J. P. Thompson, D. Donald, and R. Rafuse. 1976. (Prairie
falcon status in Saskatchewan.) *Can. Field-Nat.* 90:365–68.

Olsen, D., and J. Olsen. 1986. Why are female raptors larger than males?
Australasian Raptor Assoc. News 7:12–14.

Opdam, P., J. Thissen, P. Vershuren, and G. Müskens. 1977. (Goshawk
foraging ecology.) *J. f. Ornithologie* 118:235–51. (In German.)

Orians, G., and F. Kuhlman. 1956. (Red-tailed hawks and great horned owls in Wisconsin.) *Condor* 58:371–85.

Orians, G., and D. Paulson. 1969. Notes on Costa Rican birds. *Condor* 71:426–31.

Pache, P. H. 1974. (Biology of Harris' hawk in New Mexico.) *Wilson Bull.* 86:72–74.

Page, G., and D. F. Whitacre. 1975. Raptor predation on wintering shorebirds. *Condor* 77:73–78.

Palmer, R. 1988. *Handbook of North American birds.* Vols. 4 and 5. New Haven: Yale Univ. Press.

Parker, J. W., Jr. 1974. The breeding biology of the Mississippi kite in the Great Plains. Ph.D. diss., Univ. Kansas, Lawrence.

――――. 1975. (Mississippi Kite populations in Great Plains.) Pp. 159–72, in J. R. Murphy, C. M. White, and B. E. Havell (eds.), *Population status of raptors.* Raptor Research Reports no. 3. Vermillion, S. Dak.: Raptor Research Foundation.

――――. 1985. (Status of swallow-tailed kite in North America.) *Eyas* 7:12–13.

Parker, J. W., and J. C. Ogden. 1979. (Distribution of Mississippi kite.) *Amer. Birds* 33:119–29.

Parker, J. W., and M. Ports. 1982. (Nest-helping in Mississippi kites.) *Raptor Res.* 16:14–17.

Parker, M. A., and B. R. Tannenbaum. 1984. (Foraging and habitats of red-shouldered hawks in Missouri.) *Amer. Zool.* 24:50A (abstract).

Parkes, K. 1958. (Relationships in *Elanus.*) *Condor* 60:139–40.

Parmelee, D. F., and H. A. Stephens. 1964. Status of Harris' hawk in Kansas. *Condor* 66:443–45.

Parmelee, D. F., H. A. Stephens, and R. H. Schmidt. 1967. (Birds of Victoria Island.) *Bull. Nat. Mus. Canada* 222:1–229.

Paulson, D. R. 1983. (Flocking in hook-billed kite.) *Auk* 100:749–50.

Peck, G. K., and R. D. James. 1983. *Breeding birds of Ontario: Nidiology and distribution.* Vol. 1: Nonpasserines. Toronto: Royal Ontario Museum.

Pendleton, B. A. G., B. A. Millsap, K. W. Cline, and D. M. Bird. 1987. *Raptor management techniques manual.* Washington: National Wildlife Federation, Sci. & Tech. Series Publ. No. 10.

Peters, J. C. 1931. *Check-list of the birds of the world.* Vol. 1. Cambridge: Harvard Univ. Press.

Petersen, L. 1979. (Great horned owl and red-tailed hawk ecology in Wisconsin.) Madison: Wisconsin Dept. Nat. Res. Tech. Bull. No. 111.

Phillips, A., J. Marshall, and G. Monson. 1978. *Birds of Arizona.* Tucson: Univ. Arizona Press.

Phillips, R. L., T. P. McEneaney, and A. E. Beske. 1984. (Breeding densities of golden eagles in Wyoming.) *Wildl. Soc. Bull.* 12:269–73.

Phillips, S. R., M. A. Westall, and P. W. Zajicek. 1984. (Osprey productivity on Sanibel Island.) Pp. 61–66, in Westall, 1984.

Phipps, K. B. 1979. Hunting methods, habitat use and activity patterns of prairie falcons in the Snake River Birds of Prey Natural Area, Idaho. M.S. thesis, Western Ill. Univ., Macomb.

Picozzi, N. 1978. (Breeding dispersion of northern harriers, Scotland.) *Ibis* 120:498–508.

――――. 1984a. (Breeding biology of northern harriers, Orkney.) *Ornis. Scand.* 15:1–10.

――――. 1984b. (Territoriality of polygynous northern harriers, Orkney.) *Ibis* 126:356–65.

Platt, J. B. 1976a. (Sharp-shinned hawk nest-site selection.) *Condor* 78:102–03.

———. 1976b. (Gyrfalcon nest-site selection.) *Can. Field-Nat.* 90:338–45.

———. 1977. The breeding behavior of wild and captive gyrfalcons in relation to their environment and human disturbance. Ph.D. diss., Cornell Univ., Ithaca.

Platt, S. W. 1974. Breeding status and distribution of the prairie falcon in northern New Mexico. M.S. thesis, Okla. State Univ., Stillwater.

———. 1981. Prairie falcon: aspects of population dynamics, individual vocal identification, marking and sexual maturity. Ph.D. diss., Brigham Young Univ., Provo.

Pleasants, J. M., and B. V. Pleasants. 1988. (Reversed sexual dimorphism in raptors.) *Oikos* 52:129–35.

Poole, A. 1985. Courtship feeding and osprey reproduction. *Auk* 102:479–92.

———. 1989. *Ospreys: A natural and unnatural history.* Cambridge: Cambridge Univ. Press.

Poole, K. C. 1987. Aspects of the ecology, food habits and foraging characteristics of gyrfalcons in the central Canadian arctic. M.S. thesis, Univ. Alberta, Edmonton.

Poole, K. C., and D. A. Boag. 1988. (Diet and foraging of gyrfalcon in the Canadian arctic.) *Can. J. Zool.* 66:334–44.

Poole, K. C., and R. G. Bromley. 1988. Natural history of the gyrfalcon in the central Canadian arctic. *Arctic* 41:31–38.

Porter, R. D., and M. A. Jenkins. 1985. (Status of peregrine in Baja.) Proc. Raptor Research Found. Symp., Management of Birds of Prey, Nov. 3–6, Sacramento (abstract).

Porter, R. D., M. A. Jenkins, A. L. Gaski, and M. N. LeFranc, Jr. (eds.). 1987. *Working bibliography of the peregrine falcon.* Washington: National Wildlife Fed. Sci. & Tech. Series No. 9.

Porter, R. D., and C. M. White. 1977. (Status of hawks in western U.S.) Pp. 39–47, in Chancellor, 1977.

Porter, R. D., C. M. White, and R. J. Erwin. 1973. (Peregrine and prairie falcon in Utah.) *Brigham Young Univ. Sci. Bull. Biol. Ser.* 18(1):1–74.

Porter, R. D., and S. N. Wiemeyer. 1972. (Reproduction in captive kestrels.) *Condor* 74:46–53.

Postupalsky, S. 1977. (Nesting platforms for osprey and bald eagle.) Pp. 35–45, in S. A. Temple (ed.), *Endangered birds.* Madison: Univ. Wisconsin Press.

Powers, L. R. 1981. Nesting behavior of the ferruginous hawk (*Buteo regalis*). Ph.D. diss., Idaho State Univ., Pocatello.

Prevost, Y. A. 1983. (Osprey distribution and subspecies.) Pp. 157–74, in Bird, 1983.

Pruett-Jones, S. G., M. A. Pruett-Jones, and R. L. Knight. 1980. (Status of black-shouldered kite in North America.) *Amer. Birds* 34:682–88.

Quincy, P. A. 1976. (Caracara in Florida.) *Fla. Nat.* 49(4):7–9.

Rand, L. A. 1960. (Races of short-tailed hawk.) *Auk* 77:448–59.

Randall, P. E. 1940. (Northern harrier foods, Pennsylvania.) *Wilson Bull.* 52:165–72.

Ratcliffe, D. 1980. *The peregrine falcon.* Berkhamsted: T. & A. D. Poyser.

Rea, A. M. 1983. Cathartid affinities: a brief overview. Pp. 26–52, in S. R. Wilbur and J. J. Jackson (eds.), *Vulture biology and management.* Berkeley: Univ. Calif. Press.

Redig, P. T., and H. B. Tordoff. 1985. (Re-introducing peregrines to upper midwest.) Proc. Raptor Res. Found. Symp., Management of Birds of Prey, Nov. 3–6, Sacramento (abstract).

Reese, J. G. 1977. (Osprey reproduction, Chesapeake Bay.) *Auk* 94:202–21.

Retfalvi, L. T. 1965. Breeding behavior and feeding habits of the bald eagle (*Halieetus leucocephalus* L.) on San Juan Island, Washington. M.S. thesis, Univ. Brit. Columbia, Vancouver.

———. 1970. (Foods of bald eagles on San Juan Island.) *Condor* 72:358–61.

Reynard, G. B., L. L. Short, O. H. Garrido, and G. Alayon. 1987. (Nesting, status, and relationships of Gundlach's hawk.) *Wilson Bull.* 99:73–77.

Reynolds, R. T. 1972. (Sexual dimorphism in accipiters.) *Condor* 74:191–97.

———. 1978. Food and habitat partitioning in two groups of coexisting accipiters. Ph.D. diss., Oregon State Univ., Corvallis.

Reynolds, R. T., and E. C. Meslow. 1984. (Food and niche partitioning in accipiters.) *Auk* 101:761–79.

Reynolds, R. T., E. C. Meslow, and H. M. Wight. 1982. (Accipiter nesting habitats, Oregon.) *J. Wildl. Manage.* 46:124–38.

Reynolds, R. T., and H. M. Wight. 1978. (Accipiters in Oregon.) *Wilson Bull.* 90:182–96.

Rhodes, L. I. 1977. (Osprey nest structures.) Pp. 77–83, in Ogden, 1977.

Rice, J. N. 1969. (Decline of Pennsylvania peregrines.) Pp. 155–64, in Hickey, 1969.

Rice, W. R. 1982. (Acoustic location in northern harrier.) *Auk* 99:403–13.

Richmond, A. R. 1976. (Nestling feeding by caracara.) *Wilson Bull.* 88:667.

Risebrough, R. W., W. M. Jarman, J. E. Monk, A. M. Springer, B. J. Walton, and C. G. Thelander. 1985. (DDE in California peregrines.) Proc. Raptor Res. Found. Symp., Management of Birds of Prey, Nov. 3–6, Sacramento (abstract).

Ritchie, R. J., and J. A. Curatolo. 1982. (Golden eagle productivity, Alaska.) *Raptor Res.* 16:123–28.

Roalkvam, R. 1985. (Peregrine hunting effectiveness.) *Raptor Res.* 19:27–29.

Rodgers, J. A., Jr. 1987. (1986 snail kite survey and habitat assessment.) Unpublished annual performance report, Florida Game & Fresh Water Fish Comm.

———. 1988. (1987 snail kite survey.) Unpublished report, Florida Game & Fresh Water Fish Comm.

———. 1989. (1988 snail kite survey.) Unpublished report, Florida Game & Fresh Water Fish Comm.

Rodgers, J. A., Jr., S. T. Schwikert, and A. S. Wenner. 1988. (Snail kite status, Florida.) *Amer. Birds* 42:30–35.

Rogers, D. 1979. (Red-shouldered hawk status, Oregon.) *Oregon Birds* 5:4–8.

Root, T. 1988. *Atlas of wintering North American birds. An analysis of Christmas Bird Count data.* Chicago: Univ. Chicago Press.

Roseneau, D. G. 1972. Summer distribution, numbers and food habits of the gyrfalcon (*Falco rusticolus*) on the Seward Peninsula, Alaska. M.S. thesis, Univ. Alaska, Fairbanks.

Rosenfield, R. N. 1984. (Broad-winged hawk nesting biology, Wisconsin.) *Raptor Res.* 18:6–9.

———. 1988. (Nest-building and pre-incubation behavior of Cooper's hawk.) Proc. Annual Meeting, Raptor Res. Found., Oct. 27–29, Minneapolis (abstract).

Rosenfield, R. N., and M. W. Gratson. 1984. (Broad-winged hawk foods, Wisconsin.) *J. Field Ornith.* 55:246–47.

Roth, S. D., Jr., and J. M. Marzluff. 1985. (Ferruginous hawk status, Kansas.) Proc. Raptor Res. Found. Symp., Management of Birds of Prey, Nov. 3–6, Sacramento (abstract).

Rothfels, M., and M. R. Lein. 1983. (Sympatric red-tailed and Swainson's hawks.) *Can. J. Zool.* 61:60–64.

Runde, D. E., and S. H. Anderson. 1986. (Prairie falcon nest sites.) *Raptor Res.* 20:21–28.

Rusch, D. H., and P. D. Doerr. 1972. (Broad-winged hawk nesting biology.) *Auk* 89:139–45.

Russell, S. M. 1964. A distributional study of the birds of Honduras. Ornithol. Monogr. No. 1, American Ornith. Union.

Russock, H. I. 1979. (Wintering bald eagle behavior.) *Raptor Res.* 13:112–15.

Ryttman, H., H. Tegelstrom, K. Fredga, and J. Sondell. 1987. (Karyotype of osprey.) *Genetica* 74:143–47.

Scharf, W. C., and E. Balfour. 1971. (Northern harrier development.) *Ibis* 113:323–29.

Schempf, P. F. 1988. (Status of *Haliaeetus* eagles in Alaska.) Proc. Annual Meeting, Raptor Res. Found., Oct. 27–29, Minneapolis (abstract).

Schipper, W. J. A. 1973. (Prey selection in sympatric *Circus* species.) *Gerfaut* 63:17–20.

Schipper, W. J. A., L. S. Buurma, and P. Bossenbroek. 1975. (Hunting behavior of sympatric *Circus* species.) *Ardea* 63:1–29.

Schlorff, R. W. 1985. (Breeding status of Swainson's hawk, central valley of California.) Proc. Raptor Res. Found. Symp., Management of Birds of Prey, Nov. 3–6, Sacramento (abstract).

Schmutz, J. K. 1977. Relationships between three species of the genus *Buteo* (Aves) coexisting in the prairie-parkland ecotone of southeastern Alberta. M.S. thesis, Edmonton Univ., Alberta.

———. 1984. (Ferruginous and Swainson's hawks in Alberta.) *J. Wildl. Manage.* 48:1180–87.

———. 1985a. (Status of ferruginous hawk, Alberta.) Proc. Raptor Res. Found. Symp., Management of Birds of Prey, Nov. 3–6, Sacramento (abstract).

———. 1985b. (Status of Swainson's hawk, Alberta.) Proc. Raptor Res. Found. Symp., Management of Birds of Prey, Nov. 3–6, Sacramento (abstract).

———. 1987. (Mate choice by ferruginous hawks.) Proc. Annual Meeting, Raptor Res. Found., Oct. 28–31, Boise (abstract).

Schmutz, J. K., and R. W. Fyfe. 1987. Migration and mortality of Alberta ferruginous hawks. *Condor* 89:169–74.

Schmutz, J. K., R. W. Fyfe, D. A. Moore, and A. R. Smith. 1984. (Artificial nests for ferruginous and Swainson's hawks.) *J. Wildl. Manage.* 48:1009–13.

Schnell, G. D. 1967. (Rough-legged hawks in Illinois.) *Kans. Ornith. Soc. Bull* 18:21–28.

———. 1968. (Rough-legged and red-tailed hawk habitat utilization.) *Condor* 70:373–77.

Schnell, J. H. 1958. (Goshawk nesting and foods, California.) *Condor* 60:377–403.

———. 1979. Black hawk (*Buteogallus anthracinus*). Habitat Management Series for Unique or Endangered Species. USDI, Bur. Land Manage. Tech. Note 329.

Schönwetter, M. 1961. *Handbuch der Oologie.* Band 1. Berlin: Akademie-Verlag.

Schriver, E. C., Jr. 1969. (Status of Cooper's hawk, Pennsylvania.) Pp. 356–459, in *Peregrine falcon populations: their biology and decline* (J. J. Hickey, ed.). Madison: Univ. Wisconsin Press.

Sealy, S. G. 1966. (Rough-legged hawks in Northwest Territories.) *Blue Jay* 24:127–28.

——. 1967. (Breeding of northern harrier, Alberta and Saskatchewan.) *Blue Jay* 25:63–69.

Seidensticker, J. C., and H. V. Reynolds. 1971. (Breeding and hydrocarbon residues in red-tailed hawk and great horned owl, Montana.) *Wilson Bull.* 83:408–18.

Selander, R. K. 1966. Sexual dimorphism and differential niche utilization in birds. *Condor* 68:113–51.

Seymour, N. R., and R. P. Bancroft. 1983. (Status of ospreys, Nova Scotia.) Pp. 275–80, in Bird, 1983.

Sherrod, S. K. 1978. Diets of North American Falconiformes. *Raptor Res.* 12:49–121.

——. 1983. Behavior of fledgling peregrines. Fort Collins: Peregrine Fund.

Sherrod, S. K., C. M. White, and F. S. L. Williamson. 1976. (Biology of bald eagle, Amchitka.) *Living Bird* 15:143–82.

Shor, W. 1975. (Survival rates of prairie falcons.) *Raptor Res.* 9:46–50.

Shufeldt, R. 1891. (Osteology of North American kites.) *Ibis* 33:228–32.

Shuster, W. C. 1977. (Bibliography of goshawk.) USDI, Bur. Land Manage. Tech. Note. 309:1–11.

——. 1980. (Goshawk nest requirements.) *Western Birds* 11:89–96.

Sibley, C. G., and J. E. Ahlquist. 1972. (Egg-white proteins of non-passerines.) *Peabody Mus. Nat. Hist. Bull.* 39:1–276.

——. 1985. (African bird relationships based on DNA hybridization.) Pp. 115–61, in K.-L. Schuchmann (ed.), Proc. Intern. Symp. on African Vertebrates. Bonn: Zool. Forsch. & Museum Alex. Koenig.

Sibley, C. G., J. E. Ahlquist, and B. L. Monroe, Jr. 1988. (Classification of birds based on DNA hybridization.) *Auk* 105:409–423.

Simmons, R. 1986. (Ecological segregation of African accipiters.) *Ardea* 74:137–49.

——. 1988. (Behavior of polygynous northern harriers.) *Auk* 105:303–07.

Simmons, R., P. Barnard, R. B. MacWhirter, and G. L. Hansen. 1986. (Microtines and northern harrier breeding biology.) *Can. J. Zool.* 64:2447–56.

Simmons, R., P. Barnard, and P. C. Smith. 1987. (Comparison of N. American and European northern harrier behavior.) *Ornis. Scand.* 18:33–41.

Simmons, R., and P. C. Smith. 1985. (Northern harrier nest-site selection.) *Can. J. Zool.* 63:494–98.

Simmons, R., P. C. Smith, and R. B. MacWhirter. 1986. (Northern harrier harem hierarchies.) *J. Anim. Ecol.* 55:755–71.

Simons, T., S. K. Sherrod, M. W. Collopy, and M. A. Jenkins. 1988. Restoring the bald eagle. *American Scientist* 76:252–60.

Skinner, R. W. 1962. (Mississippi kite feeding habits.) *Auk* 79:273–74.

Skutch, A. F. 1965. (Life histories of two American kites.) *Condor* 67:235–46.

Slud, P. 1964. The birds of Costa Rica. *Am. Mus. Nat. Hist. Bull.* 128:1–430.

Smith, C. F., and C. Ricardi. 1983. (Bald eagle and osprey in New Hampshire.) Pp. 149–56, in Bird, 1983.

Smith, D. G., and J. R. Murphy. 1973. (Breeding ecology of raptors in Utah.) *Brigham Young Univ. Sci. Bull. Biol. Ser.* 18(3):1–76.

——. 1978. Biology of the ferruginous hawk. *Sociobiol.* 3:79–95.

——. 1979. (Breeding raptors and jackrabbits, Utah.) *Raptor Res.* 13:1–14.

——. 1982. (Golden eagle nesting, Utah.) *Raptor Res.* 16:129–32.

Smith, D., J. R. Murphy, and N. D. Woffinden. 1981. (Jackrabbit abundance and ferruginous hawk reproduction.) *Condor* 83:52–56.

Smith, D. G., C. R. Wilson, and H. H. Frost. 1972. (Kestrel biology in Utah.) *Southwest Nat.* 17:73–83.

Smith, N. G. 1980. (Hawk migrations in Neotropics.) Pp. 51–65, in A. Keast and E. S. Morton (eds.), *Migrant birds in the Neotropics: Ecology, behavior, distribution and conservation.* Washington: Smithsonian Inst. Press.

———. 1985. (North American hawk migrations.) Pp. 387–93, in M. Harwood (ed.), *Proceedings of the Hawk Migration Conference IV, 1983.* Moorestown, N.J.: Hawk Migration Association of N. Am.

Smith, T. B. 1982. (Hook-billed kite nests.) *Biotropica* 14:79–80.

Smith, T. B., and S. A. Temple. 1982a. (Hook-billed kite feeding habits.) *Auk* 99:197–207.

———. 1982b. (Grenada hook-billed kites.) *Condor* 84:131.

Smithe, F. B. 1966. *The birds of Tikal.* Garden City: Natural History Press.

Smithe, F. B., and R. A. Paynter, Jr. 1963. Birds of Tikal, Guatemala. *Harvard Mus. Comp. Zool. Bull.* 128:245–324.

Snelling, J. C., and B. E. Leuck. 1976. (Predatory behavior of captive aplomado falcons.) *Avic. Mag.* 82:169–71.

Snow, C. 1972. (Habitat management series for endangered species: Peregrine falcon.) USDI, Bur. Land Manage. Tech. Note 167.

———. 1973a. (Habitat management series for unique or endangered species: Golden eagle.) USDI, Bur. Land Manage. Tech. Note 239.

———. 1973b. (Habitat management series for unique or endangered species: Bald eagle.) USDI, Bur. Land Manage. Tech. Note 171.

———. 1974a. (Habitat management series for unique or endangered species: Prairie falcon.) USDI, Bur. Land Manage. Tech. Note 240.

———. 1974b. (Habitat management series for unique or endangered species: Ferruginous hawk.) USDI, Bur. Land Manage. Tech. Note 255.

———. 1974c. (Habitat management series for unique or endangered species: Gyrfalcon.) USDI, Bur. Land Manage. Tech. Note 241.

Snyder, N. F. R. 1975. (Breeding biology of swallow-tailed kites, Florida.) *Living Bird* 13:73–97.

Snyder, N. F., S. R. Beissinger, and R. C. Chandler. 1989. (Reproduction and demography of the snail kite.) *Condor* 91:300–16.

Snyder, N. F. R., and H. W. Kale II. 1983. (Snail kite predation, Colombia.) *Auk* 100:93–97.

Snyder, N. F. R., and H. A. Snyder. 1969. (Mollusk predation by limpkins, snail kites and boat-tailed grackles.) *Living Bird* 8:177–223.

———. 1970. (Feeding territories in snail kites.) *Condor* 72:492–93.

Snyder, N. F. R., H. A. Snyder, J. L. Lincer, and R. T. Reynolds. 1972. Organochlorides, heavy metals and the biology of North American accipiters. *BioSci.* 23:300–05.

Snyder, N. F. R., and J. W. Wiley. 1976. Sexual size dimorphism in hawks and owls of North America. Amer. Ornithol. Union, Ornith. Monogr. no. 20.

Southern, W. E. 1963. (Wintering bald eagles, Illinois). *Wilson Bull.* 75:42–55.

———. 1964. (Wintering bald eagles.) *Wilson Bull.* 76:121–37.

———. 1967. (Subadult bald eagle plumages.) *Jack-Pine Warbler* 45:70–80.

Speiser, R., and T. Bosakowsi. 1987. (Goshawk nest-site selection, New York.) *Condor* 89:387–94.

Spitzer, P. R. 1980. Dynamics of a discrete coastal population of ospreys in the northeastern USA, 1969–1979. Ph.D. diss., Cornell Univ., Ithaca.

Spofford, W. R. 1971. (Golden eagles in Appalachians.) *Amer. Birds* 25:3–7.

Springer, M. A. 1975. (Foods of rough-legged hawks, Alaska.) *Condor* 77:338–39.

Springer, M. A., and J. S. Kirkley. 1978. (Interactions between red-tailed hawks and great horned owls, Ohio.) *Ohio J. Sci.* 78:323–28.

Sprunt, A., W. B. Robertson, S. Postupalsky, R. J. Hensel, C. E. Knoder, and F. J. Ligas. 1973. (Productivity of six bald eagle populations.) Trans. North Amer. Wildl. & Nat. Res. Conf. 38:96–106.

Stahlecker, D. W., and H. J. Griese. 1977. (Double-brooding by kestrels, Colorado.) *Wilson Bull.* 89:618–19.

Stalmaster, M. V. 1981. Ecological energetics and foraging behavior of wintering bald eagles. Ph.D. diss., Utah State Univ., Logan.

——. 1987. *The bald eagle.* New York: Universe Books.

Steenhof, K., and M. N. Kochert. 1987. (Electrical transmission lines as raptor nest sites.) Proc. Ann. Meeting, Raptor Res. Found., Oct. 28–31, Boise (abstract).

——. 1988. (Dietary responses of raptors to changing prey densities.) *J. Anim. Ecol.* 57:37–48.

Steenhof, K., M. N. Kochert, and J. H. Doremus. 1983. (Nesting of subadult golden eagles, Idaho.) *Auk* 100:443–47.

Stendell, R. C. 1972. The occurrence, food habits and nesting strategy of white-tailed kites in relation to a fluctuating vole population. Ph.D. diss., Univ. Calif., Berkeley.

Stendell, R. C., and P. Myers. 1973. (Black-shouldered kites and voles.) *Condor* 75:359–60.

Stensrude, C. 1965. (Gray hawk observations, Arizona.) *Condor* 67:319–21.

Stevenson, J. O., and L. H. Meitzen. 1946. (White-tailed hawk foods, Texas.) *Wilson Bull.* 58:198–205.

Stewart, G. R., and B. J. Walton. 1985 (Reestablishing Harris's hawks, lower Colorado River.) Proc. Raptor Res. Found. Symp., Management of Birds of Prey, Nov. 3–6, Sacramento (abstract).

Stewart, R. E. 1949. (Ecology of red-shouldered hawk.) *Wilson Bull.* 61:26–35.

——. 1975. *Breeding birds of North Dakota.* Fargo: Tri-College Center for Environmental Studies.

Stewart, R. E., and H. A. Kantrud. 1972. (Breeding bird populations, North Dakota.) *Auk* 89:766–88.

Stewart, R. E., and C. S. Robbins. 1958. Birds of Maryland and the District of Columbia. USDI, Fish & Wildl. Serv., North American Fauna No. 63.

Stieglitz, W. O., and R. L. Thompson. 1967. (Life history of snail kite.) USDI, Fish. & Wildl. Serv. Spec. Sci. Rep.—Wildl. no. 109:1–21.

Stinson, C. H. 1977. Familial longevity in ospreys. *Bird-banding* 48:72–73.

Storer, R. W. 1966. (Sexual dimorphism and foods of accipiters.) *Auk* 83:423–36.

Stotts, V. D., and C. H. Henny. 1975. (Time of fledging in ospreys.) *Wilson Bull.* 87:277–78.

Stresemann, E., and D. Amadon. 1979. Falconiformes. Pp. 271–425, in E. Mayr and G. W. Cottrell (eds.), *Check-list of birds of the world.* 2d ed. Cambridge: Harvard Univ. Press.

Sutton, G. M. 1939. The Mississippi kite in spring. *Condor* 41:41–53.

——. 1955. Nesting of the swallow-tailed kite. *Everglades Nat. Hist.* 3:72–84.

Swenson, J. E. 1978. (Osprey prey and foraging, Wyoming.) *J. Wildl. Manage.* 42:87–90.

——. 1979. (Osprey foraging ecology.) *Auk* 96:408–12.

——. 1981. (Osprey nest sites, Yellowstone.) *J. Field Ornith.* 52:67–69.

Swenson, J. E., K. L. Alt, and R. L. Eng. 1986. Ecology of bald eagles in the Greater Yellowstone Ecosystem. *Wildl. Monog.* 95:1–46.

Sykes, P. W., Jr. 1979. (Snail kite status 1968–1978.) *Wilson Bull.* 91:494–511.

———. 1983a. (Snail kites in southern Florida.) *Fla. Field Nat.* 11:73–78.

———. 1983b. (Snail kite population trends.) *J. Field Ornithol.* 54:237–46.

———. 1985. (Snail kite evening roosts.) *Wilson Bull.* 97:57–70.

———. 1987a. (Snail kite nesting ecology.) *Fla. Field Nat.* 15:57–70.

———. 1987b. (Snail kite breeding biology.) *J. Field Ornith.* 58:171–89.

———. 1987c. (Snail kite feeding ecology.) *Colonial Waterbirds* 10:84–92.

Sykes, P. W., Jr., and H. W. Kale II. 1974. (Non-snail prey of snail kites.) *Auk* 91:818–20.

Szaro, R. C. 1978. (Foraging behavior of osprey.) *Wilson Bull.* 90:112–18.

Tarboton, W. 1978. (Energy budget of black-shouldered kite.) *Condor* 80:88–91.

———. 1984. (Peregrine incubation behavior.) *Raptor Res.* 18:131–36.

Temeles, E. J. 1985. (Prey vulnerability and raptor size dimorphism.) *Amer. Nat.* 125:485–99.

———. 1986. (Reversed sexual dimorphism in northern harriers.) *Auk* 103:70–78.

Temple, S. A. 1972a. Sex and age characteristics of North American merlins. *Bird-Banding* 43:191–96.

———. 1972b. Chlorinated hydrocarbons and reproductive success in North American merlins. *Condor* 74:105–06.

———. 1985. Recovery efforts for the peregrine falcon. *Biol. Conser.* 2:161–62.

Teresa, S. 1980. Golden eagle successfully breeding in subadult plumage. *Raptor Res.* 14:86–87.

Tewes, M. E. 1984. (White-tailed hawk foraging on burns.) *Wilson Bull.* 96:135–36.

Thiollay, J.-M. 1980. Spring hawk migration in eastern Mexico. *Raptor Res.* 14:13–19.

———. 1985a. (Ecology of rainforest falconiforms.) Pp. 167–80, in Newton and Chancellor, 1985.

———. 1985b. (Falconiform forest communities.) Pp. 181–90, in Newton and Chancellor, 1985.

Thomas, G. B. 1908. The Mexican black hawk. *Condor* 10:116–18.

Thompson, S. P., R. S. Johnstone, and C. D. Littlefield. 1982. (Golden eagle nesting biology, Oregon.) *Raptor Res.* 16:116–22.

Thompson, S. P., C. D. Littlefield, and R. S. Johnstone. 1985. (Status of Swainson's hawk in Oregon.) Proc. Raptor Res. Found. Symp., Management of Birds of Prey, Nov. 3–6, Sacramento (abstract).

Thurow, T. L., and C. M. White. 1983. (Nest sites of ferruginous and Swainson's hawks.) *J. Field Ornithol.* 54:401–06.

Tinkham, E. R. 1948. (Behavior of caracara.) *Condor* 50:274.

Titus, K., and J. A. Mosher. 1981. (Habitat selection of woodland hawks, Appalachians.) *Auk* 98:270–81.

Tjernberg, M. 1985. (Nest spacing of golden eagles.) *Ibis* 127:250–55.

Todd, C. S., L. S. Young, R. B. Owen, Jr., and F. J. Gramlich. 1982. (Foods of bald eagles in Maine.) *J. Wildl. Manage.* 46:636–45.

Toland, B. 1985a. (Harrier predation on prairie-chickens.) *Raptor Res.* 19:146–48.

———. 1985b. (Foods of Cooper's hawks, Missouri.) *J. Field Ornithol.* 56:419–22.

———. 1985c. (Double-brooding by kestrels, Missouri.) *Condor* 87:434–36.

Toups, J. A., J. A. Jackson, and E. Johnson. 1985. (Black-shouldered kites in Mississippi.) *Amer. Birds* 39:865–67.

Treleaven, D. 1980. Cartwheeling by raptors. *Brit. Birds* 73:589.

Trimble, S. A. 1975. (Habitat management series for unique or endangered species: merlin.) USDI, Bur. Land Manage. Tech. Note 271.

Ueoka, M. L., and J. R. Koplin. 1973. (Foraging behavior of ospreys.) *Raptor Res.* 7:32–38.

U.S. Dept. of Interior. 1979. Snake River Birds of Prey research project annual report. Boise: U. S. Bureau of Land Management, Boise.

———. 1985. Endangered and threatened wildlife and plants: review of vertebrate wildlife. *Federal Register* 50:37958–67.

Van Horn, D., G. McDonald, and G. Ravensfeather. 1982. (Peregrine populations in southeast Alaska.) P. 154, in Ladd and Schempf, 1982 (abstract).

Voelker, T. 1969. Mating behavior of red-tailed hawks. *Loon* 41:90–91.

Voous, K. H., and T. de Vries. 1978. (Systematics of Galapagos hawk.) *Gerfaut* 68:245–52.

Vuilleumier, F. 1970. (Generic relations of caracaras.) *Brevoria* 355:1–29.

Waian, L. B. 1973. The behavioral ecology of the North American white-tailed kite (*Elanus leucurus majusculus*) of the Santa Barbara coastal plain. Ph.D. diss., Univ. Calif, Santa Barbara.

———. 1976. A resurgence of kites. *Nat. Hist.* 85(9):40–47.

Wakeley, J. S. 1974. (Hunting methods of ferruginous hawk.) *Raptor Res.* 8:67–72.

Walker, D. G. 1987. (Post-fledging period of golden eagle.) *Ibis* 129:92–96.

Walter, H. 1979. *Eleonora's falcon: Adaptations to prey and habitat in a social raptor.* Chicago: Univ. Chicago Press.

Walls, G. L. 1942. *The vertebrate eye and its adaptive radiation.* Bloomfield Hills, Mich.: Cranbrook Inst. of Science.

Walton, B. J. 1985. (Peregrine management, Pacific coast states.) Proc. Raptor Res. Found. Symp., Management of Birds of Prey, Nov. 3–6, Sacramento (abstract).

Ward, F. P., and R. C. Laybourne. 1985. (Age differences in peregrine prey selection.) Pp. 303–09, in Newton and Chancellor, 1985.

Warkentin, I. G., and P. C. James. 1988. Nest site selection by urban merlins. *Condor* 90:734–38.

Warkentin, I. G., and L. W. Oliphant. 1985. (Winter food caching by merlins.) *Raptor Res.* 19:100–01.

Warner, J. S., and R. L. Rudd. 1975. (Hunting by black-shouldered kites.) *Condor* 77:226–30.

Waste, S. M. 1982. (Winter ecology of bald eagle.) Pp. 68–81, in Ladd and Schempf, 1982.

Watson, D. 1977. *The hen harrier.* Berkhamsted: T. & A. D. Poyser.

Wattel, J. 1973. Geographic differentiation in the genus *Accipiter. Pub. Nuttall Orn. Club* 13:1–231.

Wauer, R. H., and R. C. Russell. 1967. (Gray hawk in Utah.) *Condor* 69:420–23.

Weathers, W. W. 1983. *Birds of southern California's Deep Canyon.* Berkeley: Univ. Calif. Press.

Weaver, J. D., and T. J. Cade. 1973. Special report on the falcon breeding program at Cornell University. Ithaca: Laboratory of Ornithology. 16pp. mimeo.

Weber, W. C. 1980. (Rare and endangered bird species in British Columbia.)

Pp. 160–82, in Stace-Smith, R., L. Johns, and P. Joslin (eds.), *Threatened and endangered species and habitats in British Columbia and the Yukon.* Ministry of Environ., Fish & Wildl. Branch, Victoria, B. C.

Webster, H. M. 1944. (Prairie falcons in Colorado.) *Auk* 61:609–16.

Weller, M. W. 1964. (Buteos wintering in Iowa.) *Iowa Bird Life* 34:58–62.

Westall, M. A. 1983. (Ospreys and nest structures.) Pp. 287–91, in Bird, 1983.

Westall, M. A. (ed.). 1984. Proceedings of the Southeastern U. S. and Caribbean osprey symposium. Sanibel, Fla.: Intern. Osprey Foundation, Inc.

Weston, J. B. 1969. (Nesting ecology of ferruginous hawk.) *Brigham Young Univ. Sci. Bull. Biol. Ser.* 10(4):25–36.

Wetmore, A. 1965. *The birds of the Republic of Panama.* Pt. I. Washington: Smithsonian Institution.

Wetmore, S. P., and D. I. Gillespie. 1977. (Ospreys of Labrador and Quebec.) Pp. 87–88, in Ogden, 1977.

Whaley, W. H. 1979. The ecology and status of the Harris' hawk (*Parabuteo uncinatus*) in Arizona. M.S. thesis, Univ. Arizona, Tucson.

⸻. 1986. (Population ecology of Harris' hawk, Arizona.) *Raptor Res.* 20:1–15.

Whitacre, D., D. Ukrain, and G. Falxa. 1982. (Hunting behavior of caracara, Mexico.) *Wilson Bull.* 94:565–66.

White, C. M. 1968. (Diagnosis of tundra peregrines.) *Auk* 85:179–81.

⸻. 1975. (Peregrines in Aleutian Islands.) *Raptor Res.* 9:33–50.

White, C. M., and T. Cade. 1971. (Raptors and ravens in arctic Alaska.) *Living Bird* 10:107–50.

White, C. M., and R. B. Weeden. 1966. (Hunting behavior of gyrfalcons.) *Condor* 68:517–19.

Whitfield, D. W. A., J. M. Gerrard, W. J. Maher, and D. W. Davis. 1974. (Bald eagles in Saskatchewan and Manitoba.) *Can. Field-Nat.* 88:399–407.

Whittemore, R. E. 1984. (Ospreys at Mattamuskeet refuge.) Pp. 17–41, in Westall, 1984.

Wilbur, S. R. 1973. The red-shouldered hawk in the western United States. *Western Birds* 4:15–22.

Wiley, J. W. 1975. (Red-tailed and red-shouldered hawk biology, California.) *Condor* 77:133–39.

⸻. 1985. (Conservation of forest raptors, West Indies.) Pp. 199–203, in Newton and Chancellor, 1985.

Wilkinson, G. S., and K. R. Debban. 1980. (Raptor winter habitat, Sacramento Valley.) *Western Birds* 11:25–34.

Willis, E. O. 1963. (Zone-tailed hawk mimicry.) *Condor* 65:313–17.

⸻. 1966. (Zone-tailed hawk prey capture.) *Condor* 68:104–05.

⸻. 1980. (Ecology of birds on Barro Colorado Island.) Pp. 205–25, in *Migrant birds in the Neotropics: Ecology, behavior, distribution and conservation* (A. Keast and E. Morton, eds.). Washington, D.C.: Smithsonian Institution Press.

Willoughby, E. J., and T. J. Cade. 1964. Breeding behavior of the American kestrel. *Living Bird* 3:75–96.

Woffinden, N. D. 1975. Ecology of the ferruginous hawk (*Buteo regalis*) in central Utah: population dynamics and nest site selection. Ph.D. diss., Brigham Young Univ., Provo.

Woffinden, N. D., and J. R. Murphy. 1983. Ferruginous hawk nest site selection. *J. Wildl. Manage.* 47:216–19.

————. 1985. (Breeding status of ferruginous hawk, Utah). Proc. Raptor Res. Found. Symp., Management of Birds of Prey, Nov. 3–6, Sacramento (abstract).

————. 1988. (Status of ferruginous hawks in central Utah.) Proc. Annual Meeting, Raptor Res. Found., Oct. 27–29, Minneapolis (abstract).

Wood, C. 1917. *The fundus oculi of birds, especially as viewed by the opthalmoscope.* Chicago: Lakeside Press.

Woodbridge, B. B. 1987. (Swainson's hawks and grazing, California.) Proc. Ann. Meeting, Raptor Res. Found., Oct. 28–31, Boise (abstract).

Woodin, N. 1980. (Gyrfalcon breeding in Iceland.) *Raptor Res.* 14:97–124.

Woodrey, M. S. 1986. (Red-shouldered hawk nests in Ohio.) *Wilson Bull.* 98:466–69.

Wrege, P., and T. J. Cade. 1977. (Courtship of falcons in captivity.) *Raptor Res.* 11:1–27.

Wright, B. A. 1978. Ecology of the white-tailed kite in San Diego County. M.S. thesis, San Diego State Univ.

Wright, M., R. Green, and N. Reed. 1970. The nesting activity of the swallow-tailed kite. Privately published by the authors.

————. 1971. Swallowtails of Royal Palm. *Audubon Mag.* 73:40–49.

Zarn, M. 1974. (Habitat management series for unique or endangered species: osprey.) USDI, Bur. Land Manage. Tech. Note 254.

Zimmerman, D. A. 1965. The gray hawk in the Southwest. *Aud. Field Notes* 19:475–77.

————. 1976a. (Status of gray hawk in New Mexico.) *Auk* 93:650–55.

————. 1976b. (Zone-tailed hawk mimicry and feeding habits.) *Condor* 78:420–21.

Index

The following index includes Latin names of falconiform bird taxa (family, subfamily, tribe, genus, species, and subspecies) mentioned in the text, as well as currently accepted species-level English vernacular names (plus some commonly or previously used alternate names that have been cross-indexed). Complete indexing is limited to the entries for these accepted vernacular names; only the principal species accounts (which are indicated by italics) are indexed under the listings of scientific names. Extralimital falconiform taxa are indexed as necessary, but nonfalconiform avian and nonavian taxa are not indexed. The appendixes are also not indexed.